Applied Probability and Statistics

BAILEY · The Elements of Stochastic F
to the Natural Sciences

BARTHOLOMEW · Stochastic Models for Social Processes, *Second Edition*

BENNETT and FRANKLIN · Statistical Analysis in Chemistry and the Chemical Industry

BHAT · Elements of Applied Stochastic Processes

BLOOMFIELD · Fourier Analysis of Time Series: An Introduction

BOX and DRAPER · Evolutionary Operation: A Statistical Method for Process Improvement

BROWN and HOLLANDER · Statistics: A Biomedical Introduction

BROWNLEE · Statistical Theory and Methodology in Science and Engineering, *Second Edition*

BURY · Statistical Models in Applied Science

CHERNOFF and MOSES · Elementary Decision Theory

CHOW · Analysis and Control of Dynamic Economic Systems

CLELLAND, deCANI, BROWN, BURSK, and MURRAY · Basic Statistics with Business Applications, *Second Edition*

COCHRAN · Sampling Techniques, *Third Edition*

COCHRAN and COX · Experimental Designs, *Second Edition*

COX · Planning of Experiments

COX and MILLER · The Theory of Stochastic Processes, *Second Edition*

DANIEL · Application of Statistics to Industrial Experimentation

DANIEL and WOOD · Fitting Equations to Data

DAVID · Order Statistics

DEMING · Sample Design in Business Research

DODGE and ROMIG · Sampling Inspection Tables. *Second Edition*

DRAPER and SMITH · Applied Regression Analysis

DUNN and CLARK · Applied Statistics: Analysis of Variance and Regression

ELANDT-JOHNSON · Probability Models and Statistical Methods in Genetics

FLEISS · Statistical Methods for Rates and Proportions

GNANADESIKAN · Methods for Statistical Data Analysis of Multivariate Observations

GOLDBERGER · Econometric Theory

GROSS and CLARK · Survival Distributions

GROSS and HARRIS · Fundamentals of Queueing Theory

GUTTMAN, WILKS and HUNTER · Introductory Engineering Statistics, *Second Edition*

HAHN and SHAPIRO · Statistical Models in Engineering

HALD · Statistical Tables and Formulas

HALD · Statistical Theory with Engineering Applications

HARTIGAN · Clustering Algorithms

HILDEBRAND, LAING and ROSENTHAL · Prediction Analysis of Cross Classifications

HOEL · Elementary Statistics, *Fourth Edition*

HOLLANDER and WOLFE · Nonparametric Statistical Methods

HUANG · Regression and Econometric Methods

continued on back

RAMSEY LIBRARY UNCA
3 0509 0009629 0

NORMAN L. JOHNSON

University of North Carolina, Chapel Hill

SAMUEL KOTZ

Temple University, Philadelphia

continuous univariate

distributions in statistics

distributions-1

A WILEY-INTERSCIENCE PUBLICATION

JOHN WILEY & SONS

New York • Chichester • Brisbane • Toronto • Singapore

ISBN 0 471 44626 2

PRINTED IN THE U.S.A.

20 19 18 17 16 15 14 13 12 11

General Preface

The purpose of this compendium is to give an account of the properties and uses of statistical distributions at the present time. The idea was originally suggested by one author (Samuel Kotz) while he was visiting the University of North Carolina at Chapel Hill as a Research Associate in 1962–1963. He felt that a collection of the more important facts about the commonly used distributions in statistical theory and practice would be useful to a wide variety of scientific workers.

While preparing the book, it became apparent that "important" and "commonly" needed to be rather broadly interpreted because of the differing needs and interests of possible users. We have, therefore, included a considerable amount of detailed information, often introducing less frequently used distributions related to those other which are better-known. However, we have tried to adhere to our original intention of excluding theoretical minutiae of no apparent practical importance. We do not claim our treatment is exhaustive, but we hope we have included references to most practical items of information.

This is not an introductory textbook on general statistical theory; it is, rather, a book for persons who wish to apply statistical methods, and who have some knowledge of standard statistical techniques.

For this reason no systematic development of statistical theory is provided beyond what is necessary to make the text intelligible. We hope that the book will also prove useful as a textbook and a source book for graduate courses and seminars in statistics, and possibly for discussion or seminar groups in industrial research laboratories.

Some arbitrary choices have been unavoidable. One such choice deserves special explanation: methods of estimation for parameters occurring in the specification of distributions have been included, but tests of significance have been excluded. We felt that the form of distribution is usually well-established in situations associated with estimation, but not in cases using tests of significance. At this point we would like to emphasize the usefulness of a rough classification of distributions as "modelling" or "sampling" distributions. These classes are neither mutually exclusive nor exhaustive. However, it is true that some distributions (e.g., Wishart or correlation coefficient) are almost always encountered only as a consequence of application of sampling theory to well-established population distributions. Other distributions (e.g., Weibull or logistic) are mostly used in the construction of models of variations in real populations.

We have tried to arrange the order of discussion to produce a text which develops as naturally as possible. Although some inversions have appeared desirable in the discussion of some distributions, there are broad similarities in all the treatments and the same kind of information is provided in each case. The "historical remarks" found in the discussions are not intended to be full accounts, but are merely to help assess the part played by the relevant distribution in the development of applied statistical methods.

We have included at the end of each chapter a fairly comprehensive bibliography including papers and books mentioned in the text. This inevitably produces some repetition of titles, but is, in our opinion, important to a reader who would like a more extended treatment of a particular topic.

References are indicated by numbers in square brackets in the text e.g., [4]. References and equations are numbered separately in each chapter.

The production of a work such as the present necessarily entails the assembly of substantial amounts of information, much of which is only available from published papers and even less readily accessible sources. We would like to thank Miss Judy Allen, Miss Lorna Lansitie, Mr. K. L. Weldon, Mr. H. Spring, Mr. B. G. A. Kelly, Mr. B. Zemmel, Mr. P. Cehn, Mr. W. N. Selander and Dr. T. Sugiyama for much help in this respect.

We are particularly indebted to Professor Herman Chernoff, who read our manuscript, for valuable comments and suggestions on matters of basic importance; and to Dr. D. W. Boyd and Mr. E. E. Pickett for their assistance in compilation of the preliminary drafts of the chapters on non-central χ^2 distribution, quadratic forms, and Pareto and inverse Gaussian distributions.

We would also like to record our appreciation of Miss D. Coles, Miss V. Lewis, Mrs. G. Bem, Miss C. M. Sawyer and Miss S. Lavender for their typing work and for other technical assistance.

The assistance of our wives, especially in proofreading, cannot be overestimated.

The financial support of the U. S. Air Force Office of Scientific Research and the stimulating assistance of Mrs. Rowena Swanson of the Information Sciences Division are sincerely appreciated.

The hospitality of the Department of Statistics at Chapel Hill during the summer of 1966 and Cornell University during the summer of 1968 aided in the solution of organizational difficulties connected with this project. The help of the Librarians at the University of Toronto, the University of North Carolina at Chapel Hill, Temple University, and Cornell University is also much appreciated.

In the course of the preparation of these volumes, we were greatly assisted by existing bibliographical literature on Statistical Distributions. Among them are F. A. Haight's *Index to the Distributions of Mathematical Statistics,* J. A. Greenwood and H. O. Hartley's *Guide to Tables in Mathematical Statistics*, G. P. Patil and S. W. Joshi's *Bibliography of Classical and Contagious Discrete Distributions*, W. Buckland's *Statistical Assessment of the Life Characteristic* and L. S. Deming's *Selected Bibliographies of Statistical Literature 1930 to 1957.*

Last but not least, we would like to thank the authors from many parts of the globe, too numerous to be mentioned individually, who so generously supplied us with the reprints of their publications and thus facilitated the compilation of these volumes.

Norman L. Johnson; Samuel Kotz

Preface

The presentation of information in these two books on continuous distributions follows the same general lines as the book on discrete distributions. There are natural changes in emphasis; for example, order statistics are discussed in some detail in many of the chapters on continuous distributions, while generating functions are less frequently used. The theory of order statistics for continuous random variables is capable of simpler development than is the theory for discrete variables. Such theory is employed in the construction of many useful estimators. We give a condensed account of the theory in Chapter 12. Another change is in notation. Whereas, for discrete random variables we used bold type to indicate random variables, we now use capital letters (with a few exceptions) for this purpose.

Chapter 13 (Normal Distributions) contains considerably more detailed discussion than other chapters, reflecting the massive concentration on the normal distribution over past decades. For this reason, it may often be profitable to look at Chapter 13 to obtain information on points which are lacking in other chapters.

It is well known that models can be expected to be only approximations to real situations. This is clearly so when continuous random variables are used, since observable variables are essentially discrete due to limitations of recording apparatus. For this reason, the more formal theoretical results, such as characterizations, are less likely to have practical applications. They have been included to add detail to knowledge of the nature of the various distributions and because of the intrinsic interest such information may have for some researchers.

The final chapter (Chapter 33) includes a discussion of a number of interesting special distributions. It also includes a discussion of Mills' ratio "area to bounding ordinate" for normal distributions, which might be thought more appropriate to Chapter 13. However, the association between Mills' ratios and hazard rates, discussed in Chapter 33, combined with the very considerable length of Chapter 13, caused us to adopt the arrangement we have used.

Most of the references given at the end of each chapter are, in fact, mentioned in the text. We have included some references not so mentioned because some readers will find these references useful although their importance, or relevance, does not warrant explicit mention in the text. The titles of such references usually give a clear indication of their content. In Chapter 13 (Normal Distribution) the proportion of such references is larger than usual, and we have inserted a special note about this feature.

Our coverage of known distributions is necessarily incomplete. In some cases omission has been deliberate, on grounds of utility or importance, but there must unfortunately be cases where it is inadvertent. One such oversight, which was noted too late for correction, is the omission of a discussion of *Tukey's lambda distributions*, which are the distributions of $u^\lambda - (1 - u)^\lambda$, where u has a standard uniform distribution (Hastings, C., Mosteller, F., Tukey, J. W. and Winsor, C. P. (1947). *Annals of Mathematical Statistics*, **18,** 413–426). We hope this reference may go some way to repair the omission.

A number of recently obtained results have been excluded, as it was not possible for many additions to be made to the text after September, 1969.

In addition to the acknowledgements in the general preface, we would like once more to express our thanks to Professor Herman Chernoff for his valuable suggestions and informative comments, to Mrs. Rowena Swanson, Directorate of the Information Sciences Division, U.S. Air Force Office of Scientific Research, for her constant interest and assistance, especially during the period of technical difficulties encountered in course of preparation of these volumes, to Mrs. J. O'Fallon for her able assistance in proofreading of galleys, and to Miss Berna Rubinson, Mrs. Linda Steinbrunn and Mrs. G. Ballard for their conscientious work in typing parts of the manuscript.

Contents

17

Gamma Distribution (Including Chi-Square)

18

Exponential Distribution

19

Pareto Distribution

12

Continuous Distributions (General)

1. Introduction

Chapters 1 and 2 (of Volume 1) contain some general results and methods which we shall find useful also in the discussion of continuous distributions. In the present chapter we will supplement this information with certain techniques especially relevant to continuous distributions. Also, as in Chapter 2, some general *systems* of (continuous) distributions will be described.

Generally, continuous distributions are susceptible of more elegant mathematical treatment than are discrete distributions. This makes them especially useful as approximations to discrete distributions. Continuous distributions are used in this way in most applications, both in the construction of models and in applying statistical techniques. Continuous distributions have been used in approximating discrete distributions of discrete statistics in Volume 1. The fact that most uses of continuous distributions in model-building are as approximations to discrete distributions may be less widely appreciated but is no less true. Very rarely is it more reasonable, in an absolute sense, to represent an observed value by a continuous, rather than a discrete, random variable. Rather this representation is a convenient *approximation*, facilitating mathematical and statistical analysis.

An essential property of a continuous random variable is that there is zero probability that it takes any specified numerical value, but in general a non-zero probability, calculable as a definite integral of a *probability density function* (see Section 1.4) that it takes a value in specified (finite or infinite) intervals.

Some concepts, which have great value with discrete distributions, are much less valuable in the discussion of continuous distributions. Probability generating functions, in particular, are little used in this part of the book. Factorial moments, also, although they can be calculated, rarely offer the advantages of conciseness and simplicity that they do for discrete distributions.

On the other hand, *standardization* (use of the transformed variable

$$(X - E[X])/\sqrt{\text{var}(X)}$$

to produce a distribution with zero mean and unit standard deviation) is much more useful for continuous distributions. In particular, the shape of a distribution can be conveniently summarized by giving standardized values of a number of *quantiles* (i.e. values of the variable for which the cumulative distribution function has specified values). Care should be taken to distinguish between *standardized* and *standard* forms of distributions. The latter are usually convenient ways of writing the mathematical formulas for probability density functions. They *may* happen to be standardized, but this is not essential.

Order statistics, also, are of much greater use, and simpler in theoretical analysis, for continuous than for discrete distributions. The next section will be devoted to a general discussion of order statistics for continuous variables, with particular reference to their use in statistical analysis.

2. Order Statistics

If $X_1, X_2, \ldots, X_n$ are random variables,* and $X_1' \leq X_2' \leq \cdots \leq X_n'$ are the same variables arranged in ascending order of magnitude (so that $X_1' = \min(X_1, X_2, \ldots, X_n)$, $X_n' = \max(X_1, X_2, \ldots, X_n)$) then $X_1', X_2', \ldots, X_n'$ are called the *order-statistics* corresponding to $X_1, X_2, \ldots, X_n$. (See also Chapter 1, Section 7.2.) If it is necessary to indicate the total number of variables explicitly, the symbols $X_{1;n}'$, $X_{2;n}'$, etc. will be used.

If the differences $\{X_i - X_j\}$ are continuous random variables then the events $\{X_i = X_j\}$ all have zero probability. Being finite in number, they can be neglected in probability calculations. We will suppose, from now on, that this is the case, so we can assume that $X_1' < X_2' \cdots < X_n'$, without altering any probabilities which relate to the joint distribution of $X_1', X_2', \ldots, X_n'$.

The cumulative distribution function of X_n' is defined by

$$\Pr[X_n' \leq x] = \Pr\left[\bigcap_{j=1}^{n} (X_j \leq x)\right]. \tag{1}$$

If $X_1, X_2, \ldots, X_n$ are mutually independent, then

$$\Pr[X_n' \leq x] = \prod_{j=1}^{n} \Pr[X_j \leq x] \tag{2}$$

whence the probability density function of X_n' is equal to

$$\text{(cumulative distribution function of } X_n'\text{)} \times \sum_{j=1}^{n} \left(\frac{\text{probability density function of } X_j}{\text{cumulative distribution function of } X_j}\right). \tag{3}$$

*Continuous random variables will generally be indicated by capital letters in this volume.

If all X_j's have identical distributions with $\Pr[X_j \le x] = F(x)$, and $dF(x)/dx = p(x)$, then the probability density function of X'_n is

$$n[F(x)]^{n-1}p(x). \tag{4}$$

Similarly (again assuming all X_j's independent and identically distributed) the probability density function of X'_n is

$$n[1 - F(x)]^{n-1}p(x). \tag{5}$$

More generally, in this case, with $1 \le a_1 < a_2 < \cdots < a_s \le n$ (and putting $a_0 = 0$; $a_{s+1} = n$; $F(x_{a_0}) = 0$; $F(x_{a_{s+1}}) = 1$), the joint probability density function of $X'_{a_1}, X'_{a_2}, \ldots, X'_{a_s}$ is (in an obvious notation)

$$\frac{n!}{\prod_{j=1}^{s+1}(a_j - a_{j-1})!}\left[\prod_{j=1}^{s+1}\{F(x_{a_j}) - F(x_{a_{j-1}})\}^{a_j - a_{j-1}}\right]\prod_{j=1}^{s} p(x'_{a_j}), \tag{6}$$

$$(x_1 \le x_2 \le \cdots \le x_s).$$

In particular, the joint probability density function of X'_1 and X'_n is

$$n(n-1)p(x_1)p(x_n)[F(x_n) - F(x_1)]^{n-2}, \qquad (x_1 \le x_n). \tag{7}$$

From this joint distribution it is possible to evaluate the cumulative distribution function of the range ($W = X'_n - X'_1$). The formulas

$$\Pr[W \le w] = n\int_{-\infty}^{\infty} p(x)[F(x) - F(x-w)]^{n-1}\,dx \tag{8}$$

and

$$E[W] = \int_{-\infty}^{\infty}\{1 - [F(x)]^n - [1 - F(x)]^n\}\,dx \tag{9}$$

are of interest.

If $n = 2m + 1$ is odd (i.e. m is an integer) then X'_{m+1} represents the (*sample*) *median* of $X_1, X_2, \ldots, X_n$. Its probability density function is

$$\frac{(2m+1)!}{(m!)^2}[F(x)\{1 - F(x)\}]^m p(x)\,. \tag{10}$$

Generally the $100p$-% sample percentile is represented by $X'_{(n+1)p}$ and is defined only if $(n+1)p$ is an integer. The median corresponds to $p = \frac{1}{2}$; we have the *lower* and *upper quartile* for $p = \frac{1}{4}, \frac{3}{4}$ respectively.

Under certain conditions of regularity, which are satisfied in many commonly encountered cases, it is possible to obtain useful approximations to the moments of order statistics in terms of the common probability density function of the X's.

This makes use of the fact that the statistics $Y_1 = F(X_1)$, $Y_2 = F(X_2), \ldots,$ $Y_n = F(X_n)$ are independently distributed with common rectangular distribution (see Chapter 25) over the range 0 to 1. The corresponding order statistics

$Y'_1, Y'_2, \ldots, Y'_n$ have the joint probability density function

$$p_{Y'_1 \ldots, Y'_n}(y_1, \ldots, y_n) = n! \qquad (0 \le y_1 \le y_2 \le \cdots \le y_n \le 1) \cdot$$

The joint probability density function of any subset $Y'_{a_1}, \ldots, Y'_{a_s}$ $1 \le a_1 < a_2 < \cdots < a_s \le n$ is (using (6)).

$$(11) \qquad p_{\{Y'_a\}}(y_{a_1}, \ldots, y_{a_s}) = \frac{n!}{\prod_{j=1}^{s+1} (a_j - a_{j-1})!} \prod_{j=1}^{s+1} (y_{a_j} - y_{a_{j-1}})^{a_j - a_{j-1} - 1}$$

with $a_0 = 0$, $a_{s+1} = n$; $y_{a_0} = 0$; $y_{a_{s+1}} = 1$, and denoting $Y'_{a_1}, \ldots, Y'_{a_s}$ by $\{Y'_a\}$. The moments and product moments of the Y''s are given by the formula

$$(12) \qquad E\left[\prod_{j=1}^{s} Y'^{\,r_j}_{a_0}\right] = \frac{n!}{\left(n + \sum_{j=1}^{s} r_j\right)!} \prod_{j=1}^{s} \frac{\left(a_j + \sum_{i=1}^{j} r_i - 1\right)!}{\left(a_j + \sum_{i=1}^{j-1} r_i - 1\right)!} \cdot$$

We now expand X'_r, as a function of Y'_r, about the value $E(Y'_r) = r/(n+1)$; thus

$$(13) \qquad X'_r = F^{-1}\left(\frac{r}{n+1}\right) + \left(Y'_r - \frac{r}{n+1}\right) \cdot \left[\left.\frac{dF^{-1}(y)}{dy}\right|_{y=r/(n+1)}\right]$$
$$+ \frac{1}{2}\left(Y'_r - \frac{r}{n+1}\right)^2 \cdot \left[\left.\frac{d^2F^{-1}(y)}{dy^2}\right|_{y=r/(n+1)}\right] + \cdots$$

and take expected values of each side of (13) (using the method of statistical differentials, described in Chapter 1).

Note that since

$$y = F(x) = \int_{-\infty}^{x} p(t)\, dt,$$

$$\frac{dF^{-1}}{dy} = \frac{dx}{dy} = \frac{1}{dy/dx} = \frac{1}{p(x)}$$

and

$$\left.\frac{dF^{-1}(y)}{dy}\right|_{y=r/(n+1)} = \frac{1}{p(\xi'_r)}$$

where ξ'_r satisfies the equation

$$(14) \qquad \frac{r}{n+1} = \int_{-\infty}^{\xi'_r} p(x)\, dx.$$

Similarly $\dfrac{d^2F^{-1}}{dy^2} = -[p(x)]^{-2} \dfrac{dp(x)}{dy} = -[p(x)]^{-3} \dfrac{dp(x)}{dx}$, and so on.

David and Johnson [26] found it convenient to arrange the series so obtained in descending powers of $(n+2)$. Some of their results follow. (In these formulas, $p_s = s/(n+1)$; $q_s = 1 - p_s$;

$$(F^{-1})'_r = \left.\frac{dF^{-1}}{dy}\right|_{y=r/(n+1)}; \; (F^{-1})''_r = \left.\frac{d^2F^{-1}}{dy^2}\right|_{y=r/(n+1)} \text{ etc.}).$$

(15.1)
$$E[X'_r] = X'_r + \frac{p_r q_r}{2(n+2)}(F^{-1})''_r + \frac{p_r q_r}{(n+2)^2}[\tfrac{1}{3}(q_r - p_r)(F^{-1})'''_r + \tfrac{1}{8}p_r q_r (F^{-1})^{iv}_r] + \cdots$$

(15.2)
$$\operatorname{var}(X'_r) = \frac{p_r q_r}{n+2}\{(F^{-1})'_r\}^2 + \frac{p_r q_r}{(n+2)^2}[2(q_r p_r)(F^{-1})'_r(F^{-1})''_r + p_r q_r\{(F^{-1})'_r(F^{-1})'''_r + \tfrac{1}{2}[(F^{-1})''_r]^2\}] + \cdots$$

(15.3)
$$\operatorname{cov}(X'_r, X'_s) = \frac{p_r q_r}{n+2}\{(F^{-1})'_r(F^{-1})'_s\} \qquad (r < s)$$
$$+ \frac{p_r q_r}{(n+2)^2}[(q_r - p_r)(F^{-1})''_r(F^{-1})_s + (q_s - p_s)(F^{-1})'_r(F^{-1})''_s + \tfrac{1}{2}p_r q_r(F^{-1})'''_r(F^{-1})'_s + \tfrac{1}{2}p_s q_s(F^{-1})'_r(F^{-1})'''_s + \tfrac{1}{2}p_r q_s(F^{-1})''_r(F^{-1})''_s] + \cdots$$

(15.4)
$$\mu_3(X'_r) = \frac{p_r q_r}{(n+2)^2}[2(q_r - p_r)\{(F^{-1})'_r\}^3 + 3p_r q_r\{(F^{-1})'_r\}^2(F^{-1})''_r\}] + \cdots$$

(15.5)
$$\mu_4(X'_r) = \frac{3p_r^2 q_r^2}{(n+2)^2}\{(F^{-1})'_r\}^4 + \frac{p_r q_r}{(n+2)^3}[6\{(q_r - p_r)^2 - p_r q_r\}\{(F^{-1})'_r\}^4]$$
$$+ 36 p_r q_r(q_r - p_r)\{(F^{-1})'_r\}^3(F^{-1})'''_r + 5p_r^2 q_r^2[2\{(F^{-1})'_r\}^3(F^{-1})''_r + 3\{(F^{-1})'_r(F^{-1})''_r\}^2] + \cdots$$

By inserting the values of ξ'_r, $(F^{-1})'_r$, $(F^{-1})''_r$, etc. appropriate to the particular distribution, approximate formulas can be obtained corresponding to any absolutely continuous common distribution of the original independent variables. These formulas generally tend to be more accurate for larger n, and for larger $\min(p_r, q_r)$ (with $\operatorname{cov}(X'_r, X'_s)$, for larger $\min(p_r, p_s, q_r, q_s)$).

If the distribution of X is such that $\Pr[X \leq x]$ is a function of $(x - \theta)/\phi$ only, so that θ and ϕ (> 0) are *location* and *scale* parameters, then it is easy to see that $Z = (X - \theta)/\phi$ has a distribution which does not depend on θ or ϕ. Denoting the order statistics corresponding to independent random variables $Z_1, Z_2, \ldots, Z_n$, each distributed as Z, by $Z'_{1;n}, Z'_{2;n}, \ldots, Z'_{n;n}$ it is easy to see that

(16.1)
$$E[X'_{r;n}) = \theta + \phi E[Z'_{r;n}]$$

and, further that

(16.2)
$$\operatorname{var}(X'_{r;n}) = \phi^2 \operatorname{var}(Z'_{r;n})$$
$$\operatorname{cov}(X'_{r;n}, X'_{s;n}) = \phi^2 \operatorname{cov}(Z'_{r;n}, Z'_{s;n}).$$

Hence it is possible to obtain *best linear unbiased estimators* of θ and ϕ, based on the order statistics $X'_{1;n}, X'_{2;n}, \ldots, X'_{n;n}$, by minimizing the quadratic form:

$$\sum_r \sum_s c_{rs}(X'_{r;n} - \theta - \phi E[Z'_{r;n}])(X'_{s;n} - \theta - \phi E[Z'_{s;n}])$$

where the matrix (c_{rs}) is the inverse of the matrix of variances and covariances of the $Z'_{r;n}$'s. (Lloyd [51].)

In later chapters a number of results obtained by this method will be presented. The method is of particular value when not all the $X_{r;n}$'s are used. For example, when data are *censored* (as described in Section 1.7.2) not all the order statistics are available. Even if they are available we may wish to use only a limited number. It is useful, in such cases, to know which sets of a fixed number of order statistics will minimize the variance of the best linear unbiased estimator of θ or ϕ (or perhaps some function of these parameters). Exact calculation is usually tedious, but approximate calculation, using only the first terms of formulas (15.1), (15.2) is less troublesome.

In using these results, it is desirable to bear in mind that (*a*) there may be nonlinear estimators which are (in some sense) more accurate, (*b*) 'best' is defined in terms of variance, which is not always appropriate, and (*c*) the constraint of unbiasedness may exclude some good estimators. However, it does appear that the best linear unbiased estimators of location and scale parameters, based on order statistics, usually offer accuracy close to the utmost attainable from the data.

Bennett [2] has developed a general method for determining "asymptotically efficient linear unbiased estimators." These are estimators of the form

$$L_n = \sum_{j=1}^{n} J(j/(n+1))X'_{j;n} \tag{17}$$

where $J(\cdot)$ is a 'well-behaved' function; that is the limiting distribution of $\sqrt{n}(L_n - \theta)$ is normal with expected value zero. Bennett's thesis is not easily available. The following results are quoted from Chernoff *et al.* [20] who have also demonstrated the asymptotic normality (as $n \to \infty$) of these estimators.

Relatively simple formulas for $J(\cdot)$ are available for the special case when the parameters θ_1, θ_2 are location and scale parameters so that (for each unordered X)

$$\Pr[X \leq x] = g\left(\frac{x - \theta_1}{\theta_2}\right) \qquad (\theta_2 > 0).$$

The corresponding density function is $\theta_2^{-1} g'\left(\frac{x - \theta_1}{\theta_2}\right)$ and the Fisher information matrix is

$$\begin{pmatrix} I_{11} & I_{12} \\ I_{21} & I_{22} \end{pmatrix} = \begin{pmatrix} \int_{-\infty}^{\infty} \frac{dL_1}{dy} g(y)\,dy & \int_{-\infty}^{\infty} \frac{dL_2}{dy} g(y)\,dy \\ \int_{-\infty}^{\infty} y\frac{dL_1}{dy} g(y)\,dy & \int_{-\infty}^{\infty} y\frac{dL_2}{dy} g(y)\,dy \end{pmatrix} \theta_2^{-2}$$

where
$$L_1(y) = -g'(y)/g(y)$$
$$L_2(y) = -1 - y \cdot g'(y)/g(y).$$

(Note that $I_{21} = I_{12}$ provided $f''(y)$ exists and $\lim_{y\to\pm\infty} yf'(y) = 0$.) Then for estimating θ_1, with θ_2 known we can use

$$J(u) = I_{11}^{-1}L_1'(F^{-1}(u)). \tag{18}$$

To make the estimator unbiased, $I_{11}^{-1}I_{12}\theta_2$ must be subtracted. For estimating θ_2, with θ_1 known, we can use

$$J(u) = I_{22}^{-1}L_2'(F^{-1}(u)).$$

To make the estimator unbiased, $I_{22}^{-1}I_{12}\theta_1$ must be subtracted.

If neither θ_1, nor θ_2 is known then, for estimating θ_1

$$J(u) = I^{11}L_1'(F^{-1}(u)) + I^{12}L_2'(F^{-1}(u))$$

and for estimating θ_2

$$J(u) = I^{12}L_1'(F^{-1}(u)) + I^{22}L_2'(F^{-1}(u))$$

where $\begin{pmatrix} I^{11} & I^{12} \\ I^{12} & I^{22} \end{pmatrix}$ is the inverse of the matrix I. These estimators are unbiased.

Chernoff *et al.* [20] also obtain formulas to use when the data are censored.

The *limiting distributions* of order statistics as n tends to infinity have been studied by a number of workers. It is not difficult to establish that if $r - n\omega$ tends to zero as n tends to infinity the limiting distribution of $n(X'_{r;n} - X_\omega)$ (where $\Pr[X \leq X_\omega] = \omega$) is normal with expected value zero and standard deviation $\sqrt{\omega(1-\omega)}/p(X_\omega)$. However, other limiting distributions are possible. Wu [71] has shown that a lognormal limiting distribution may be obtained.

Chan [17] has shown that the distribution function is characterized by either of the sets of values $\{E[X'_{1;n}]\}$ or $\{E[X'_{n;n}]\}$ (for all n) provided the expected value of the distribution is finite.

3. Calculus of Probability Density Functions

The reader may have noted that many of the results of the preceding section were expressed in terms of probability density functions. Although these are only auxiliary quantities — actual probabilities being the items of real importance — they are convenient in the analysis of continuous distributions. In this section we briefly describe techniques for working with probability density functions, which will be employed in later chapters. More detailed discussions, and proofs, can be found in textbooks.

If $X_1, X_2, \ldots, X_n$ are independent random variables with probability density functions $p_{X_1}(x_1), p_{X_2}(x_2), \ldots, p_{X_n}(x_n)$ then the joint probability density function may be taken as

$$p_{X_1,X_2,\ldots,X_n}(x_1,x_2,\ldots,x_n) = \prod_{j=1}^{n} p_{X_j}(x_j). \tag{19}$$

If the variables are not independent, conditional probability density functions must be used, and in place of (19) we have

(20)
$$p(x_1,x_2,\ldots,x_n) = p(x_1)p(x_2 \mid x_1)p(x_3 \mid x_1,x_2),\ldots, p(x_n \mid x_{n-1},\ldots,x_1).^*$$

Of course (20) includes (19) since if $X_1, \ldots, X_n$ are a mutually independent set of variables then

$$p(x_2 \mid x_1) = p(x_2)$$
$$p(x_3 \mid x_1,x_2) = p(x_3), \text{ etc.}$$

If $p(x_1,\ldots,x_n)$ is known, then the joint probability density function of any subset of the n random variables can be obtained by repeated use of the formula

$$\int_{-\infty}^{\infty} p(x_1,\ldots,x_n)\,dx_p = p(x_1,\ldots,x_{p-1},x_{p+1},\ldots,x_n). \tag{21}$$

If it is desired to find the joint distribution of n functions of $X_1, \ldots, X_n$ (*statistics*), $T_1 \equiv T_1(X_1,\ldots,X_n), \ldots, T_n = T_n(X_1,\ldots,X_n)$, and the transformation from $(X_1,\ldots,X_n)$ to $(T_1,\ldots,T_n)$ is one-to-one then the formula

$$p_{T_1,\ldots,T_n}(t_1,\ldots,t_n) = p_{X_1,\ldots,X_n}(x_1(\mathbf{t}),\ldots,x_n(\mathbf{t}))\left|\frac{\partial(x_1,\ldots,x_n)}{\partial(t_1,\ldots,t_n)}\right| \tag{22}$$

may be used. ($t_1, \ldots, t_n$ and $x_1, \ldots, x_n$ are related in the same way as $T_1, \ldots, T_n$ and $X_1, \ldots, X_n$; $x_j(\mathbf{t})$ means x_j expressed in terms of $t_1, \ldots, t_n$ and $\partial(x_1,\ldots,x_n)/\partial(t_1,\ldots,t_n)$ is the Jacobian of $(x_1,\ldots,x_n)$ with respect to $(t_1, \ldots, t_n)$ — it is a determinant of n rows and n columns with the element in the ith row and the jth column equal to $\partial x_i/\partial t_j$.)

If the transformation is not one-to-one the simple formula (22) cannot be used. In particular cases, however, straightforward modifications of (22) can be employed. For example, if, in general, k different sets of values of the x's produce the *same* set of values of the t's, it may be possible to split up the transformation into k separate transformations. Then (22) is applied to each, and the results added together.

Having obtained the joint distribution of $T_1, T_2, \ldots, T_n$, the joint distribution of any subset thereof can be obtained by using (21) repeatedly.

The conditional distribution of X_1, given $X_2, \ldots, X_n$ is sometimes called the *array distribution* of X_1 (given $X_2,\ldots,X_n$). The expected value of this conditional distribution (a function of $X_2,\ldots,X_n$) is called the *regression* of

*In the remainder of this and succeeding sections of this chapter, subscripts following p will usually be omitted for convenience. In succeeding chapters, the subscript will sometimes appear.

X_1 on $X_2, \ldots, X_n$. The variance is called the *array variance* (of X_1, given $X_2, \ldots, X_n$); if it does not depend on $X_2, \ldots, X_n$ the variation is said to be *homoscedastic*.

4. Systems of Distributions

Some families of distributions have been constructed which are intended to provide approximations to as wide a variety of observed distributions as is possible. Such families are often called *systems* of distributions, or, more often, *systems of frequency curves*. Although theoretical arguments may indicate the relevance of a particular system, their value should be judged primarily on practical, *ad hoc* considerations. Particular requirements, are ease of computation and facility of algebraic manipulation. Such requirements make it desirable to use as few parameters as is possible in defining an individual member of the system. How few we may use, without prejudicing the variety of distributions included, is a major criterion in judging the utility of systems of distributions.

For most practical purposes it is sufficient to use four parameters. There is no doubt that at least three parameters are needed; for some purposes this is enough. Inclusion of a fourth parameter does effect noticeable improvement, but it is doubtful whether the improvement obtained by including a 5th or 6th parameter is commensurate with the extra labor involved.

Here we will describe some systems of frequency curves. Among these systems there should be at least one which suffices for practical needs and possibilities in most situations.

4.1 *Pearson System*

This system was originated by Pearson [58] between 1890 and 1900. For every member of the system, the probability density function $p(x)$ satisfies a differential equation of form

$$\frac{1}{p}\frac{dp}{dx} = -\frac{a + x}{c_0 + c_1 x + c_2 x^2}. \tag{23}$$

The shape of the distribution depends on the values of the parameters a, c_0, c_1 and c_2. Provided a is not a root of the equation

$$c_0 + c_1 x + c_2 x^2 = 0$$

p is finite when $x = a$, and $\dfrac{dp}{dx}$ is zero when $x = a$. The slope (dp/dx) is also zero when $y = 0$; but if $x \neq a$ and $y \neq 0$ then $dp(x)/dx \neq 0$. Since the conditions $p(x) \geq 0$ and

$$\int_{-\infty}^{\infty} p(x)\, dx = 1$$

must be satisfied, it follows from (23) that $p(x)$ must tend to zero as x tends to infinity; and so, also must dp/dx. This may not be true of *formal* solutions of (23), but in such cases the condition $p(x) \geq 0$ is not satisfied and it is necessary to restrict the range of values of x to those for which $p(x) > 0$, and assign the value $p(x) = 0$ when x is outside this range.

The shape of the curve representing the probability density function varies considerably with a, c_0, c_1 and c_2.

Pearson classified the different shapes into a number of types. We will give a resumé of his classification. We follow his system of numbering because it is well-established, but it does not have a clear systematic basis.

The form of solution of (23) evidently depends on the nature of the roots of the equation

$$c_0 + c_1 x + c_2 x^2 = 0 \tag{24}$$

and the various types correspond to these different forms of solution.

We first note that if $c_1 = c_2 = 0$, equation (23) becomes

$$\frac{d \log p(x)}{dx} = -\frac{x + a}{c_0}$$

whence

$$p(x) = K \exp\left[-\frac{(x + a)^2}{2c_0}\right]$$

where K is a constant, chosen to make

$$\int_{-\infty}^{\infty} p(x)\, dx = 1.$$

In fact it is clear that c_0 must be positive, and $K = \sqrt{2\pi c_0}$, so that the corresponding distribution is *normal* with expected value $-a$ and standard deviation $\sqrt{c_0}$. The next chapter is devoted to this distribution.

The normal curve is not assigned to particular type. It is, in fact, a limiting distribution of all types. From now on we will suppose that the origin of the scale of X has been so chosen that $E(X) = 0$.

Type I corresponds to both roots of (23) being real, and of opposite signs. Denoting the roots by a_1, a_2, with

$$a_1 < 0 < a_2$$

we have

$$c_0 + c_1 x + c_2 x^2 = -c_2(x - a_1)(a_2 - x)$$

and Equation (23) can be written

$$\frac{d \log p(x)}{dx} = \frac{x + a}{c_2(x - a_1)(a_2 - x)} = \frac{1}{c_2(a_2 - a_1)}\left[\frac{a + a_1}{x - a_1} + \frac{a + a_2}{a_2 - x}\right]$$

whence

$$p(x) = K(x - a_1)^{m_1}(a_2 - x)^{m_2} \tag{25}$$

with $m_1 = \dfrac{a + a_1}{c_2(a_2 - a_1)}$; $m_2 = -\dfrac{a + a_2}{c_2(a_2 - a_1)}$.

In order for both $(x - a_1)$ and $(a_2 - x)$ to be positive we must have $a_1 < x < a_2$, and so we limit the range of variation to these values of x. Equation (25) can represent a proper probability density function provided $m_1 > -1$ and $m_2 > -1$.

This is a general form of *beta distribution*, which will be discussed further in Chapter 24. Here we briefly note a few points relating to the function $p(x)$.

The limiting value of $p(x)$, as x tends to a_j is zero or infinite according as m_j is positive or negative (for $j = 1,2$). If m_1 and m_2 have the same sign, $p(x)$ has a single mode or antimode (according as the m's are positive or negative, respectively). Type I distributions can be subdivided according to the appearance of the graph of $p(x)$ against x. Thus we have

Type I(U) if $m_1 < 0$ and $m_2 < 0$.
Type I(J) if $m_1 < 0$ and $m_2 > 0$, or if
$m_1 > 0$ and $m_2 < 0$.

If m_j is zero then $p(x)$ tends to a non-zero limit as x tends to $a_j (j = 1$ or $2)$.

The symmetrical form of (25), with $m_1 = m_2$, is called a *Type II* distribution. If the common value is negative the distribution is U-shaped, and is sometimes described as Type II(U).

Type III corresponds to the case $c_2 = 0$ (and $c_1 \neq 0$). In this case (23) becomes

$$\frac{d \log p(x)}{dx} = -\frac{x + a}{c_0 + c_1 x} = -\frac{1}{c_1} - \frac{a - c_0/c_1}{c_0 + c_1 x}$$

whence

$$p(x) = K(c_0 + c_1 x)^m \exp(-x/c_1) \tag{26}$$

with $m = c_1^{-1}(c_0 c_1^{-1} - a)$.

If $c_1 > 0$ we take the range of x as $x > -c_0/c_1$; if $c_1 < 0$, the range is taken to be $x < -c_0/c_1$.

Type III distributions are *gamma distributions* and are discussed further in Chapter 17.

Type IV distributions correspond to the case when the equation

$$c_0 + c_1 x + c_2 x^2 = 0$$

does not have real roots. In this case we use the identity

$$c_0 + c_1 x + c_2 x^2 = C_0 + c_2(x + C_1)^2$$

with $C_0 = c_0 - \frac{1}{4}c_1^2 c_2^{-1}$; $C_1 = \frac{1}{2}c_1 c_2^{-1}$, and write (23) as

$$\frac{d \log p(x)}{dx} = \frac{-(x + C_1) - (a - C_1)}{C_0 + c_2(x + C_1)^2}.$$

From this it follows that

$$(27) \qquad p(x) = K[C_0 + c_2(x + C_1)^2]^{-(2c_2)^{-1}} \exp\left[-\frac{a - C_1}{\sqrt{c_2 C_0}} \tan^{-1} \frac{x + C_1}{\sqrt{C_0/c_2}}\right].$$

(Note that, since $c_0 + c_1 x + c_2 x^2 = 0$ has no real roots, $c_1^2 < 4c_0 c_2$ and so $c_2 C_0 = c_0 c_2 - \frac{1}{4}c_1^2$ is positive.)

No common statistical distributions are of Type IV form, therefore it will not be discussed in a later chapter, so we denote a little space here to this type. Formula (27) leads to intractable mathematics if one attempts to calculate values of the cumulative distribution function. As we shall see later, it is possible to express the parameters a, c_0, c_1, c_2 in terms of the first four moments of the distribution, so it is possible to fit by equating actual and fitted moments. Fitting by maximum likelihood is so difficult (with unknown accuracy in finite-sized samples) that it is practically never attempted.

The tables of Johnson *et al.* [42] give standardized quantiles of Type IV distributions to three decimal places for $\sqrt{\beta_1} = 0.0(0.1)2.0$ and for β_2 increasing by intervals of 0.2. With some interpolation, these tables can provide approximate values of the cumulative distribution function, without the need to evaluate K in formula (27). To calculate K, special tables must be used, or a special quadrature of $K^{-1}p(x)$, (according to (27)) carried out.

On account of technical difficulties associated with the use of Type IV distributions, efforts have been made to find other distributions with simpler mathematical form, which, according to circumstances, are close enough to Type IV distributions to replace them. Tables 2 and 3, later in this chapter, contain some information on this point.

Type V corresponds to the case when $c_0 + c_1 x + c_2 x^2$ is a perfect square ($c_1^2 = 4c_0 c_2$). Equation (23) can now be written

$$\frac{d(\log p(x))}{dx} = -\frac{x + a}{c_2(x + C_1)^2}$$
$$= -\frac{1}{c_2(x + C_1)} - \frac{a - C_1}{c_2(x + C_1)^2}$$

whence

$$(28) \qquad p(x) = K(x + C_1)^{-1/c_2} \exp\left[\frac{a - C_1}{c_2(x + C_1)}\right].$$

If $(a - C_1)/c_2 < 0$ then $x > -C_1$; if $(a - C_1)/c_2 > 0$ then $x < -C_1$. (The inverse Gaussian distribution (Chapter 15) belongs to this family.) If $a = C_1$ and $|c_2| < 1$ then we have the special case

$$p(x) = K(x + C_1)^{-1/c_2}$$

which are sometimes called Type VIII and IX according as $c_2 > 0$ or $c_2 < 0$. From (28) it can be seen that $(x + C_1)^{-1}$ has a Type III distribution.

Type VI corresponds to the case when the roots of $c_0 + c_1 x + c_2 x^2 = 0$ are real and of the same sign. If they are both negative — ($a_1 < a_2 < 0$ say) —

then an analysis similar to that leading to equation (25) can be carried out, with the result written in the form

(29) $$p(x) = K(x - a_1)^{m_1}(x - a_2)^{m_2}.$$

Since the expected value is greater than a_2 it is clear that the range of variation of x must be $x > a_2$. (Formula (29) can represent a proper probability density function provided $m_2 < -1$ and $m_1 + m_2 < 0$).

Finally, *Type VII* corresponds to the case when $c_1 = a = 0$, $c_0 > 0$, $c_2 > 0$. In this case equation (23) becomes

(30) $$\frac{d \log p(x)}{dx} = -\frac{x}{c_0 + c_2x^2}$$

whence

(30)′ $$p(x) = K(c_0 + c_2x^2)^{-(2c_2)^{-1}}$$

A particularly important distribution belonging to this family is the (*central*) *t distribution*, which will be discussed further in Chapter 27. Distribution (30)′ can be obtained by a simple multiplicative transformation from a t distribution with 'degrees of freedom' (possibly fractional) equal to $(c_2^{-1} - 1)$. The parameters a, c_0, c_1, and c_2 in (23) can be expressed in terms of the moments of the distribution. Equation (23) may be written (after multiplying both sides by x^r).

(31) $$x^r(c_0 + c_1 + c_2x^2)\frac{dp(x)}{d(x)} + x^r(a + x)p(x) = 0\cdot$$

Integrating both sides of (23) between $-\infty$ and $+\infty$ and assuming $x^rp(x) \to 0$ as $x \to \pm\infty$ for $r \leq 5$ we obtain the equation

(32) $$-rc_0\mu'_{r-1} + [-(r + 1)c_1 + a]\mu'_r + [-(r + 2)c_2 + 1]\mu'_{r+1} = 0.$$

Putting $r = 0, 1, 2, 3$ in (32), and noting that $\mu'_0 = 1$ and (in the present context) $\mu'_{-1} = 0$, we obtain four simultaneous linear equations for a, c_0, c_1 and c_2 with coefficients which are functions of μ'_1, μ'_2, μ'_3 and μ'_4. It can always be arranged (as we have done above) that the expected value of the variable is zero. If this be done than $\mu'_1 = 0$ and $\mu'_r = \mu_r$ for $r \geq 2$. The formulas for a, c_0, c_1 and c_2, are then

(33.1) $$c_0 = (4\beta_2 - 3\beta_1)(10\beta_2 - 12\beta_1 - 18)^{-1}\mu_2$$

(33.2) $$a = c_1 = \sqrt{\beta_1}\,(\beta_2 + 3)(10\beta_2 - 12\beta_1 - 18)^{-1}\sqrt{\mu_2}$$

(33.3) $$\begin{aligned} c_2 &= (2\mu_4\mu_2 - 3\mu_3^2 - 6\mu_2^3)(10\mu_4\mu_2 - 12\mu_3^2 - 18\mu_2^3)^{-1} \\ &= (2\beta_2 - 3\beta_1 - 6)(10\beta_2 - 12\beta_1 - 18)^{-1}. \end{aligned}$$

Remembering the definitions of the various types it is clear from equations (33) that for:

Type I: $$\begin{aligned} \kappa &= \tfrac{1}{4}c_1^2(c_0c_2)^{-1} \\ &= \tfrac{1}{4}\beta_1(\beta_2 + 3)^2(4\beta_2 - 3\beta_1)^{-1}(2\beta_2 - 3\beta_1 - 6)^{-1} < 0 \end{aligned}$$

Type II: $\beta_1 = 0, \beta_2 < 3$
Type III: $2\beta_2 - 3\beta_1 - 6 = 0$
Type IV: $0 < \kappa < 1$
Type V: $\kappa = 1$
Type VI: $\kappa > 1$
Type VII: $\beta_1 = 0, \beta_2 > 3$

The division of the (β_1,β_2) plane among the various types is exhibited in Figure 1. (Note that it is impossible to have $\beta_2 - \beta_1 - 1 < 0$.)

The upside-down presentation of this Figure is in accordance with well-established convention. Note that only Types I, VI and IV correspond to areas in the (β_1,β_2) diagram. The remaining types correspond to lines and are sometimes called *transition types*. Other forms of diagrams have been proposed by Boetti [7] and Craig [23]. The latter uses $(2\beta_2 - 3\beta_1 - 6)/(\beta_2 + 3)$ in place of β_2 for one axis.

Examples of fitting Pearson curves to numerical data are given in [34].

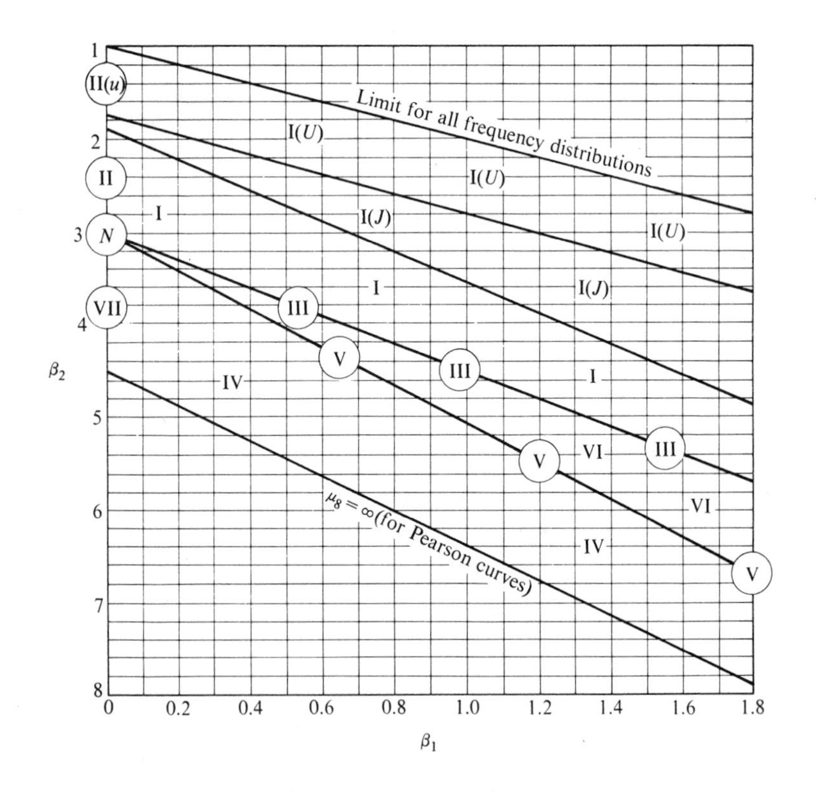

FIGURE 1

A Chart Relating the Type of Pearson Frequency Curve to the Values of β_1, β_2

A computer program for producing values of random variables having Pearson type distributions has been described by Cooper *et al.* [21].

We conclude this subsection by noting a few general properties of Pearson type distributions.

By an analysis similar to that leading to Equations (33) it can be shown that

$$\text{mean deviation} = 2\left(\frac{1-3c_2}{1-2c_2}\right)\mu_2 p(\mu_1') \tag{34}$$

for all Pearson type distributions (Pearson [59], Kamat [43], Suzuki [64]). (Note that

$$\frac{1-3c_2}{1-2c_2} = \frac{4\beta_2 - 3\beta_1}{6(\beta_2 - \beta_1 - 1)} > 0.)$$

As mentioned earlier, the derivative $dp(x)/dx$ equals zero at $x = -a$. There is a mode, or an antimode, of the distribution at this value of x, so

$$\text{Mode (or Antimode)} - \text{Expected Value} = -a. \tag{35}$$

Finally we have,

$$\begin{aligned}\frac{d^2p(x)}{dx^2} &= -\frac{x+a}{c_0+c_1x+c_2x^2}\frac{dp(x)}{dx} + \frac{c_2x^2+2ac_2x+ac_1-c_0}{(c_0+c_1x+c_2x^2)^2}p(x)\\ &= \frac{p(x)}{(c_0+c_1x+c_2x^2)^2}[(x+a)^2+c_2(x+a)^2+a^2(1-c_2)-c_0]\\ &\qquad\qquad (\text{since } c_1 = a).\end{aligned}$$

Provided $a^2(1 - c_2) < c_0$ there are points of inflexion, equidistant from the mode, at

$$x = \text{Mode} \pm \sqrt{\{c_0 - a^2(1-c_2)\}(1+c_2)^{-1}}. \tag{36}$$

4.2 *Expansions*

For a wide class of continuous distributions it is possible to change the values of the cumulants by a simple application of an operator to the probability density function.

If $f(x)$ is a probability density function with cumulants $\kappa_1, \kappa_2, \ldots$, then the function

$$g(x) = \exp\left[\sum_{j=1}^{\infty} \epsilon_j\{(-D)^j/j!\}\right] f(x) \tag{37}$$

will have cumulants $\kappa_1 + \epsilon_1, \kappa_2 + \epsilon_2, \ldots$. It is necessary to explain the meaning of (37) rather carefully. The operator

$$\exp\left[\sum_{j=1}^{\infty} \epsilon_j\{(-D)^j/j!\}\right]$$

is to be understood in the sense described in Chapter 1. That is, the exponential must be formally expanded as

$$\sum_{i=0}^{\infty}\left[\sum_{j=1}^{\infty}\epsilon_j\{(-D)^j/j!\}\right]^i \Big/ i!$$

and then applied to $f(x)$. (As in Chapter 1 (Section 1.2) D is the differentiation operator, and $D^j f(x) = d^j f(x)/dx^j$.) It should be clearly understood that $g(x)$ may not satisfy the condition $g(x) \geq 0$ for all x. (The cumulants of $g(x)$ are defined as coefficients of $t^r/r!$ in the expansion of

$$\log\left[\int_{-\infty}^{\infty} e^{tx} g(x)\, dx\right]$$

whether $g(x) \geq 0$ or not.)

Despite this limitation it is possible to obtain useful approximate representation of a distribution with known moments (and so, known cumulants) in terms of a known distribution $f(x)$. By far the most commonly used initial family of distributions is the normal distribution. The representations arising from this choice of initial distribution are called *Gram-Charlier series*. From (37) we find (formally)

$$\begin{aligned}(38)\qquad g(x) = f(x) &- \epsilon_1 Df(x) + \tfrac{1}{2}(\epsilon_1^2 + \epsilon_2)D^2 f(x)\\ &- \tfrac{1}{6}(\epsilon_1^3 + 3\epsilon_1\epsilon_2 + \epsilon_3)D^3 f(x)\\ &+ \tfrac{1}{24}(\epsilon_1^4 + 6\epsilon_1^2\epsilon_2 + 4\epsilon_1\epsilon_3 + \epsilon_4)D^4 f(x) + \cdots.\end{aligned}$$

For the approximation to the cumulative distribution function, we have

$$(39)\qquad \int_{-\infty}^{x} g(t)\,dt = \int_{-\infty}^{x} f(t)\,dt - \epsilon_1 f(x) + \tfrac{1}{2}(\epsilon_1^2 + \epsilon_2)Df(x)\ldots \text{etc.}$$

In many cases (including the case when $f(x)$ is a normal probability density function)

$$D^j f(x) = P_j(x) f(x)$$

where $P_j(x)$ is a polynomial of degree j in x. Then (38) can be written in the form

$$\begin{aligned}(38)'\qquad g(x) = [1 &- \epsilon_1 P_1(x) + \tfrac{1}{2}(\epsilon_1^2 + \epsilon_2)P_2(x) - \tfrac{1}{6}(\epsilon_1^3 + 3\epsilon_1\epsilon_2 + \epsilon_3)P_3(x)\\ &+ \tfrac{1}{24}(\epsilon_1^4 + 6\epsilon_1^2\epsilon_2 + 4\epsilon_1\epsilon_3 + \epsilon_4)P_4(x) - \cdots]f(x)\end{aligned}$$

with a corresponding form for (39).

If the expected values and standard deviations of $f(x)$ and $g(x)$ have been made to agree then $\epsilon_1 = \epsilon_2 = 0$ and (38)′ becomes

$$(40)\qquad g(x) = [1 - \tfrac{1}{6}\epsilon_3 P_3(x) + \tfrac{1}{24}\epsilon_4 P_4(x) - \cdots]f(x)$$

and also

$$(41)\qquad \int_{-\infty}^{x} g(t)\,dt = \int_{-\infty}^{x} f(t)\,dt - [\tfrac{1}{6}\epsilon_3 P_2(x) - \tfrac{1}{24}\epsilon_4 P_3(x) + \cdots]f(x)$$

assuming $P_j(x)f(x) \to 0$ at the extremes of the range of variation of x.
A common way of ensuring this agreement in regard to the expected value and standard deviation is to use standardized variables, and to choose $f(x)$ so that the corresponding distribution is standardized. If desired, the actual expected value and standard deviation can be restored by an appropriate linear transformation.

Supposing that we are using a standardized variable, and taking $f(x) = (\sqrt{2\pi})^{-1}e^{-\frac{1}{2}x^2}$ (normal) then $(-1)^jP_j(x)$ is the Hermite polynomial $H_j(x)$ described in Chapter 1 and since $\kappa_r = 0$ when r greater than 2 for the normal distribution, $\epsilon_3, \epsilon_4, \ldots$ are equal to the corresponding cumulants of the distribution we desire to approximate. Further, since this function is standardized we have

$$\epsilon_3 = \alpha_3 = \sqrt{\beta_1}\,;\ \epsilon_4 = \alpha_4 - 3 = \beta_2 - 3,$$

where the shape factors refer to this distribution. Thus we have

$$g(x) = [1 + \tfrac{1}{6}\sqrt{\beta_1}\,H_3(x) + \tfrac{1}{24}(\beta_2 - 3)H_4(x) + \cdots](\sqrt{2\pi})^{-1}e^{-\frac{1}{2}x^2} \tag{42}$$

and integrating both sides of (42),

$$\begin{aligned}\int_{-\infty}^{x} g(t)\,dt &= \Phi(x) - [-\tfrac{1}{6}\sqrt{\beta_1}\,H_2(x) + \tfrac{1}{24}(\beta_2 - 3)H_3(x) + \cdots]Z(x) \\ &= \Phi(x) - \tfrac{1}{6}\sqrt{\beta_1}\,(x^2 - 1)Z(x) \\ &\quad - \tfrac{1}{24}(\beta_2 - 3)(x^3 - 3x)Z(x) + \cdots,\end{aligned} \tag{43}$$

where

$$\Phi(x) = (\sqrt{2\pi})^{-1}\int_{-\infty}^{x} e^{-\frac{1}{2}t^2}\,dt;\ Z(x) = (\sqrt{2\pi})^{-1}e^{-\frac{1}{2}x^2}.$$

Equations (42) and (43) are known as *Gram-Charlier expansions* [18] (some earlier writers refer to them as Bruns-Charlier expansions [11]). In these expansions the terms occur in sequence determined by the successive derivatives of $Z(x)$. This is not necessarily in decreasing order of importance, and a different ordering is sometimes used. This is based on the fact that for a sum of n independent, identically distributed standardized random variables, the rth cumulant is proportional to $n^{1-r/2}(r \geq 2)$. This means that, in our notation, $\epsilon_r \propto n^{1-r/2}$. Collecting terms of equal order in $n^{-1/2}$, and arranging in ascending order, gives an *Edgeworth expansion* [29] [30] the leading terms of which are

$$\begin{aligned}g(x) = [1 &+ \tfrac{1}{6}\sqrt{\beta_1}\,H_3(x) + \tfrac{1}{24}(\beta_2 - 3)H_4(x) + \tfrac{1}{72}\beta_1 H_6(x) + \cdots] \\ &\times \frac{e^{-\frac{1}{2}x^2}}{\sqrt{2\pi}}\end{aligned} \tag{44}$$

from which we obtain

$$\begin{aligned}\int_{-\infty}^{x} g(t)\,dt &= \Phi(x) - \tfrac{1}{6}\sqrt{\beta_1}\,(x^2 - 1)Z(x) - \tfrac{1}{24}(\beta_2 - 3)(x^3 - 3x)Z(x) \\ &\quad - \tfrac{1}{72}\beta_1(x^5 - 10x^3 + 15x)Z(x) + \cdots.\end{aligned} \tag{44'}$$

FIGURE 2

β_1, β_2 Plane Showing Regions of Unimodal Curves and Regions of Curves Composed Entirely of Non-negative Ordinates

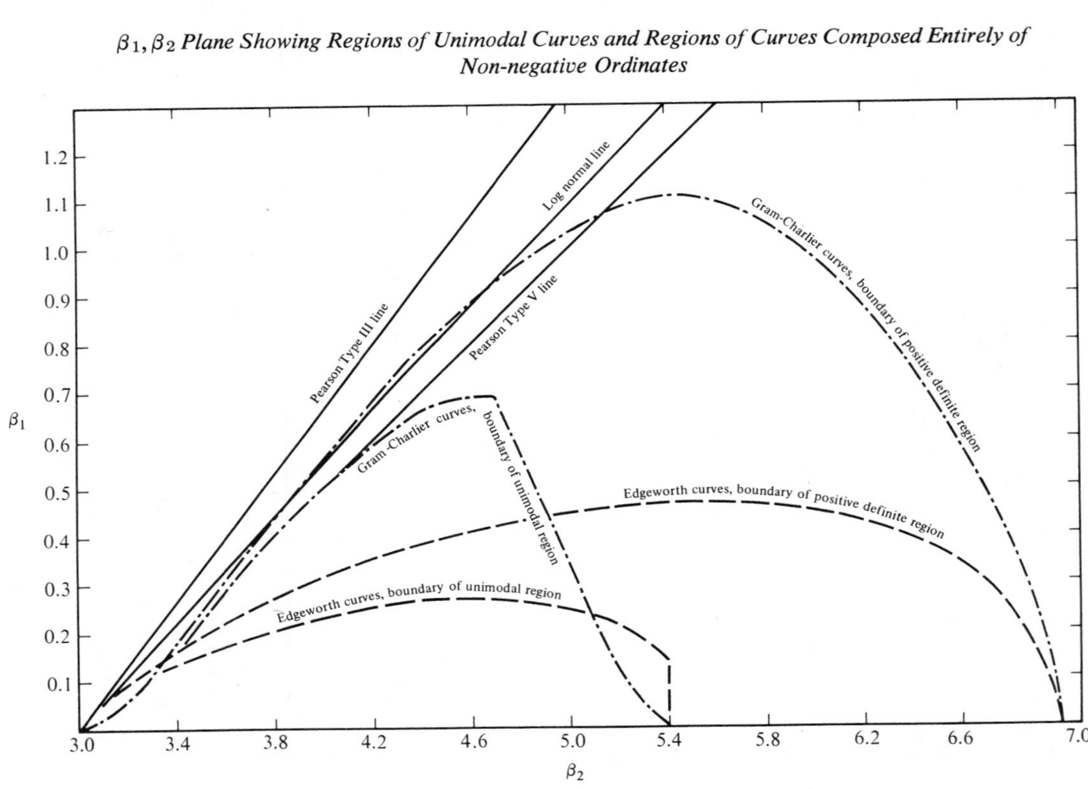

We noted, at the beginning of this subsection, that the mathematical expression obtained by applying a cumulant modifying function, in general, will not represent a proper probability density function, because there are many intervals throughout which it is negative.

This is also true when only a finite number of terms of the expansion is used. Figure 2 presents the results of an investigation by Barton and Dennis [1] and shows the regions in the (β_1,β_2) plane where the expressions (42) and (44) are never negative.

Figure 2 also shows the region where the curves corresponding to (42) and (44) are unimodal. Multimodality in expansions like (42) or (44), fitted to empirical data, often indicates an unnecessary fidelity to more or less accidental features of the data, in the form of 'humps' in the tails. This kind of phenomenon is more likely to be encountered, the more terms used in the expansion.

In most applications only the first four moments are used, and the following terminating expressions are used:

$$g(x) = [1 - \tfrac{1}{6}\sqrt{\beta_1}\,(x^3 - 3x) + \tfrac{1}{24}(\beta_2 - 3)(x^4 - 6x^2 + 3)]Z(x) \tag{45}$$

(Gram-Charlier)

or

$$\begin{aligned} g(x) = [1 &- \tfrac{1}{6}\sqrt{\beta_1}\,(x^3 - 3x) + \tfrac{1}{24}(\beta_2 - 3)(x^4 - 6x^2 + 3) \\ &+ \tfrac{1}{72}\beta_1(x^5 - 10x^3 + 15x)]Z(x). \end{aligned} \tag{46}$$

(Edgeworth)

(Note that the Edgeworth has no *general* theoretical superiority over the Gram-Charlier expansion — it depends on a particular assumption about the orders of magnitude of successive cumulants which may, or may not, be a good approximation to actual conditions.)

Although the expansions (45) and (46) terminate, and so the general theory at the beginning of this paragraph does not apply, it can be seen, from the orthogonality (with normal weight function) of Hermite polynomials that they are functions with the correct values for the first four moments, and also satisfy the condition $\int_{-\infty}^{\infty} g(x)\,dx = 1$.

Since

$$\frac{1}{\sqrt{2\pi}}\int_{-\infty}^{\infty} |x|x^j e^{-\frac{1}{2}x^2}\,dx = 0 \text{ for } j \text{ odd}$$

and for j even

$$\begin{aligned} \frac{1}{\sqrt{2\pi}}\int_{-\infty}^{\infty} |x|x^j e^{-\frac{1}{2}x^2}\,dx &= \sqrt{\frac{2}{\pi}}\int_{\infty}^{\infty} x^{j+1}e^{-\frac{1}{2}x^2}\,dx \\ &= \sqrt{\frac{2}{\pi}}\,2^{\frac{1}{2}(j+1)-\frac{1}{2}}\int_0^{\infty} t^{\frac{1}{2}j}e^{-t}\,dt, \end{aligned}$$

i.e.

$$\frac{1}{\sqrt{2\pi}}\int_{\infty}^{\infty} |x|x^j e^{-\frac{1}{2}x^2}\,dx = \sqrt{\frac{2}{\pi}}\,2^{\frac{1}{2}j}(\tfrac{1}{2}j)! \qquad (j \text{ even})$$

it follows that for the Gram-Charlier (finite term) distribution as given by (45) the mean deviation is

$$\sqrt{\frac{2}{\pi}}\left[1 - \tfrac{1}{24}(\beta_2 - 3)\right] = \sqrt{\frac{2}{\pi}}\,\frac{27 - \beta_2}{24}.$$

This is also the ratio of mean deviation to standard deviation for this Gram-Charlier expansion, with general values for the expected value and variance. Note that for $\beta_2 > 27$ the mean deviation is negative. This is because the probability density function is negative for some values of x.

However, for $1 < \beta_2 < 7$; $\frac{5}{6}\sqrt{\frac{2}{\pi}} < \frac{\text{m.d.}}{\text{s.d.}} < \frac{13}{12}\sqrt{\frac{2}{\pi}}$.

Similar results can be obtained for the Edgeworth expansion. For the latter (46), Bhattarcharjee [4] and Singh [61] have obtained the distributions of extreme values and ranges, and have given numerical values for expected values and variances in random samples of size up to 12. Subrahmaniam [63] has obtained the distributions of linear functions of independent sample values, and of sample variance for random samples from this distribution.

It is possible to derive expansions of Gram-Charlier form by arguments similar to those used in deriving certain central limit theorems (Chapter 13). Cramér [24] gives a general discussion; Longuet-Higgins [52] gives an analysis from a physicist's point of view.

Some theoretical results of Bol'shev [8] are relevant here. Starting from the normalizing transformation

$$y(x) = \Phi^{-1}[\Pr[S \leq x]] \tag{47}$$

one can expand the argument $\Pr[S \leq x]$ about $\Phi(x)$, as a power series in $(\Pr[S \leq x] - \Phi(x))$. If this difference, in turn, be expanded as a power series in x then (47) gives a power series (in x) expression for $y(x)$. (See also Equation (13) *et seq.*, of this chapter.)

In the particular case when $S \equiv S_n$ is the standardized sum of n independent identically distributed variables $X_1, X_2, \ldots, X_n$ with finite expected value ξ, and standard deviation σ respectively (and finite cumulants of all orders) (i.e., $S_n = (n\sigma^2)^{-\frac{1}{2}} \sum_{j=1}^{n} (X_j - \xi)$) there is an expansion

$$\Pr[S_n \leq x] - \Phi(x) = Q(x) \tag{48}$$

where $Q(x)$ is a polynomial with coefficient depending on the common moment ratios of each $X_i S$. Inserting (47) in (48), we obtain

$$y(x) = x + \sum P_{3j-1}(x) n^{-j/2} + O(n^{-(r-2)/2}) \tag{49}$$

with

$$P_2(x) = \tfrac{1}{6}\sqrt{\beta_1}\,(x^2 - 1)$$

$$P_5(x) = \tfrac{1}{36}\beta_1(4x^3 - 7x) - \tfrac{1}{24}(\beta_2 - 3)(x^3 - 3x).$$

Bol'shev shows that of all functions $u(x,n)$ satisfying the conditions

(i) $\partial^{r-2}u/\partial n^{r-2}$ exists and is continuous with respect to n on the line $n = 0$
(ii) $\partial u/\partial x$ exists in a domain

$$|x| < Cn^{-(r-2)}(r - 1) \qquad (C > 0, r \geq 3)$$

the *only* one for which

$$\Pr[u(x,n) \leq u_0] = \Phi(u_0) + O(n^{-(r-2)/2})$$

is the function given by (49).

Bol'shev has applied this result in a number of special cases.

It will be appreciated that it is not essential that $f(x)$ in (38)′ must be a normal probability density function. In particular, if $f(x)$ is a standard gamma probability density function, expansions in terms of Laguerre polynomials are obtained. Such expansions have been discussed by Khamis [46] and applied to approximate the distribution of non-central F (Chapter 30) by Tiku [67].

If $f(x)$ is a standard beta distribution, expansions in terms of Jacobi polynomials are obtained, but these have not been much used.

Woods and Posten [70] have made a systematic study of the use of Fourier series expansions in calculating cumulative distribution functions. Their methods are based on the following theorem:

If X is a random variable with $F_X(x) = 0, 1$ for $x < 0, x > 1$ respectively then, for $0 \leq x \leq 1$,

$$F_X(x) = 1 - \theta\pi^{-1} - \sum_{j=1}^{\infty} b_j \sin j\theta$$

where

$$\theta = \cos^{-1}(2x - 1)$$

and

$$b_j = 2(j\pi)^{-1}E[\cos(j\cos^{-1}(2X - 1))].$$

(The function of which the expected value is to be taken is the j-th Chebyshev polynomial, $T_j(X)$. See Chapter 1, Section 3.)

Woods and Posten also use a generalized form of this theorem which expresses $F_X(x)$ in terms, of any conveniently chosen "distribution function" $G(x)$, with $G(x) = 0, 1$ for $x < 0, x > 1$ respectively. For $0 \leq x \leq 1$,

$$F_X(x) = G(x) - \sum_{j=1}^{\infty} d_j \sin j\theta$$

with

$$d_j = b_j - 2(j\pi)^{-1}\int_0^1 T_j(x)\,dG(x).$$

Appropriate choice of $G(x)$ — usually close to $F_X(x)$ — can increase the rate of convergence of the infinite series, though the d's are not so easily computed as the b's.

There are similar results for the case when $F_X(x) = 0, 1$ for $x < -1, x > 1$.

Computer programs based on these theorems are given in [70] for evaluating the cumulative distribution and percentage points of the beta F and chi-square distributions (Chapters 24, 26 and 17, respectively) and of the non-central forms of these distributions (Chapters 30 and 29). In Chapter 24 (Section 6) some further details are given in regard to the application of these series to the beta distributions.

4.3 *Transformed Distributions*

If the distribution of a random variable X is such that a simple explicit function $f(X)$ has a well-known distribution, it becomes possible to use the results of research on the latter — including, in particular, published tables — in studying the former distribution.

The best known of such distributions is the *lognormal* distribution (Chapter 14) where $\log (X - \xi)$ has a normal distribution. Other well known families of distributions correspond to cases in which $(X - \xi)^c$ or $e^{-(X-\xi)}$ have exponential distributions (Type II (or Weibull) and Type III extreme value distributions, respectively (see Chapters 19 and 20)).

Edgeworth [32] [33] considered the possibility of polynomial transformations to normality. To make sure the transformation is monotonic, it is necessary to impose restrictions on the coefficients in the polynomial. This analysis is rather complicated and this kind of transformation is not often used at present.

Plotting on probability paper* will indicate the form of the transformation. One of the earliest papers on this method was published in 1928 by Kameda [44], one of the most recent is Flapper [38]. Sets of 'model plots' of quantiles of various distributions against those of the unit normal distribution contained in Chambers and Fowlkes [16] can be helpful in deciding on suitable transformations.

By analogy with the Pearson system of distributions, it would be convenient if a simple transformation to a normally distributed variable could be found such that, for any possible pair of values $\sqrt{\beta_1}$, β_2 there is just one member of the corresponding family of distributions. No such single simple transformation is available, but Johnson [40] has described a set of three such transformations which, when combined, do provide one distribution corresponding to each pair of values $\sqrt{\beta_1}$ and β_2.

One of the three transformations is, simply

$$Z = \gamma + \delta \log (X - \xi) \qquad (X \geq \xi) \tag{50.1}$$

— corresponding to the family of lognormal distributions.

*Probability paper is graph paper designed so that a plot of the cumulative frequency against the variable value would give a linear relation for a specified distribution — often the normal distribution.

The others are

$$Z = \gamma + \delta \log \{(X - \xi)/(\xi + \lambda - X)\} \qquad (\xi \leq X \leq \xi + \lambda) \tag{50.2}$$

$$Z = \gamma + \delta \sinh^{-1} \{(X - \xi)/\lambda\}. \tag{50.3}$$

The distribution of Z is, in each case, unit normal. The symbols γ, δ, ξ and λ represent parameters. The value of λ must be positive, and we conventionally make the sign of δ positive also.

The range of variation of X in (50.2) is bounded and the corresponding family of distributions is denoted by S_B; in (50.3) the range is unbounded, and the symbol S_U is used. For lognormal distributions, the range is bounded below (if $\delta < 0$, it would be bounded above).

It is clear that the shapes of the distribution of X depends only on the parameters γ and δ (δ only, for lognormal). For, writing $Y = (X - \xi)/\lambda$ we have

$$Z = \gamma' + \delta \log Y \qquad \text{for lognormal } (\gamma' = \gamma - \delta \log \lambda) \tag{51.1}$$

$$Z = \gamma + \delta \log \{Y/(1 - Y)\} \qquad \text{for } S_B \tag{51.2}$$

$$Z = \gamma + \delta \sinh^{-1} Y \qquad \text{for } S_U \tag{51.3}$$

and Y must have a distribution of the same shape as X. The moments of Y in (51.1) are given in Chapter 14.

For S_B, from (51.2)

$$\mu_r'(Y) = (\sqrt{2\pi})^{-1} \int_{-\infty}^{\infty} [1 + e^{-(z-\gamma)/\delta}]^{-r} e^{-\frac{1}{2}z^2}\, dz.$$

Although it is possible to give explicit expressions for $\mu_r'(Y)$ (not involving integral signs) (see Johnson [40], Equations (56) and (57)), they are very complicated.

It is interesting to note that, for S_B distributions,

$$\frac{\partial \mu_r'}{\partial \gamma} = \frac{r}{\delta} (\mu_{r+1}' - \mu_r') \tag{52.1}$$

$$\frac{\partial \mu_r'}{\partial \delta} = \frac{r}{\delta^3} (\gamma\delta - r)(\mu_r' - \mu_{r+1}') + \frac{r(r+1)}{\delta^3} (\mu_{r+1}' - \mu_{r+2}'). \tag{52.2}$$

For S_U, however, we obtain from (51.3),

$$\mu_r'(Y) = (\sqrt{2\pi})^{-1} 2^{-r} \int_{-\infty}^{\infty} [e^{(z-\gamma)/\delta} - e^{-(z-\gamma)/\delta}]^r e^{-\frac{1}{2}z^2}\, dz \tag{53}$$

and this can be evaluated in a straightforward manner, yielding the following values for the expected value, and lower central moments of Y:

$$\begin{aligned} \mu_1'(Y) &= \omega^{\frac{1}{2}} \sinh \Omega \\ \mu_2(Y) &= \tfrac{1}{2}(\omega - 1)(\omega \cosh 2\Omega + 1) \end{aligned} \tag{54}$$

$$\mu_3(Y) = -\tfrac{1}{4}\omega^{\frac{1}{2}}(\omega - 1)^2\{\omega(\omega + 2)\sinh 3\Omega + 3\sinh\Omega\}$$
$$\mu_4(Y) = \tfrac{1}{8}(\omega - 1)^2\{\omega^2(\omega^4 + 2\omega^3 + 3\omega^2 - 3)\cosh 4\Omega + 4\omega^2(\omega + 2) \times \cosh 2\Omega + 3(2\omega + 1)\}$$

where $\omega = \exp(\delta^{-2})$, $\Omega = \gamma/\delta$.

(Note that $\omega > 1$, and that μ_3, and so $\alpha_3(= \sqrt{\beta_1})$ has the *opposite* sign to γ.)

For $\gamma = 0$, the shape factors (for both X and Y) are

$$\alpha_3^2 = \beta_1 = 0;\ \alpha_4 = \beta_2 = \tfrac{1}{2}(\omega^4 + 2\omega^2 + 3).$$

As γ increases both β_1 and β_2 increase, and the (β_1,β_2) point approaches the point with coordinates $((\omega - 1)(\omega + 2)^2,\ \omega^4 + 2\omega^3 + 3\omega^2 - 3)$ as $\gamma \to \infty$. The latter point is on the 'lognormal line' (see Chapter 14). It corresponds to a lognormal distribution defined by (50.1).

The variation of (β_1,β_2) with γ and δ is shown diagrammatically in Figure 3. This Figure can be used to obtain approximate values of γ and δ, for given β_1 and β_2. (Note that the *sign* of γ must be *opposite* to that of $\sqrt{\beta_1}$ — see above.) More accurate values can be obtained using tables in Johnson [41], possibly combined with methods of iterative calculation described in this reference.

It can be seen from Figure 3 (and proved analytically) that for any (β_1,β_2) point "below" the lognormal line, there is an appropriate S_U distribution. Similarly, for any possible point "above" the lognormal line there is an appropriate S_B distribution. In fact the lognormal, S_U and S_B families (or systems)

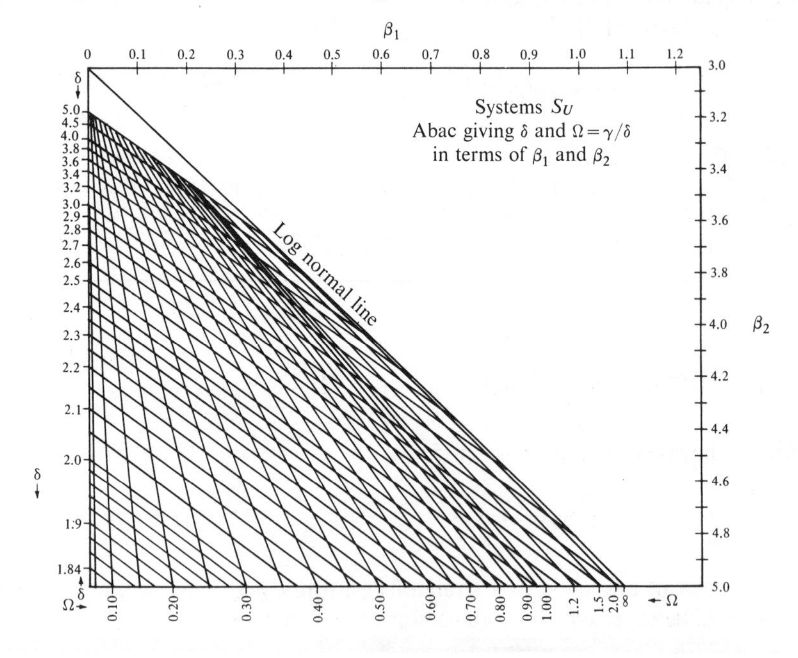

FIGURE 3

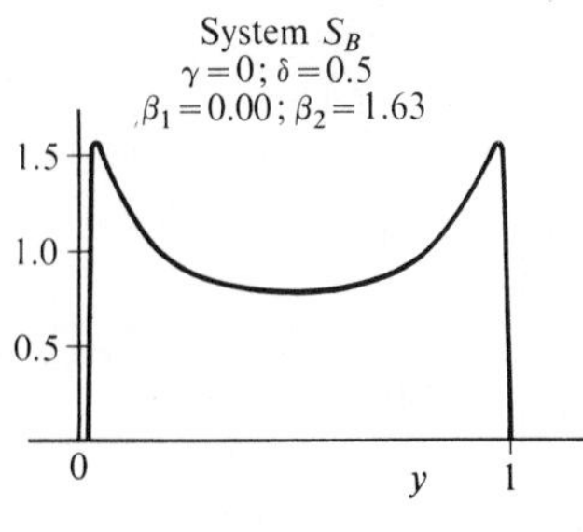

FIGURE 4a

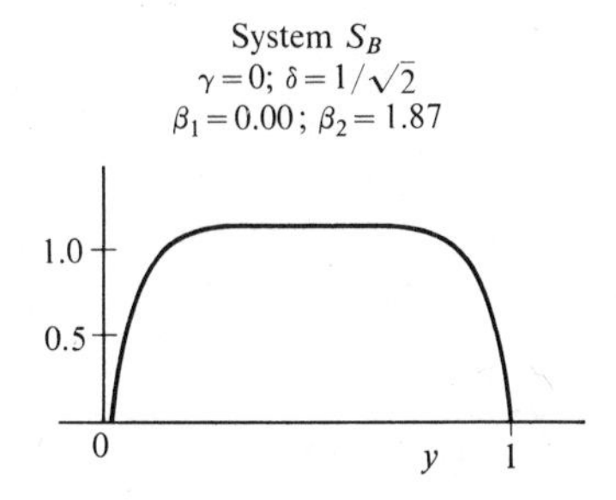

FIGURE 4b

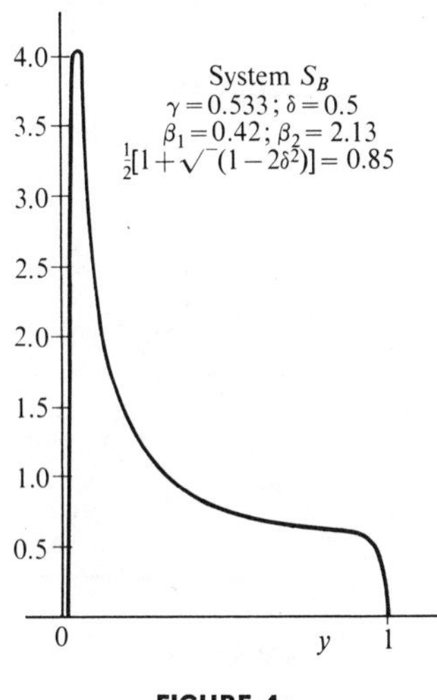

FIGURE 4c

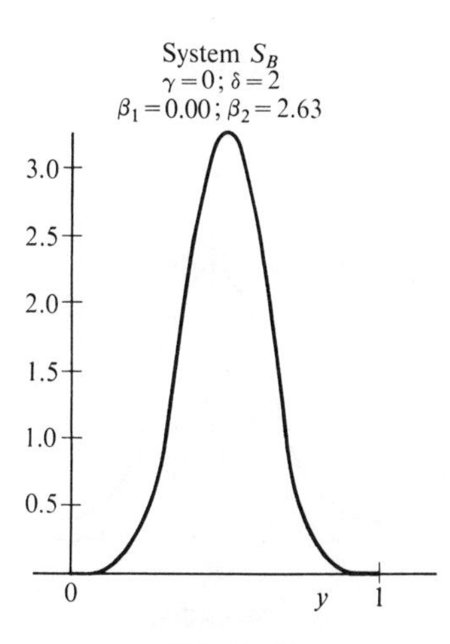

FIGURE 4d

cover the whole possible (β_1,β_2) plane uniquely — i.e. there is just one appropriate distribution corresponding to each (β_1,β_2) point. The normal distribution, corresponding to the point (0,3), has to be considered as the limiting form of all three families as $\delta \to \infty$.

Some typical probability density functions belonging to the S_B and S_U families are shown in Figures 4a-d(S_B) and 5a-e(S_U) (taken from Johnson [40]). All S_U curves are unimodal; S_B curves may be unimodal, or they may have two modes, with an antimode between them. The latter case occurs if

$$\delta < 1/\sqrt{2} \text{ and } |\gamma| < \delta^{-1}\sqrt{1-2\delta^2} - 2\delta \tanh^{-1}\sqrt{1-2\delta^2}. \tag{55}$$

There are transition cases in which one mode and the antimode coalesce into a point of inflection (see Figures 4a, c). If $\gamma = 0$ and $\delta = 1/\sqrt{2}$ (Figure 4a) a flat-topped distribution is obtained.

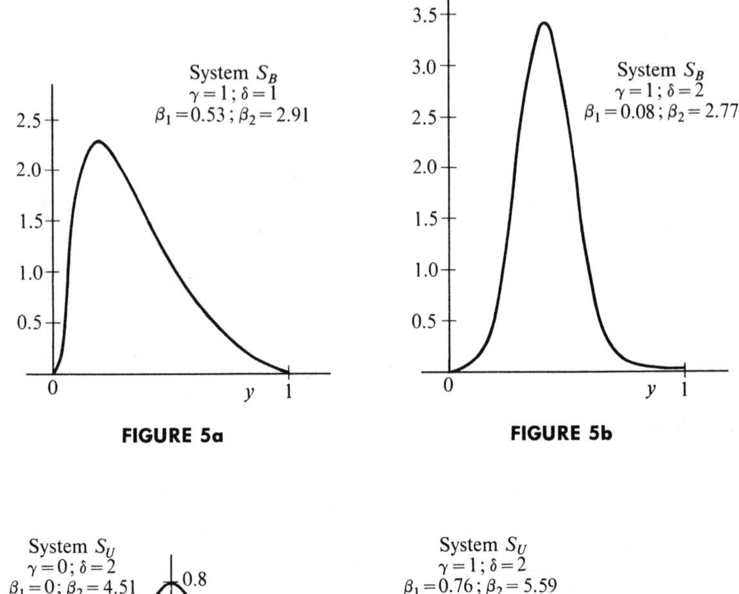

FIGURE 5a

FIGURE 5b

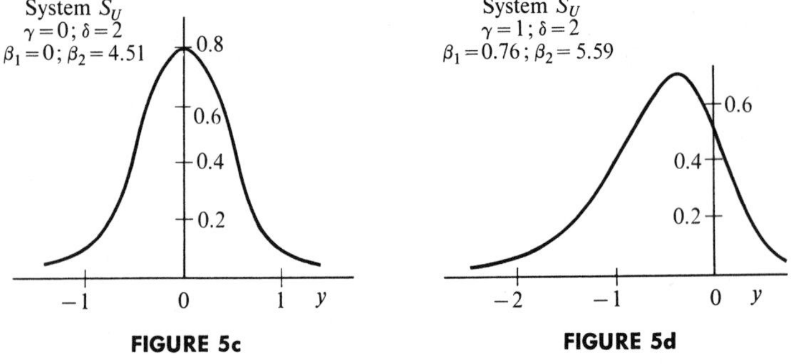

FIGURE 5c

FIGURE 5d

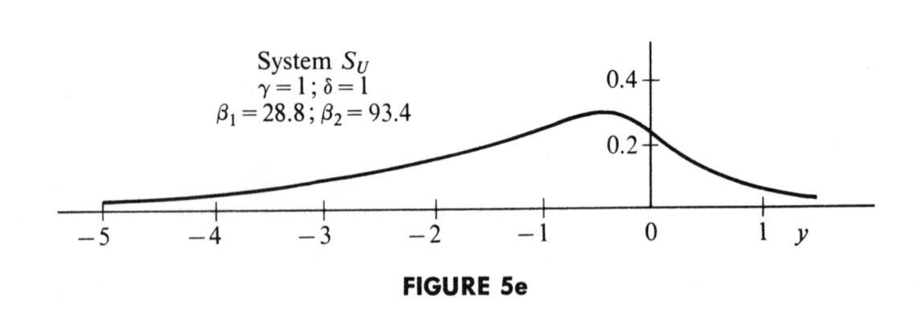

FIGURE 5e

For all S_B and S_U (and also lognormal) distributions there is 'high contact' at the extremities of the range of variation. That is to say, not only does the probability density function tend to zero as the extremity is approached, but so do all derivatives. (This applies as $Y \to \pm\infty$, as well as when the extremities are finite.) This property is not shared by all Pearson system distributions.

TABLE 1 *Comparison of Type IV (or I) and S_U Distributions*

		Cumulative Distribution Function								
$\sqrt{\beta_1}$	β_2	0.01	0.05	0.1	0.25	0.5	0.75	0.9	0.95	0.99
0	4	−2.472	−1.621	−1.227	−0.626	0.000	0.626	1.227	1.621	2.472
		−2.486	−1.623	−1.225	−0.622	0.000	0.622	1.225	1.623	2.486
0	6	−2.566	−1.587	−1.176	−0.586	0.000	0.586	1.176	1.587	2.566
		−2.626	−1.585	−1.158	−0.566	0.000	0.566	1.158	1.585	2.626
0	8	−2.598	−1.567	−1.151	−0.568	0.000	0.568	1.151	1.567	2.598
		−2.693	−1.556	−1.116	−0.533	0.000	0.533	1.116	1.556	2.693
0.5	4	−2.100	−1.506	−1.197	−0.675	−0.068	0.597	1.278	1.737	2.741
		−2.109	−1.502	−1.191	−0.672	−0.071	0.594	1.278	1.742	2.753
0.5	6	−2.300	−1.508	−1.155	−0.619	−0.049	0.561	1.206	1.668	2.789
		−2.349	−1.498	−1.133	−0.602	−0.052	0.541	1.194	1.674	2.851
0.5	8	−2.373	−1.502	−1.133	−0.596	−0.041	0.546	1.174	1.635	2.796
		−2.460	−1.484	−1.095	−0.563	−0.044	0.511	1.144	1.630	2.892
1	6	−1.889	−1.389	−1.126	−0.672	−0.118	0.535	1.259	1.782	3.030
		−1.931	−1.380	−1.109	−0.660	−0.121	0.522	1.252	1.788	3.065
1	8	−2.064	−1.414	−1.112	−0.634	−0.093	0.522	1.209	1.720	3.009
		−2.147	−1.397	−1.077	−0.606	−0.097	0.493	1.186	1.720	3.081
1.5	8	−1.552	−1.245	−1.056	−0.690	−0.181	0.481	1.264	1.850	3.278
		−1.614	−1.244	−1.043	−0.675	−0.179	0.470	1.255	1.848	3.304
1.5	10	−1.754	−1.300	−1.064	−0.657	−0.147	0.481	1.221	1.785	3.225
		−1.874	−1.291	−1.030	−0.624	−0.145	0.454	1.193	1.776	3.288
2	12	−1.356	−1.134	−0.986	−0.679	−0.218	0.425	1.231	1.860	3.467
		−1.473	−1.147	−0.974	−0.655	−0.209	0.412	1.212	1.848	3.493
2	14	−1.508	−1.188	−1.005	−0.661	−0.190	0.433	1.207	1.815	3.408
		−1.694	−1.195	−0.973	−0.618	−0.179	0.407	1.169	1.789	3.456

Note: Upper figures are all Type IV except the case $\sqrt{\beta_1} = 2$, $\beta_2 = 12$ which is Type I. Lower figures correspond to S_U.

Numerical comparisons of the cumulative distribution functions of Pearson Type IV and S_U distributions corresponding to the same pair of values of β_1 and β_2 indicate remarkably close agreement. Distributions of either system may often be used in place of their counterpart in the other system with low risk of errors of practical importance being incurred. Agreement is poorest in the lower tails for large β_1 and/or β_2.

Table 1, above, compares standardized deviates of (a) Type IV (or I) and (b) S_U distributions for various values of β_1 and β_2, and for various values of the cumulative distribution function.

4.4 *Bessel Function Distributions*

McKay [56] described a system of distributions which would provide (at least) one distribution corresponding to any pair of values (β_1,β_2) for which $(\beta_2 - 3)/\beta_1 > 1.5$ (i.e., 'below' the Type III line). In fact for the narrow strip

$$1.5 < (\beta_2 - 3)/\beta_1 \leq 1.57735$$

there are *three* possible distributions of this system corresponding to a single (β_1,β_2) point. Although most of the formulas presented below were derived by McKay, we will first mention a possible genesis of the system constructed by Bhattacharyya [5].

The distributions can, in fact, be obtained as distributions of $X_1\sigma_1^2 \pm X_2\sigma_2^2$ where X_1, X_2 are mutually independent random variables, each distributed as χ^2 with ν degrees of freedom (see Chapter 17).

The distribution of $Y = X_1\sigma_1^2 + X_2\sigma_2^2$ is the first of McKay's forms. It has probability density function

$$p(y) = \frac{|1 - c^2|^{m+\frac{1}{2}}|y|^m}{\pi^{\frac{1}{2}}2^m b^{m+1}\Gamma(m + \frac{1}{2})} e^{-cy/b} I_m(|y/b|) \qquad (y > 0) \tag{56}$$

with

$$b = 4\sigma_1^2\sigma_2^2(\sigma_1^2 - \sigma_2^2)^{-1}$$
$$c = (\sigma_1^2 + \sigma_2^2)(\sigma_1^2 - \sigma_2^2)^{-1} > 1$$

and

$$m = 2\nu + 1.$$

The distribution of $Z = X_1\sigma_1^2 - X_2\sigma_2^2$ is the second of McKay's forms. It has probability density function.

$$p_Z(z) = \frac{|1 - c^2|^{m+\frac{1}{2}}|z|^m}{\pi^{\frac{1}{2}}2^m b^{m+1}\Gamma(m + \frac{1}{2})} e^{-cz/b} K_m(|z/b|) \tag{57}$$

with

$$b = 4\sigma_1^2\sigma_2^2(\sigma_1^2 + \sigma_2^2)^{-1}$$
$$c = -(\sigma_1^2 - \sigma_2^2)(\sigma_1^2 + \sigma_2^2)^{-1} \qquad (|c| < 1)$$

and

$$m = 2\nu + 1.$$

In (56) and (57), $I_m(\cdot)$ and $K_m(\cdot)$ are modified Bessel functions (of the second kind) of order m. (See Section 3, Chapter 1.)

For *both* kinds of distribution (56) and (57) the moment generating function is

$$[(1 - c^2)\{1 - (c - tb)^2\}^{-1}]^{m+\frac{1}{2}}$$

(with, of course, appropriate values of b and c, depending on which kind of distribution is being considered).

It follows that the rth cumulant is

$$\kappa_r = (r-1)!(m+\tfrac{1}{2})b^r(c^2-1)^{-r}[(c-1)^r+(c+1)^r]. \tag{58}$$

(This can be established directly from Bhattacharyya's approach.)

In particular

$$\begin{cases} \mu_1' = (2m+1)bc(c^2-1)^{-1} \\ \mu_2 = (2m+1)b^2(c^2+1)(c^2-1)^{-2} \\ \beta_1 = 4c^2(c^2+3)^2(2m+1)^{-1}(c^2+1)^{-3} \\ \beta_2 = 3+6(c^4+6c^2+1)(2m+1)^{-1}(c^2+1)^{-2} \end{cases} \tag{59}$$

From the last two equations of (59) it can be shown that

$$2c^2(c^2+3)^2\left(\frac{\beta_2-3}{\beta_1}\right) - 3(c^2+1)(c^4+6c^2+1) = 0. \tag{60}$$

Regarded as a cubic equation in c^2, (60) has a single positive root for

$$(\beta_2-3)/\beta_1 > 1.57735,$$

and three positive roots for $(\beta_2-3)/\beta_1 > 1.5$. (The three positive roots are given by McKay [56] to 5 significant figures for $(\beta_2-3)/\beta_1 = 1.502(0.002)1.576.$) For the region between the line

$$\beta_2 - 1.57735\beta_1 - 3 = 0 \tag{61}$$

(termed the *Bessel line* by McKay) and the axis of β_2 there is a unique 'Bessel distribution' corresponding to any given (β_1,β_2) point. Figure 6 on page 30 (taken from Bhattacharyya [5]) shows how the values of m and c^2 vary over this region — called the K-region by McKay (because only distributions of form (57) can be used here). This diagram also shows how narrow is the K and I *region*, where there are two distributions of form (56) and one of form (57) for each pair of values (β_1,β_2).

It would seem that the K-form (57) is likely to be more generally useful than the I-form (56). Indeed it has been suggested that (57) would lead to less troublesome computation than the Pearson Type IV distributions which have the same (β_1,β_2) values. However, the I-form (56) has been used (Bose [10]) for graduating an observed frequency distribution. Also the family of I-distributions includes the noncentral χ^2 distributions (see Laha [48] and Chapter 28) and the distribution of the Mahalanobis D^2-statistic.

A generalization of the K-form, with probability density function proportional to $|z|^{m'}K_m(|z/b|)$, $(m' \neq m)$, has been studied by Sastry [60].

McNolty [57] has described the application of Bessel function I-distributions to the distribution of signal or noise in output processed by a radar receiver under various sets of conditions. This paper contains interesting accounts of how the Bessel distribution might be expected to arise in each case.

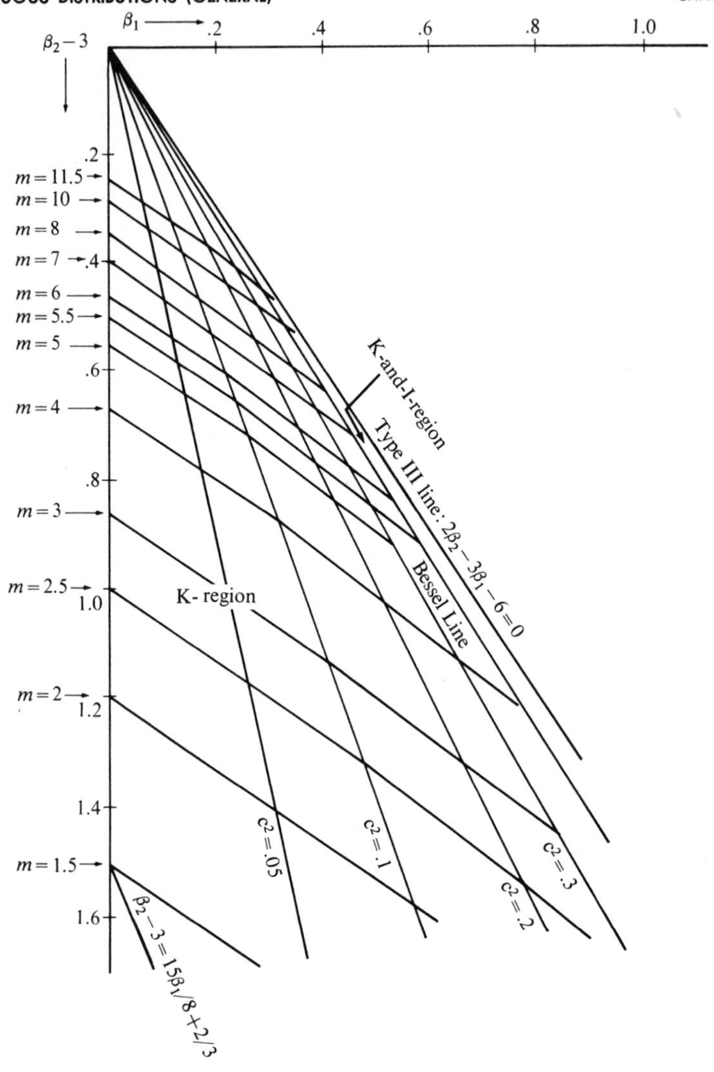

FIGURE 6

Bessel Distribution Regions in (β_1, β_2) *Plane*

4.5 *Miscellaneous*

Burr [12] has suggested a number of forms of cumulative distribution function which might be useful for purposes of graduation. The principal aim, in choosing one of these forms of distribution, is to facilitate the mathematical analysis to which it is to be subjected, while attaining a reasonable approximation.

The forms for the cumulative distribution function listed by Burr are shown below. (The first corresponds to a uniform distribution; it is included because it is in the original list.)

(I)	y	$(0 < y < 1)$
(II)	$(e^{-y} + 1)^{-k}$	
(III)	$(y^{-c} + 1)^{-k}$	$(0 < y)$
(IV)	$\left[\left(\frac{c-y}{y}\right)^{1/c} + 1\right]^{-k}$	$(0 < y < c)$
(V)	$(ce^{-\tan y} + 1)^{-k}$	$(-\pi/2 < y < \pi/2)$
(VI)	$(ce^{-k \sinh y} + 1)^{-k}$	
(VII)	$2^{-k}(1 + \tanh y)^k$	
(VIII)	$\left(\frac{2}{\pi}\tan^{-1} e^y\right)^k$	
(IX)	$1 - \frac{2}{c[(1 + e^y)^k - 1] + 2}$	
(X)	$(1 - e^{-y^2})^k$	$(0 < y)$
(XI)	$\left(y - \frac{1}{2\pi}\sin 2\pi y\right)^k$	$(0 < y < 1)$
(XII)	$1 - (1 + y^c)^{-k}$	$(0 < y)$

k and c are positive parameters. Putting $y = (x - \xi)/\lambda$, we can introduce two extra parameters.

Burr [12] devoted special attention to the family (XII) of distributions. The probability density function of $Z = Y^c$ is

$$k(1 + z)^{-(k+1)} \qquad (z > 0)$$

and the rth moment of X is

$$E[Z^{r/c}] = kB\left(\frac{r}{c} + 1, k - \frac{r}{c}\right) \qquad (r < ck).$$

In [12] there are given, for $c = 1(1)10$ and $k = 1(1)11$, tables of mean and standard deviation to 5 decimal places, and of $\sqrt{\beta_1}$ and β_2 to 3 decimal places or 4 significant figures. In later papers [13] [14] there is some further discussion of properties of this family, with special reference to the distributions of sample median and range.

It should be noticed that it is possible to relate some of the above forms of distribution by simple transformations. Thus (III) can be obtained from (II) by replacing y by $c \log y$.

Ferreri [35] has described a system of distributions with probability density functions of form

$$(62) \qquad p_X(x) = \sqrt{b}\,[G(-\tfrac{1}{2},a)]^{-1}[\exp\{a + b(x - \xi)^2\} + c]^{-1} \qquad (c = \pm 1)$$

where $$G(p,a) = \Gamma(p+1) \sum_{j=1}^{\infty} (-c)^{j-1} e^{-ja} j^{-(p+1)}.$$

This distribution depends on four parameters a, b, c and ξ. The rth absolute moment about ξ (the mean) is

$$b^{-\frac{1}{2}r} G(\tfrac{1}{2}(r-1), a)/G(-\tfrac{1}{2},a). \tag{63}$$

For $c = 1$, $\beta_2 < 3$; for $c = -1$, $\beta_2 > 3$. As a increases the distribution approaches normality.

Among other systems of distributions we note those described by Toranzos [69] and Laha [48]. Toranzos describes a class of bell-shaped frequency distributions with probability density functions of form

$$(\text{constant}) \cdot x^c \exp[-\tfrac{1}{2}(\alpha + \beta x)^2] \qquad (x > 0). \tag{64}$$

Laha considers distributions for which there is a standard form having characteristic function $(1 + |t|^\alpha)^{-1}$ for some value of α in the range $0 < \alpha \leq 2$.

There are some very broad classes of distributions, including most of the common distributions. We will not devote much attention to these, as we are concerned more with properties of specific distributions than with broad classification. The *exponential-type class* includes all density functions which can be written in the form

$$p(x) = \exp[A(x)B(\theta) + C(x) + D(\theta)] \tag{65}$$

where $A(\cdot)$, $B(\cdot)$, $C(\cdot)$, $D(\cdot)$ are arbitrary functions. This class was obtained, nearly simultaneously, by Darmois [25] and Koopman [47] (as the form taken by the density function if a single sufficient statistic for θ exists, given values of n independent identically distributed random variables). It is often called the *Darmois-Koopman* (or *Koopman-Darmois*) class.

An even broader class is that of the *Pólya-type* distributions, introduced by Karlin [45].

A frequency function $\Lambda(x)$ is said to be a *Pólya-type* frequency function if for every positive integer n and every pair of sets of increasing numbers $x_1 < x_2 < \cdots < x_n$, $y_1 < y_2 < \cdots < y_n$ the determinant $\|\Lambda(x_i - y_k)\| \geq 0$.

A characterization of Pólya-type frequency functions by means of the structure of their characteristic functions was recently given by Lukacs [53].

Mathai and Saxena [54] have pointed out that the formula for the density function

$$p_X(x) = \frac{da^{c/d}\Gamma(\alpha)\Gamma(\beta)\Gamma(r - c/d)}{\Gamma(c/d)\Gamma(r)\Gamma(\alpha - c/d)\Gamma(\beta - c/d)} x^{c-1} F(\alpha,\beta; r; -ax^d) \tag{66}$$

(with $x > 0, c > 0, \alpha - c/d > 0, \beta - c/d > 0$) and the limiting form obtained by letting α tend to infinity, and a to zero in such a way that a^α tends to a':

$$p_X(x) = \frac{da'^{c/d}\Gamma(\beta)\Gamma(r - c/d)}{\Gamma(c/d)\Gamma(r)\Gamma(\beta - c/d)} x^{c-1} M(\beta; r; -a'x^d) \tag{67}$$

can be made to represent a considerable variety of commonly used distributions by appropriate choice of values for the parameters.

5. Cornish-Fisher Expansions

If any distribution is fitted by making the first s moments of the fitted and actual distributions agree, it is (in principle) possible to calculate quantiles of the fitted distribution, and to regard these as approximations to the corresponding quantiles of the actual distribution. In fact, the fitted quantiles are functions of the s fitted moments, and so we have estimators of the actual quantiles which are functions of these s moments.

Usually these functions are very complicated and not easily expressible in explicit form. However, in the case of the Gram-Charlier and Edgeworth expansions, described in Section 4.2, it is possible to obtain explicit expansions for standardized quantiles as functions of corresponding quantiles of the unit normal distributions. In these expansions, the terms are polynomial functions of the appropriate unit normal quantile, with coefficients which are functions of the moment-ratios of the distribution. We now outline the method of derivation of these expansions, using an argument propounded by Cornish and Fisher [22]. (The argument has been reformulated in Fisher and Cornish [37] and extended by Finney [36].)

From (37), we have formally

$$\int_{-\infty}^{x} g(t)\,dt = D^{-1}\exp\left[\sum_{j=1}^{\infty} \epsilon_j(-D)^j/j!\right] f(x) \tag{68}$$

and if $f(x) = Z(x) = (\sqrt{2\pi})^{-1}e^{-\frac{1}{2}x^2}$, then $D^j f(x) = (-1)^j H_j(x)Z(x)$.

Now suppose X_α and U_α are defined by

$$\int_{-\infty}^{X_\alpha} g(x)\,dx = \alpha = \int_{-\infty}^{U_\alpha} Z(x)\,dx.$$

Using the expansion (68),

(69)

$$\int_{-\infty}^{X_\alpha} Z(x)\,dx + \left[\sum_{i=0}^{\infty} D^{-1}\left\{\sum_{j=1}^{\infty} \epsilon_j(-D)^j/j!\right\}^i \Big/ i!\right] Z(X_\alpha) = \int_{-\infty}^{U_\alpha} Z(x)\,dx.$$

We now expand the right-hand side as

$$\begin{aligned} \int_{-\infty}^{U_\alpha} Z(x)\,dx &= \int_{0}^{X_\alpha} Z(x)\,dx + \sum_{j=1}^{\infty} \{(U_\alpha - X_\alpha)^j/j!\}\, D^j Z(X_\alpha) \\ &= \int_{0}^{X_\alpha} Z(x)\,dx + \sum_{j=1}^{\infty} \{(X_\alpha - U_\alpha)^j/j!\} H_j(X_\alpha) Z(X_\alpha). \end{aligned} \tag{70}$$

Inserting this in (69) gives the *identity*

$$\text{(71)}\quad \left[\sum_{i=0}^{\infty} D^{-1}\left\{\sum_{j=1}^{\infty} \epsilon_j(-D)^j/j!\right\}^i \Big/ i!\right] Z(X_\alpha) = \left[\sum_{j=1}^{\infty} \{(X_\alpha - U_\alpha)^j/j!\} H_j(X_\alpha)\right] Z(X_\alpha).$$

Expanding the left-hand side and dividing both sides by $Z(X_\alpha)$ gives an identity, of polynomial form, between $(X_\alpha - U_\alpha)$ and X_α.

By straightforward (though tedious) algebra it is possible to rearrange (71) to give: (a) U_α as a function of X_α, i.e. $U_\alpha = U(X_\alpha)$ or (b) X_α as a function of U_α, i.e. $X_\alpha = X(U_\alpha)$.

Cornish and Fisher [22] gave detailed formulas for $U(X_\alpha)$ and $X(U_\alpha)$, and extended these in [37]. They collected terms according to Edgeworth's system (see Section 4.3). Based on their formula we have (to order n^{-1} if κ_r is of order $n^{1-\frac{1}{2}r}$ with the distribution standardized).

$$\begin{aligned}
\text{(72)}\quad X(U_\alpha) = {} & U_\alpha + \tfrac{1}{6}(U_\alpha^2 - 1)\kappa_3 \\
& + \tfrac{1}{24}(U_\alpha^3 - 3U_\alpha)\kappa_4 - \tfrac{1}{36}(2U_\alpha^3 - 5U_\alpha)\kappa_3^2 \\
& + \tfrac{1}{120}(U_\alpha^4 - 6U_\alpha^2 + 3)\kappa_5 - \tfrac{1}{24}(U_\alpha^4 - 5U_\alpha^2 + 2)\kappa_3\kappa_4 \\
& + \tfrac{1}{324}(12U_\alpha^4 - 53U_\alpha^2 + 17)\kappa_3^3 \\
& + \tfrac{1}{720}(U_\alpha^5 - 10U_\alpha^3 + 15U)\kappa_6 \\
& - \tfrac{1}{180}(2U_\alpha^5 - 17U_\alpha^3 + 21U_\alpha)\kappa_3\kappa_5 \\
& - \tfrac{1}{384}(3U_\alpha^5 - 24U_\alpha^3 + 29U_\alpha)\kappa_4^2 \\
& + \tfrac{1}{288}(14U_\alpha^5 - 103U_\alpha^3 + 107U_\alpha)\kappa_3^2\kappa_4 \\
& - \tfrac{1}{7776}(252U_\alpha^5 - 1688U_\alpha^3 + 1511U_\alpha)\kappa_3^4 \\
& + \cdots .
\end{aligned}$$

Also

$$\begin{aligned}
\text{(73)}\quad U(X_\alpha) = {} & X_\alpha - \tfrac{1}{6}(X_\alpha^2 - 1)\kappa_3 \\
& - \tfrac{1}{24}(X_\alpha^3 - 3X_\alpha)\kappa_4 + \tfrac{1}{36}(4X_\alpha^3 - 7X_\alpha)\kappa_3^2 \\
& - \tfrac{1}{120}(X_\alpha^4 - 6X_\alpha^2 + 3)\kappa_5 + \tfrac{1}{144}(11X_\alpha^4 - 42X_\alpha^2 + 15)\kappa_3\kappa_4 \\
& - \tfrac{1}{648}(69X_\alpha^4 - 187X_\alpha^2 + 52)\kappa_3^3 \\
& - \tfrac{1}{720}(X_\alpha^5 - 10X_\alpha^3 + 15X_\alpha)\kappa_6 \\
& + \tfrac{1}{360}(7X_\alpha^5 - 48X_\alpha^3 + 51X_\alpha)\kappa_3\kappa_5 \\
& + \tfrac{1}{384}(5X_\alpha^5 - 32X_\alpha^3 + 35X_\alpha)\kappa_4^2 \\
& - \tfrac{1}{864}(111X_\alpha^5 - 547X_\alpha^3 + 456X_\alpha)\kappa_3^2\kappa_4 \\
& + \tfrac{1}{7776}(948X_\alpha^5 - 3628X_\alpha^3 + 2473X_\alpha)\kappa_3^4 + \cdots .
\end{aligned}$$

The formulas given in [22] and [37] include terms adjusting the mean and variance. In the above formulas (72) and (73), however, it has been assumed that the distribution to be fitted has been standardized, so that no correction is needed.

Numerical values of the coefficients in formula (72) are given in [37] to 5 decimal places for $\alpha = 0.5, 0.75, 0.9, 0.95, 0.975, 0.99, 0.995, 0.999$ and 0.9995. This paper also gives the values of the first seven Hermite polynomials, to 12 decimal places, for the same values of α.

It is especially to be noted that the functional forms $U(\cdot)$ and $X(\cdot)$ do *not* depend on the value of α. The function $U(X)$ may be regarded as a *normalizing transformation* of the random variable X. The function $X(\cdot)$ expresses the quantiles of the (standardized) distribution of X as a function of corresponding quantiles of the unit normal distribution.

In practice only a finite number of terms of the expansions $U(\cdot)$ or $X(\cdot)$ are used. It is important to recognize that the results obtained are not equivalent to those obtained by retaining a similar number (or, indeed, any specified number) of terms in the Gram-Charlier expansions. They may be (in favorable cases) good approximations to the quantiles of the 'distributions' represented by the complete expansions.

Although the Cornish-Fisher expansions are directly related to the Edgeworth form of distribution, there is a difference in the way these two are used. It is unusual to use moments higher than the fourth in fitting an Edgeworth (or Gram-Charlier) expansion. This is mainly because the possibility of negative values (and multimodality) becomes more serious as further terms are added, but also because, with observed data, estimation of higher moments is often of low accuracy. Cornish-Fisher expansions, on the other hand, are more usually applied to theoretically determined distributions (with known moments), and it is quite usual to use moments of order as high as six, or even greater.

As Finney [36] has pointed out, it would be possible to obtain analogues of Cornish-Fisher expansions by operating on Laguerre series (or other) forms of distribution in the same way as described (for Edgeworth series) at the beginning of this section. Details of such analyses have not (so far as we know) yet been published.

REFERENCES

[1] Barton, D. E. and Dennis, K. E. R. (1952). The conditions under which Gram-Charlier and Edgeworth curves are positive definite and unimodal, *Biometrika*, **39,** 425–427.

[2] Bennett, C. A. (1952). *Asymptotic properties of ideal linear estimators*, Unpublished thesis, University of Michigan.

[3] Benson, F. (1949). A note on the estimation of mean and standard deviation from quantiles, *Journal of the Royal Statistical Society, Series B*, **11,** 91–100.

[4] Bhattarcharjee, G. P. (1965). Distribution of range in non-normal samples, *Australian Journal of Statistics*, **7,** 127–141.

[5] Bhattacharyya, B. C. (1942). The use of McKay's Bessel function curves for graduating frequency distributions, *Sankhyā*, **6,** 175–182.

[6] Bhattacharyya, B. C. (1943). On an aspect of Pearsonian system of curves and a few analogies, *Sankhyā*, **6,** 415–418.

[7] Boetti, G. (1964). Una rappresentazione grafica per la determinazione del tipo di curva, tra le curve del sistema di Pearson, corrispondente a dati valori numerici dei momenti 3° e 4° delle distribuzioni, *Giornale dell' Istituto Italiano degli Attuari*, **27,** 99–121.

[8] Bol'shev, L. N. (1963). Asymptotic Pearsonian transformations, *Teoriya Veroyatnostei i ee Primeneniya*, **8,** 129–155. (In Russian)

[9] Bortolotti, G. (1965). Nuove vedute sulle distribuzioni di frequenze e sui loro rapporti con la legge normale delle probabilita. I & II, *Statistica, Bologna*, **25,** 197–288 and 329–362.

[10] Bose, S. S. (1938). On a Bessel function population, *Sankhyā*, **3,** 253–261.

[11] Bruns, H. (1906). *Wahrscheinlichkeitsrechnung und Kollektivmasslehre*, Leipzig: Teubner.

[12] Burr, I. W. (1942). Cumulative frequency functions, *Annals of Mathematical Statistics*, **13,** 215–232.

[13] Burr, I. W. (1968). On a general system of distributions III. The sample range, *Journal of the American Statistical Association*, **63,** 636–643.

[14] Burr, I. W. and Cislak, P. J. (1968). On a general system of distributions, I. Its curve-shape characteristics, II. The sample median, *Journal of the American Statistical Association*, **63,** 627–635.

[15] Castoldi, L. (1963). A continuous analogon of Poisson's distribution, *Rendiconti del Seminario della Facoltà di Scienze della Università di Cagliari*, **33,** 1–5.

[16] Chambers, E. and Fowlkes, E. B. (1966). *A Dictionary of Distributions: Comparisons with the Standard Normal*, Bell Telephone Laboratories, Murray Hill, New Jersey.

[17] Chan, L. K. (1967). On a characterization of distributions by expected values of extreme order statistics, *American Mathematical Monthly*, **74,** 950–951.

[18] Charlier, C. V. L. (1905). Über die Darstellung willkürlicher Funktionen, *Arkiv för Matematik, Astronomi och Fysik*, **2,** No. 20, 1–35.

[19] Charlier, C. V. L. (1914). Contributions to the mathematical theory of statistics, *Arkiv för Matematik, Astronomi och Fysik*, **9,** 1–18.

[20] Chernoff, H., Gastwirth, J. L. and Johns, M. V. (1967). Asymptotic distribution of linear combinations of functions of order statistics with applications to estimation, *Annals of Mathematical Statistics*, **38,** 52–72.

[21] Cooper, J. D., Davis, S. A. and Dono, N. R. (1965). Pearson Universal Distribution Generator (PURGE), *Proceedings of the 19th Annual Conference, American Society for Quality Control*, 402–411.

[22] Cornish, E. A. and Fisher, R. A. (1937). Moments and cumulants in the specification of distributions, *Review of the International Statistical Institute*, **5,** 307–320.

[23] Craig, C. C. (1936). A new exposition and chart for the Pearson system of curves, *Annals of Mathematical Statistics*, **7,** 16–28.

[24] Cramér, H. (1928). On the composition of elementary errors, *Skandinavisk Aktuarietidskrift*, **11,** 13–74 and 141–180.

[25] Darmois, G. (1935). Sur les lois de probabilités à estimation exhaustif, *Comptes Rendus de l'Académie des Sciences, Paris*, **200,** 1265–1267.

[26] David, F. N. and Johnson, N. L. (1954). Statistical treatment of censored data, Part I: Fundamental formulae, *Biometrika*, **41,** 228–240.

[27] de Fériet, J. K. (1966). *The Gram-Charlier approximation of the normal law etc.* Report **2013,** Applied Mathematics Laboratory, David Taylor Model Basin, Department of the Navy, Washington, D. C.

[28] Dershowitz, A. F. (1966). *Polynomial fit to percentiles of Pearson curves*, T. I. S. Report 66-Ch-SD-511, General Electric Company.

[29] Edgeworth, F. Y. (1896). The asymmetrical probability curve, *Philosophical Magazine, 5th Series*, **41,** 90–99.

[30] Edgeworth, F. Y. (1907). On the representation of statistical frequency by a series, *Journal of the Royal Statistical Society, Series A*, **70,** 102–106.

[31] Edgeworth, F. Y. (1914). On the use of analytic geometry to represent certain kinds of statistics, *Journal of the Royal Statistical Society, Series A*, **77,** 838–852.

[32] Edgeworth, F. Y. (1916). On the mathematical representation of statistical data, *Journal of the Royal Statistical Society, Series A*, **79,** 455–500.

[33] Edgeworth, F. Y. (1917). On the mathematical representation of statistical data, *Journal of the Royal Statistical Society, Series A*, **80,** 65–83; **80,** 266–288; **80,** 411–437.

[34] Elderton, W. P. and Johnson, N. L. (1969). *Systems of Frequency Curves*, London: Cambridge University Press.

[35] Ferreri, C. (1964). A new frequency distribution for single variate analysis (In Italian), *Statistica (Bologna)*, **24,** 223–251.

[36] Finney, D. J. (1963). Some properties of a distribution specified by its cumulants, *Technometrics*, **5,** 63–69.

[37] Fisher, R. A. and Cornish, E. A. (1960). The percentile points of distributions having known cumulants, *Technometrics*, **2,** 209–226.

[38] Flapper, P. (1967). Transformatie van niet-normale verdelingen, *Statistica Neerlandica*, **21,** 151–155.

[39] Harter, H. L. (1968). The use of order statistics in estimation, *Operations Research*, **16,** 783–798.

[40] Johnson, N. L. (1949). Systems of frequency curves generated by methods of translation, *Biometrika*, **36,** 149–176.

[41] Johnson, N. L. (1965). Tables to facilitate fitting S_U frequency curves, *Biometrika*, **52,** 547–558.

[42] Johnson, N. L., Nixon, E., Amos, D. E. and Pearson, E. S. (1963). Table of percentage points of Pearson curves, for given $\sqrt{\beta_1}$ and β_2, expressed in standard measure, *Biometrika*, **50,** 459–498.

[43] Kamat, A. R. (1966). A property of the mean deviation for the Pearson type distributions, *Biometrika*, **53,** 287–289.

[44] Kameda, T. (1928). On the reduction of frequency curves, *Skandinavisk Aktuarietidskrift*, **11,** 112–118.

[45] Karlin, S. (1957). Pólya-type distributions II, *Annals of Mathematical Statistics*, **28,** 281–308.

[46] Khamis, S. H. (1958). Incomplete gamma functions, *Bulletin of the International Institute of Statistics*, **37,** 385–396.

[47] Koopman, B. O. (1936). On distributions admitting a sufficient statistic, *Transactions of the American Mathematical Society*, **39,** 399–409.

[48] Laha, R. G. (1953). On some properties of the Bessel function distributions, *Bulletin of the Calcutta Mathematical Society*, **46,** 59–72.

[49] Laha, R. G. (1961). On a class of unimodal distributions, *Proceedings of the American Mathematical Society*, **12,** 181–184.

[50] Linnik, Yu. V. (1953). Linear forms and statistical criteria. II, *Ukrainskiĭ Matematicheskiĭ Zhurnal*, **5,** 247–290. (In Russian)

[51] Lloyd, E. H. (1952). Least squares estimation of location and scale parameters using order statistics, *Biometrika*, **39,** 88–95.

[52] Longuet-Higgins, M. S. (1964). Modified Gaussian distribution for slightly nonlinear variables, *Journal of Research of The National Bureau of Standards*, **68 D,** 1049–1062.

[53] Lukacs, E. (1968). Contributions to a problem of D. van Dantzig, *Teoriya Veroyatnostei i ee Primeneniya*, **13,** 114–125.

[54] Mathai, A. M. and Saxena, R. K. (1966). On a generalized hypergeometric distribution, *Metrika*, **11,** 127–132.

[55] Mardia, K. V. (1965). Tippett's formulas and other results on sample range and extreme, *Annals of the Institute of Statistical Mathematics, Tokyo*, **17,** 85–91.

[56] McKay, A. T. (1932). A Bessel function distribution, *Biometrika*, **24,** 39–44.

[57] McNolty, F. (1967). Applications of Bessel function distributions, *Sankhyā, Series B*, **29,** 235–248.

[58] Pearson, K. (1895). Contributions to the mathematical theory of evolution. II. Skew variations in homogeneous material, *Philosophical Transactions of the Royal Society of London, Series A*, **186,** 343–414.

[59] Pearson, K. (1924). On the mean error of frequency distributions, *Biometrika*, **16,** 198–200.

[60] Sastry, K. V. K. (1948). On a Bessel function of the second kind and Wilks' Z-distribution, *Proceedings of the Indian Academy of Sciences, Series A*, **28,** 532–536.

[61] Singh, C. (1967). On the extreme values and range of samples from non-normal populations, *Biometrika*, **54,** 541–550.

[62] Stok, J. P. van der (1908). On the analysis of frequency curves according to a general method, *Koninklijke Nederlandse Akademie van Wetenschappen, Series A. Mathematical Sciences*, **10,** 799–817.

[63] Subrahmaniam, K. (1966). Some contributions to the theory of non-normality. I. (univariate case), *Sankhyā, Series A*, **28,** 389–406.

[64] Suzuki, G. (1965). A consistent estimator for the mean deviation of the Pearson type distribution, *Annals of the Institute of Statistical Mathematics, Tokyo*, **17,** 271–285.

[65] Thiele, T. N. (1903). *Theory of Observations*, London: Layton.

[66] Thionet, P. (1966). Note sur les mélanges de certaines distributions de probabilités, *Publications de l'Institut de Statistique de l'Université de Paris*, **15,** 61–80.

[67] Tiku, M. L. (1965). Laguerre series forms of non-central χ^2 and F distributions, *Biometrika*, **52,** 415–427.

[68] Tippett, L. H. C. (1925). On the extreme individuals and the range of the samples taken from the normal population, *Biometrika*, **17,** 364–387.

[69] Toranzos, F. I. (1952). An asymmetric bell-shaped frequency curve, *Annals of Mathematical Statistics*, **23,** 467–469.

[70] Woods, J. D. and Posten, H. O. (1968). *Fourier series and Chebyshev polynomials in statistical distribution theory*, Research Report No. 37, Department of Statistics, University of Connecticut, Storrs, Connecticut.

[71] Wu, Chuan-yi (1966). The types of limit distributions for some terms of variational series, *Scientia Sinica*, **15,** 749–762.

13

Normal Distributions

1. Definition and Tables

A random variable X is *normally* distributed if it has the probability density function:

(1)
$$\frac{1}{\sqrt{2\pi}\,\sigma}\exp\left[-\frac{1}{2}\left(\frac{x-\xi}{\sigma}\right)^2\right] \qquad (\sigma > 0)\cdot$$

The probability density function of $U = (X - \xi)/\sigma$ is:

(2)
$$p_U(u) = (\sqrt{2\pi})^{-1}\exp(-\tfrac{1}{2}u^2),$$

which does not depend on the parameters ξ, σ. This is called the *standard form* of normal distribution. (It is also, in fact, the *standardized form*.) The random variable U is called a *standard*, or *unit*, normal variable.

Since

(3)
$$\Pr[X \leq x] = \Pr[U \leq (x - \xi)/\sigma],$$

such probabilities can be evaluated from tables of the cumulative distribution function of U, which is:

(4)
$$\Phi(u) = \Pr[U \leq u] = (\sqrt{2\pi})^{-1}\int_{-\infty}^{u} e^{-\frac{1}{2}x^2}\,dx.$$

The notation $\Phi(\cdot)$ is widely used, and will be used in this book. Further, it is convenient to have a systematic notation for the quantiles of the distribution of U. We use the system defined by

$$\Phi(U_\alpha) = \alpha$$

so that $U_{1-\alpha}$ is the upper $100\alpha\%$ point, and U_α $(= -U_{1-\alpha})$ is the lower $100\alpha\%$ point of the distribution.

There are other forms of notation which are much less frequently encountered in statistical work. The parameter in (1) is sometimes replaced by the '*precision modulus*'

$$h = (\sigma\sqrt{2})^{-1}.$$

Other functions are:

$$\text{(5.1)} \qquad \operatorname{erf}(x) = 2\pi^{-\frac{1}{2}}\int_0^x e^{-t^2}\,dt = 2\Phi(x\sqrt{2}) - 1$$

$$\text{(5.2)} \qquad \operatorname{erfc}(x) = 1 - \operatorname{erf}(x)$$

(erf: "*error function*" or "*Cramp function*"; erfc: "*error function complement*"). Other names for the distribution are *Second Law of Laplace*, *Laplace*, *Gaussian*, *Laplace-Gauss*, *de Moivre* $\Phi(\cdot)$ is also called the *Laplace-Gauss integral*, or simply the *probability integral;* erf $(\cdot)$ is also known by this last name, and is sometimes called the *error integral*.

Tables relating to the unit normal distribution are a necessary ingredient of any textbook in statistical theory or its applications. This, because for many decades the normal distribution held a central position in statistics. As pointed out, tables of the *unit* normal distribution suffice for calculations relating to *all* normal distributions. Some care is necessary in using these tables — for example, putting

$$\text{(6)} \qquad (\sqrt{2\pi})^{-1}e^{-\frac{1}{2}x^2} = Z(x)$$

it is necessary to remember the multiplier σ^{-1} in

$$\text{(7)} \qquad (\sqrt{2\pi}\,\sigma)^{-1}\exp[-\tfrac{1}{2}\{(x-\xi)/\sigma\}^2] = \sigma^{-1}Z((x-\xi)/\sigma)$$

— but no real difficulties are presented by the extended use of tables of the unit normal distribution. (The symbols $\varphi(x)$, $\phi(x)$ are often used in place of $Z(x)$.)

In most of the tables only positive values of the variable are given. This is all that is necessary, since

$$\text{(8)} \qquad Z(x) = Z(-x) \qquad \text{and} \qquad \Phi(x) = 1 - \Phi(-x).$$

Here we give a list of only the more easily available tables. Fuller lists are given in the National Bureau of Standards [220] (up to 1952) and Greenwood and Hartley [110] (up to 1958). The functions most usually tabulated are $\Phi(x)$, $Z(x)$ and U_α, but there are many variants for special uses.

Pearson and Hartley [240] give tables based on values originally computed by Sheppard [280][281]. These contain:

(a) $\Phi(x)$ and $Z(x)$ to 7 decimal places for
$x = 0.00(0.01)4.50$; and to 10 decimal places for
$x = 4.50(0.01)6.00$.
(b) U_α to 4 decimal places for
$\alpha = 0.501(0.001)0.980(0.0001)0.9999$.
(c) $Z(U_\alpha)$ to 5 decimal places for $\alpha = 0.500(0.001)0.999$.

Fisher and Yates [86] give U_α to 6 decimal places for $\alpha = 0.505(0.005)0.995$, and to 5 decimal places for $1 - \alpha = 0.0^r1$ ($r = 2(1)8$).

These tables also include values of 'probits'—$(5 + U_\alpha)$—to 4 decimal places for $\alpha = 0.001(0.001)0.980(0.0001)0.9999$ and of $Z(u)$ to 4 decimal places for $u = 0.00(0.01)3.00(0.1)3.9$.

Owen [225] gives $Z(x)$ and $\Phi(x)$ to 6 decimal places, $Z^{(1)}(x)$, $Z^{(2)}(x)$, $Z^{(3)}(x)$, and $\{1 - \Phi(x)\}/Z(x)$ to 5 decimal places, and $\Phi(x)/Z(x)$ to 4 decimal places, for $x = 0.00(0.01)3.99$; also $\{1 - \Phi(x)\}$ to 5 significance figures for $x = 3.0(0.1)6.0(0.2)10.0(1)20(10)100(25)200(50)500$, and U_α and $Z(U_\alpha)$ to 5 decimal places for $\alpha = 0.500(0.001)0.900(0.005)0.990$.

Kelley [164] gives U_α to 8 decimal places for $\alpha = 0.5000(0.0001)0.9999$.

Hald [119] gives $Z(x)$ and $\Phi(x)$ to 4 significant figures for

$$x = \pm 0.00(0.01)4.99,$$

and probits $(5 + U_\alpha)$ for 3 decimal places for

$$\alpha = 0.0001(0.0001)0.0250(0.001)0.9750(0.0001)0.9999.$$

We now describe some tables containing larger numbers of decimal places, useful for special calculations.

In Zelen and Severo [327] there are tables of $Z(x)$, $\Phi(x)$ and $Z^{(1)}(x)$ to 15 decimal places, $Z^{(2)}(x)$ to 10, and $Z^{(r)}(x)$ ($r = 3,4,5,6$) to 8 decimal places for $x = 0.00(0.02)3.00$. For the values $x = 3.00(0.05)5.00$, $\Phi(x)$ is given to 10 decimal places, $Z(x)$ to 10 significant figures, and $Z^{(r)}(x)$ ($r = 2,\ldots,6$) to 8 significant figures for $x = 3.00(0.05)$. A further table gives $Z^{(r)}(x)$ ($r = 7,\ldots,12$) to 8 significant figures for $x = 0.0(0.1)5.0$. There are also tables (based on Kelley [164]) of U_α and $Z(U_\alpha)$ to 5 decimal places for $\alpha = 0.500(0.001)0.99$, and of U_α to 5 decimal places for $\alpha = 0.9750(0.0001)0.9999$.

In the National Bureau of Standards tables [219] there are given tables of $Z(x)$ and $2\Phi(x) - 1$ $(= \text{erf}(x/\sqrt{2}))$ to 15 decimal places for $x = 0(0.0001)1.0000(0.001)7.800$; and also of $2(1 - \Phi(x))$ to 7 significant figures for $x = 6.00(0.01)10.00$.

Pearson [246] gives tables (calculated by Kondo and Elderton [168], and Mills and Camp (Mills [207])) of U_α, $Z(U_\alpha)$, $\alpha/Z(U_\alpha)$, $(1 - \alpha)/Z(U_\alpha)$, $Z(U_\alpha)/\alpha$ and $Z(U_\alpha)/(1 - \alpha)$ to 10 decimal places for $\alpha = 0.500(0.001)0.999$ and of $(1 - \Phi(x))/Z(x)$ to 5 decimal places for $x = 0.00(0.01)4.00(0.05)5.00(0.1)10.0$. (This last quantity "ratio of area to bounding ordinate" is known as *Mills ratio*.

Such ratios are discussed in Chapter 33.)

In Emersleben [75] there are tables of erf x, and $2^{\frac{3}{2}}Z(x\sqrt{2})$ to 10 decimal places for $x = 0(0.01)2.00$, and also of xe^{x^2} erfc(x) to 7 decimal places for $x = 0.000(0.005)0.250$ and erfc($\sqrt{n\pi}$) to 15 decimal places for $n = 1(1)10$.

Bol'shev and Smirnov [19] give tables of $\Phi(x)$ to six decimal places for $x = 0.000(0.001)3.000$ and to five decimal places for $x = 3.00(0.01)5.00$; $Z(x)$, $Z^{(r)}(x)$ (r = 1,2,3,4,5) for $x = 0.000(0.004)3.00(0.02)4.00(0.04)5.0(0.1)6.0$ and U_α to 6 decimal places for $x = 0.500(0.001)0.9700(0.0001)0.9990$.

Harvard University Computation Laboratory Tables [292] give values of $\Phi(x)$ to six decimal places for $x = 0.000(0.004)4.892$; of $Z(x)$ to six decimal places for $x = 0.000(0.004)5.216$ and of $Z^{(r)}(x)$ ($r = 1,2,\ldots,20$) for

$r = 1(1)4$ to six decimal places for $x = 0.000(0.004)6.468$;
$r = 5(1)10$ to six decimal places for $x = 0.000(0.004)8.236$;
$r = 11(1)15$ to six decimal places for $x = 0.000(0.002)6.198$ and seven significant figures for $x = 6.2(0.002)9.61$;
$r = 16(1)20$ to seven significant figures for $x = 0(0.002)8.398$ and to six decimal places for $x = 8.4(0.002)10.902$.

The set of tables of the Advanced Series of Mathematics and Engineering [293] contains values of $Z(x)$ for

$x = 0.0000(0.0001)2.7000$ to 8 decimal places
$2.7000(0.0001)3.4500$ to 9 decimal places
$3.4500(0.0001)4.1000$ to 10 decimal places
$4.1000(0.0001)4.6500$ to 11 decimal places
$4.6500(0.0001)5.0000$ to 12 decimal places.

Values of $\Phi(x)$ are given according to the same scheme, except that only 7 decimal places are given for $0.0000 \leq x < 1.0000$, while 11 places are given for $4.0500 \leq x < 4.1000$.

There are also values of $Z^{(2)}(x)$, $Z^{(3)}(x)$ and $Z^{(4)}(x)$, to six decimal places, for $x = 0.000(0.001)5.000$.

Extremely detailed tables of the error function and its first 20 derivatives have been published recently in the U.S.S.R. [286], [287] and [294].

There are many other publications containing various forms of tables of the normal distribution. Further tables of special functions associated with the normal distribution are used in connection with probit analysis.

There is no need for extensive tables of the normal distribution to be given here. We confine ourselves, in Table 1, to a few commonly used values of U.

Tables of random unit normal deviates (representing values of a random variable having a unit normal distribution) have been constructed from tables of random numbers (representing values of a random variable having a discrete rectangular distribution over the integers 0–9). In 1948, Wold [321] published

TABLE 1

Percentile Points of Normal Distribution, as Standardized Deviates (Values of U_α)

α	U_α
0.5	0.000000
0.6	0.253347
0.7	0.524401
0.75	0.674490*
0.8	0.841621
0.9	1.281552
0.95	1.644854
0.975	1.959964
0.99	2.326348
0.995	2.575829
0.9975	2.807034
0.999	3.090232

*The value of $U_{0.75}$ (= 0.6745), the upper quartile of the unit normal distribution, is occasionally called the *probable error* of the distribution, though this nomenclature is seldom used at present. The probable error of distribution (1) is, of course, $U_{0.75}\sigma$.

a set of 25,000 random unit normal deviates (to 3 decimal places), based on Kendall and Babington Smith's [165] table of random numbers. A set of 10,400 random unit normal deviates (also to 3 decimal places), based on Tippett's [305] table of random numbers, was published by Sengupta and Bhattacharya [276]. These replaced an earlier set of tables, first appearing in 1936 (Mahalanobis *et al.* [198]) which were found to contain a number of errors.

A set of 100,000 random unit normal deviates, to 3 decimal places, based on the first half million random numbers produced in 1947, was published by RAND [258] in 1955. In Buslenko *et al.* [26] there is a table of 1,000 random unit normal deviates, to 4 decimal places. These were calculated from the values of five independent random variables $R_1, \ldots, R_5$ each randomly distributed over the range 0 to 1 (see Chapter 25), using the formulas

$$U = X - 0.01(3X - X^3)$$

where

$$X = (1/\sqrt{5}) \sum_{j=1}^{5} [\sqrt{3}(2R_j - 1)].$$

(This formula was suggested by Bol'shev [18]. Note that $\sqrt{3}(2R_j - 1)$ has a *standardized* rectangular distribution.)

Box and Muller [23] described a method of constructing pairs of independent unit normal variables from pairs of independent uniformly distributed random variables R_1, R_2 using the formulas

$$(9) \qquad \begin{aligned} U_1 &= (-2 \log R_1) \sin (2\pi R_2); \\ U_2 &= (-2 \log R_1) \cos (2\pi R_2). \end{aligned}$$

Muller [213] has summarized methods of generating unit normal distributions available in 1959. A method described by Marsaglia and Bray [201] in 1964 appears to combine speed and accuracy in a satisfactory manner.

2. Historical Remarks

Because of the importance of the normal distribution, considerable attention has been paid to its historical development. The earliest workers regarded the distribution only as a convenient approximation to the binomial distribution. At the beginning of the nineteenth century appreciation of its broader theoretical importance spread with the work of Laplace and Gauss. The normal distribution became widely and uncritically accepted as the basis of much practical statistical work, particularly in astronomy. Around the beginning of the present century, a more critical spirit developed with more attention being paid to systems of "skew (non-normal) frequency curves" (see Chapter 12). This critical spirit has persisted, but it is offset by developments in both theory and practice. The normal distribution has a unique position in probability theory, and can be used as an approximation to other distributions. In practice, 'normal theory' can frequently be applied, with small risk of serious error, when substantially non-normal distributions correspond more closely to observed values. This allows us to take advantage of the elegant nature and extensive supporting numerical tables of normal theory.

The earliest published derivative of the normal distribution (as an approximation to a binomial distribution) seems to be that in a pamphlet of de Moivre [208], dated 12 November 1733. This pamphlet was in Latin; in 1738, de Moivre [209] published an English translation, with some additions. (See also Archibald [3] and Daw [63].)

In 1774, Laplace [176] obtained the normal distribution as an approximation to hypergeometric distribution, and four years later, in [177], he advocated tabulation of the probability integral ($\Phi(x)$, in our notation). The work of Gauss [89] [90] in 1809 and 1816 respectively established techniques based on the normal distribution, which became standard methods used during the nineteenth century.

Most theoretical arguments for the use of the normal distribution are based on forms of *central limit theorems*. These theorems state conditions under which the distribution of standardized sums of random variables tends to a unit normal distribution as the number of variables in the sum increases — i.e. with conditions sufficient to ensure an *asymptotic* unit normal distribution. Gauss' [90] derivation of the normal distribution, as the resultant of a large number of additive independent errors, may be regarded as one of the earliest results of this kind.

Formal rigorous mathematical discussion of central limit theorems (for independent random variables) may be said to start with the work of Lyapunov

[196]. A useful theorem associated with his name states that if $X_1, X_2, \ldots, X_n$ are independent, identically distributed random variables with finite mean and standard deviation then the distribution of the standardized sum

$$\left(\sum_{j=1}^{n} X_j - nE[X]\right) \Big/ \sqrt{n \operatorname{var}(X)}$$

tends to the unit normal distribution as n tends to infinity. Lyapunov also obtained an upper bound for the magnitude of the difference between the cumulative distribution functions of the standardized sum and the unit normal. This upper bound was of the form $Cn^{-\frac{1}{2}} \log n$, where C is a constant depending on the variances and third moments of the X_i's. It has subsequently been considerably improved by Cramér [51], Berry [10], Esseen [76], Zahl [325] and Zolotarev [330]. For the case when the variables $\{X_i\}$ are identically distributed the best upper bound so far obtained is (Zolotarev [330].):

$$0.82(\nu_3/\sigma^3)n^{-\frac{1}{2}}$$

where

$$\sigma^2 = \operatorname{var}(X_i); \qquad \nu_3 = E[|X_i - E[X_i]|^3].$$

This result was an improvement on an earlier result of Wallace [313] (correcting a result of Berry [10]). Zahl [325] has shown that the upper bound

$$0.65(\nu_3/\sigma^3)n^{-\frac{1}{2}}$$

can be obtained, provided $\nu_3/\sigma^3 \geq 3/\sqrt{2} = 2.22$.

It can be shown by consideration of particular cases that the upper bound must be at least

$$C(\nu_3/\sigma^3)n^{-\frac{1}{2}} \quad \text{with} \quad C = \frac{\sqrt{13}}{6\sqrt{2\pi}} = 0.40974.$$

Zolotarev [330] has shown that if the variance and absolute third central moment of X_j are σ_j^2, ν_{3j} respectively ($j = 1, 2, \ldots, n$) then an upper bound for the magnitude of the difference between cumulative distribution functions is

$$0.9051 \left(\sum_{j=1}^{n} \nu_{3j}\right)\left(\sum_{j=1}^{n} \sigma_j^2\right)^{-\frac{3}{2}}.$$

For the general case of independent (but not necessarily identically distributed) variables, Lindeberg [183] showed that, putting $\operatorname{var}(X_i) = \sigma_i^2$ and

$$\sigma_{(n)}^2 = \sum_{i=1}^{n} \sigma_i^2,$$

then if

$$\lim_{n\to\infty} \sigma_{(n)}^{-2} \sum_{i=1}^{n} (\Pr\{|X_i - E[X_i]| \geq t\sigma_{(n)}\} \\ \times E[\{X_i - E[X_i]\}^2 \,|\, |X_i - E[X_i]| \geq t\sigma_{(n)}]) = 0$$

for all $t > 0$, the distribution of the standardized sum

$$\sigma_{(n)}^{-1} \sum_{i=1}^{n} (X_i - E[X_i])$$

tends to the unit normal distribution as n tends to infinity.

The *necessity* of Lindeberg's condition was established by Feller [80].

More recently, attention has moved to consideration of conditions under which a limiting normal distribution applies to sums of non-independent random variables. An account of some such conditions can be found in a book by Loève [188].

A comprehensive account of the central limit theorem and related problems (up to the early 1950's) has been given by Gnedenko and Kolmogorov [101]. Multidimensional extensions of central limit theorems have been investigated by Bergström [9], Esseen [78], Sadikova [267], and Sazanov [273] among others.

3. Moments and Other Properties

If U has the unit normal distribution then, since the distribution is symmetrical about $U = 0$

$$E(U) = 0 \tag{10}$$

and so

$$\mu_r = \mu_r' = E(U^r) = (\sqrt{2\pi})^{-1} \int_{-\infty}^{\infty} x^r e^{-\frac{1}{2}x^2}\, dx. \tag{11}$$

If r is odd,

$$\mu_r = 0.$$

If r is even,

$$\begin{aligned} \mu_r &= (\sqrt{2/\pi}) \int_0^{\infty} x^r e^{-\frac{1}{2}x^2}\, dx \\ &= (\sqrt{2/\pi})\, 2^{\frac{1}{2}(r+1)} \int_0^{\infty} t^{\frac{1}{2}(r-1)} e^{-t}\, dt \\ &= 2^{\frac{1}{2}r}\Gamma(\tfrac{1}{2}(r+1))/\sqrt{\pi} \\ &= (r-1)(r-3)\ldots 3.1. \end{aligned} \tag{12}$$

Hence

$$\begin{aligned} \mathrm{var}(U) &= \mu_2 = 1 \\ \alpha_3(U) &= 0 \\ \beta_2(U) &= \alpha_4(U) = 3. \end{aligned}$$

Thus, as pointed out in Section 1, the unit normal is also the standardized normal distribution. If X has the general normal distribution (1), then

$$X = \xi + \sigma U \tag{13}$$

where U is a unit normal variable.

Some normal probability density functions are shown in Figure 1. The nine curves shown correspond to all possible combinations of $\xi = -1, 0, 1$ and

FIGURE 1

Normal Density Functions

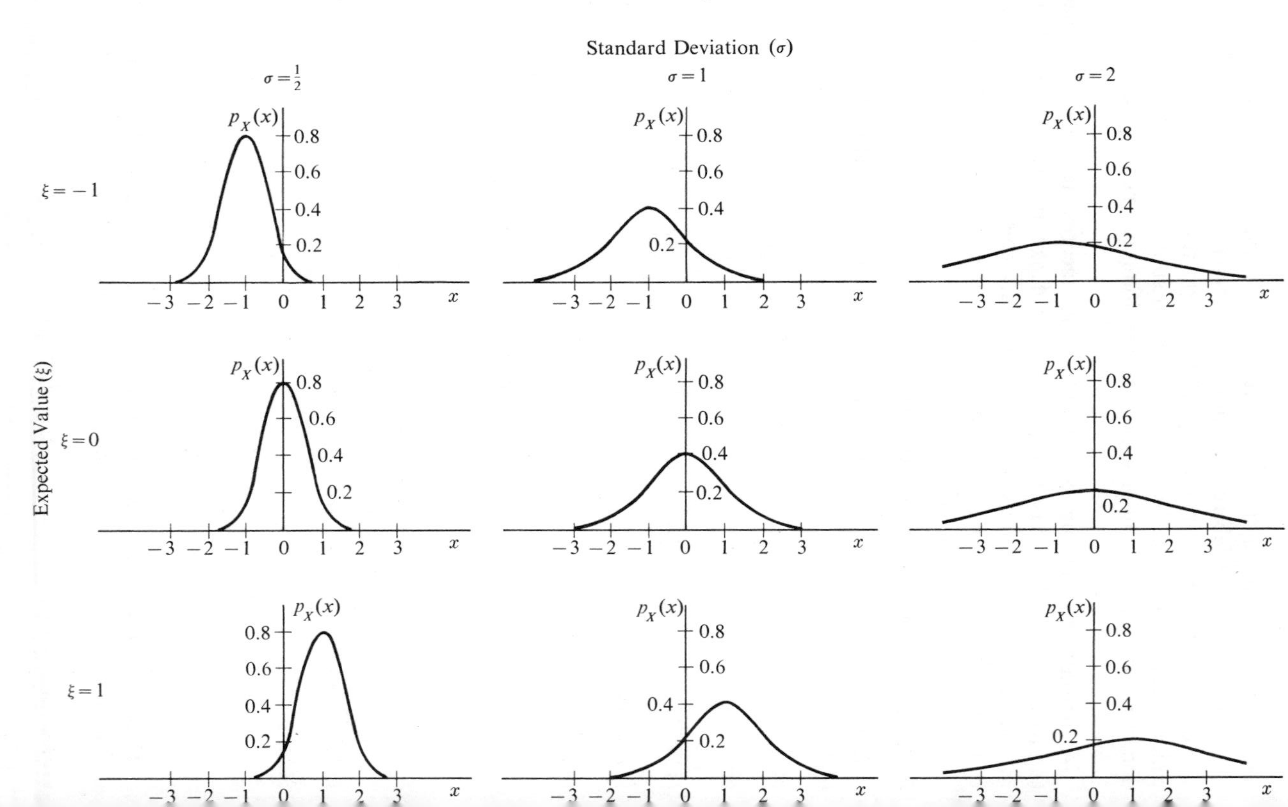

$\sigma = \frac{1}{2}$, 1, 2. The curve in the center represents the unit normal distribution ($\xi = 0, \sigma = 1$). The distribution is symmetrical about $X = \xi$; the probability density function has points of inflexion at $X = \xi \pm \sigma$. The distribution is unimodal with mode at $X = \xi$ (which is also, of course, the median of the distribution). The modal value of the probability density function is $(\sqrt{2\pi})^{-1} =$ 0.3979.

The moment generating function of $X (= \xi + \sigma U)$ is

$$E[e^{tX}] = e^{t\xi + \frac{1}{2}t^2\sigma^2} \tag{14}$$

and the characteristic function is $e^{it\xi - \frac{1}{2}t^2\sigma^2}$. For all $r > 2$, the cumulants κ_r are zero. This property characterizes normal distributions.

The mean deviation of X is $\sigma\sqrt{2/\pi} = 0.798\sigma$. For all normal distributions

$$\frac{\text{mean deviation}}{\text{standard deviation}} = \sqrt{\frac{2}{\pi}} = 0.798. \tag{15}$$

The information generating function of X is

$$(\sqrt{2\pi}\,\sigma)^{-u}(\sqrt{2\pi}\,\sigma/\sqrt{u}) = (\sqrt{2\pi}\,\sigma)^{-(u-1)}u^{-\frac{1}{2}}. \tag{16}$$

The entropy is

$$+\log(\sqrt{2\pi}\,\sigma) - \tfrac{1}{2}. \tag{17}$$

It is of some interest to note that the probability density function (1) can be expressed in the numerical form

$$0.3989\,\sigma^{-1}(0.6065)^{[(x-\xi)/\sigma]^2}. \tag{18}$$

The derivatives of the function $Z(\cdot)$ are also of some interest. They are used, for example, in the Gram-Charlier expansion (see Chapter 12). We have already discussed them in Chapters 1 and 12, and Section 1 of the latter chapter contains some references to tables of their numerical values.

If $X_1, X_2, \ldots, X_n$ are independent, normally distributed random variables, then any linear function of these variables is also normally distributed.

It is of interest to note that if X_1 and X_2 are independent, and each is normally distributed with zero expected value, then $X_1X_2(X_1^2 + X_2^2)^{-\frac{1}{2}}$ is also normally distributed. If, further, $\text{var}(X_1) = \text{var}(X_2)$ then $(X_1^2 - X_2^2)/(X_1^2 + X_2^2)$ is also normally distributed (Shepp [279]).

If $X_1, X_2, \ldots, X_n$ are independent random variables each distributed as (1), then by applying the transformation

$$\begin{cases} X_1 = \overline{X} + (1\cdot2)^{-\frac{1}{2}}U_2\sigma + (2\cdot3)^{-\frac{1}{2}}U_3\sigma + \cdots + [(n-1)n]^{-\frac{1}{2}}U_n\sigma \\ X_2 = \overline{X} - (1\cdot2)^{-\frac{1}{2}}U_2\sigma + (2\cdot3)^{-\frac{1}{2}}U_3\sigma + \cdots + [(n-1)n]^{-\frac{1}{2}}U_n\sigma \\ X_3 = \overline{X} \qquad\qquad\qquad - 2(2\cdot3)^{-\frac{1}{2}}U_3\sigma + \cdots + [(n-1)n]^{-\frac{1}{2}}U_n\sigma \\ \vdots \\ X_n = \overline{X} \qquad\qquad\qquad\qquad\qquad\qquad - n[(n-1)n]^{-\frac{1}{2}}U_n\sigma \end{cases} \tag{19}$$

it can be shown that

(i) $\overline{X}\,(= n^{-1}\sum_{j=1}^{n} X_j)$ has a normal distribution with expected value ξ and standard deviation $\sigma/\sqrt{n}$.

(ii) Each U_j $(j = 2,\ldots,n)$ is a unit normal variable.

(iii) $\overline{X}, U_2, \ldots, U_n$ are a mutually independent set of variables, and hence

(iv) $\sum_{j=1}^{n} (X_j - \overline{X})^2 = \sigma^2 \sum_{j=2}^{n} U_j^2$ is distributed as $\sigma^2(\chi^2$ with $(n-1)$ degrees of freedom).

This last result was obtained by Helmert [141], [142] in 1875–6. The transformation (19) is called *Helmert's transformation.*

We further note that, from (19) and (iii), since any function

$$g(X_1 - \overline{X},\ldots,X_n - \overline{X})$$

of the deviations $\{X_j - \overline{X}\}$ alone is a function of $\{U_j\}$ alone, that

(v) $\overline{X}$ and any function $g(X_1 - \overline{X},\ldots,X_n - \overline{X})$ are mutually independent.

This result is helpful in calculating moments and distributions of statistics like $\overline{X}[\text{Range}\,(X_1,\ldots,X_n)]^{-1}$; $\overline{X}[n^{-1}\sum_{j=1}^{n} |X_j - \overline{X}|]^{-1}$; etc.

It also can be shown that

(vi) $\sum_{j=1}^{n} (X_j - \overline{X})^2$ and any function of the ratios

$$\left\{(X_i - \overline{X})\left[\sum_{j=1}^{n} (X_j - \overline{X})^2\right]^{-\frac{1}{2}}\right\}$$

are mutually independent.

4. Characterizations

We have already noted, in Section 3, that the property $\kappa_r = 0$, for all $r > 2$, characterizes normal distributions.

It is of some interest to consider how far various combinations of the properties (i)–(vi) stated at the end of Section 3 (or similar) properties characterize the normal distribution (given that $X_1, \ldots, X_n$ are independent and have a common distribution). It is clear that (i) does so; and, in 1942, Lukacs [190] showed that if $\overline{X}$ and $\sum_{j=1}^{n} (X_j - \overline{X})^2$ are independent, then the common distribution of the X_j's must be normal. In 1943, Kaplansky [163] showed that this is also ensured by the weaker condition that the joint probability density function of $\overline{X}$ and $S = [n^{-1}\sum_{j=1}^{n} (X_j - \overline{X})^2]^{\frac{1}{2}}$ is of form $g(\overline{X},S)S^{n-2}$ (provided $g(\cdot)$ possesses partial derivatives of first order). Basu and Laha [6] and Lukacs [191] in 1954 and 1955 respectively, extended Lukacs' 1942 result by showing that S could be replaced by any k-statistic (symmetrical polynomial in the X's with expected value equal to a cumulant).

Shimizu [283] showed that if the common distribution has a finite variance σ^2, and if $T \equiv T(X_1,\ldots,X_n)$ is an unbiased estimator of σ^2, i.e. $E[T \mid \sigma^2] = \sigma^2$ for all σ^2, then independence of $\overline{X}$ and T ensures normality of the common distribution. Shimizu [284] also showed that, under the same conditions, existence of a linear function $\sum_{j=1}^{n} a_j X_j$, with no a_j's equal to zero, which has the same distribution as each of the X_j's, ensures normality. Shimizu states that his result can be derived from a result of Linnik [185].

Similar results had been derived earlier by Darmois [56] and Skitovich [285]. Without assuming finiteness of variance (or, indeed, the existence of a common distribution), independence of $\sum_{j=1}^{n} a_j X_j$ and $\sum_{j=1}^{n} b_j X_j$, with no a_j or b_j equal to zero, implies normality of the distribution of each of the X_j's. An alternative proof of this result was given later (in 1966) by Tranquilli [306].

Starting with the assumption that finite moments of *every* order exist, Cacoullos [27] has shown that normality is ensured if $\sum_{j=1}^{n} a_j b_j = 0$, and $a_j \neq 0$ $(j = 1,\ldots,n)$ implies that

$$E[(\textstyle\sum b_j X_j)^2 \mid \textstyle\sum a_j X_j] = E[(\textstyle\sum b_j X_j)^2]. \tag{20}$$

Zeigler [326] showed that if Y is a continuous random variable, and has the same variance as a random variable X with probability density function

$$p_X(x) = e^{tx} p_Y(x) \bigg/ \int_{-\infty}^{\infty} e^{ty} p_Y(y)\, dy$$

for all t, then Y is normally distributed.

Marcinkiewicz [199] has shown that if the common distribution of X_1, X_2, $\ldots$, X_n has finite moments of all orders, then the existence of two non-identical linear functions $\sum_{j=1}^{n} a_j X_j$, $\sum_{j=1}^{n} b_j X_j$ (i.e. such that $\{a_j\}$ is not just a rearrangement of $\{b_j\}$) with identical distributions, is sufficient to ensure normality of the common distribution.

Linnik [185] has shown that if some moments are infinite, this result is no longer valid.

Cramér [52], in 1936, showed that if the sum of two independent random variables (not necessarily identically distributed) has a normal distribution, then each of the two variables has a normal distribution. A similar result is true when there is a sum of n independent random variables which is normally distributed (e.g. Lukacs [192]).

Paskevich [227] and Rao [261] (see also Zinger [328]) have investigated conditions under which properties like (v) above ensure normality. With fairly light restrictions, independence of $\overline{X}$ and *any* continuous, non-constant, function $g(X_1 - \overline{X},\ldots,X_n - \overline{X})$ of $\{X_j - \overline{X}\}$ satisfying the conditions

(a) $g(X_1 - \overline{X},\ldots,X_n - \overline{X}) = 0$ if and only if $X_j = \overline{X}$ for all j;
(b) $g(c(X_1 - \overline{X}),\ldots,c(X_n - \overline{X})) = |c| g(X_1 - \overline{X},\ldots,X_n - \overline{X})$

ensures normality of the common distribution of the X_j's.

Kagan *et al.* [157] have shown that if it be assumed that the common distribution of the X_j's has a finite expected value ξ, and (with $n > 2$):

$$E[\overline{X} \mid X_2 - X_1, X_3 - X_1, \ldots, X_n - X_1] = \xi, \tag{21}$$

then the common distribution must be normal. It is interesting to note that if $n = 2$, not only is the stated condition insufficient to ensure normality, but also it is satisfied by *any* symmetrical distribution with finite expected value. In a more recent paper Rao [260] has shown that if $X_1 \ldots, X_n$ ($n \geq 3$) are independent identically distributed variables with zero mean and finite variance then the condition $E[X \mid X_j - \overline{X}] = 0$ is sufficient to establish normality of the common distribution of the X's.

Even if identity of distribution and finiteness of variance are not required, the existence of n linearly independent functions $Y_j = \sum_{i=1}^{n} a_{ji}X_i$ ($j = 1,2,\ldots, n$) with $a_{ji} \neq 0$ for all $i = 1, 2, \ldots, n$; such that

$$E[Y_1 \mid Y_2, \ldots, Y_n] = 0$$

is sufficient to ensure normality of each X_i.

Recently Cacoullos [28] has investigated characterization under conditions similar to Rao's. He shows that if $X_1, X_2, \ldots, X_n$ are independent and identically distributed with zero expected value and finite variance, then the existence of $(n - 1)$ linearly independent statistics $Y_j = \sum_{i=1}^{n} a_{ji}X_i$ ($j = 1,2,\ldots, n - 1$) such that there is a statistic $Z = \sum_{i=1}^{n} b_i X_i$ for which $E[Y_j \mid Z] = 0$ for all j is sufficient to ensure the normality of the common distribution of the X_i's.

Tamhankar [296] has obtained characterizations based essentially on the property that (for $X_1, \ldots, X_n$ independent unit normal variables) the sum of squares $\left(\sum_{i=1}^{n} X_i^2\right)$ and the set of ratios $\left\{X_j \Big/ \left(\sum_{i=1}^{n} X_i^2\right)^{\frac{1}{2}}\right\}$ are mutually independent.

Kotlarski [170] has shown that if $X_1, \ldots, X_n$ are independent random variables satisfying the conditions

(a) $\Pr[X_j = 0] = 0 \qquad (j = 1, \ldots, n)$

(b) each X_j is symmetrically distributed about zero, then the further condition that the variables

$$Y_j = \sqrt{j-1}\, X_j \Big/ \left\{\sum_{i=1}^{j-1} X_i^2\right\}^{\frac{1}{2}} \qquad (j = 2, \ldots, n)$$

are independent and distributed as t with $1, 2, \ldots, (n - 1)$ degrees of freedom (i.e. Y_j is distributed as t with $(j - 1)$ degrees of freedom) is necessary and sufficient to ensure that the X's are each normally distributed with expected value zero and common standard deviation.

Characterization conditions of quite another kind have been described by

Govindarajulu [109]. If the order statistics corresponding to the independent random variables $X_1, X_2, \ldots, X_n$ be denoted by $X'_{1;n} \leq X'_{2;n} \leq \cdots \leq X'_{n;n}$, then, provided the common distribution of the X's has a finite variance, the set of conditions

$$(22) \qquad E[X'^2_{n;n} - X'_{n-1;n}X'_{n;n}] = \sigma^2 \qquad \text{for all } n = 2, 3, \ldots$$

ensures that the common distribution is *either* a normal distribution with expected value zero and standard deviation σ, *or* this same distribution truncated from above (as described in Section 7).

If it be *assumed* that the common distribution has expected value zero, then the set of conditions

$$(23) \qquad nE[X'_{i,n}\overline{X}] = \sigma^2 \qquad \begin{pmatrix} i = 1,2,\ldots,n; \\ n = 2,3,\ldots \end{pmatrix}$$

ensures that the common distribution is normal with expected value zero and standard deviation σ.

If, on the other hand, it be assumed that each X_j is a *positive random variable* (i.e. $\Pr[X_j \leq 0] = 0$) then the set of conditions

$$(24) \qquad nE[X'_{1,n}\overline{X}] = \sigma^2 \qquad (n = 2,3,\ldots)$$

ensures that the common distribution is a *half-normal distribution* (see Section 7).

Kagan and Shalayevskii [158] have shown that the noncentral χ^2 distribution (see Chapter 28) may be used to characterize the normal distribution. They show that if $X_1, X_2, \ldots, X_n$ are independent, identically distributed random variables and $a_1, a_2, \ldots, a_n$ are constants then the distribution function of $\sum_{i=1}^{n} (X_i + a_i)^2$ will depend on the a's as a function of $\sum_{i=1}^{n} a_i{}^2$ if and only if the common distribution of the X's is normal.

It may be of some interest to note properties which do *not* characterize the normal distribution. For example, the distribution of the ratio X_1/X_2 of two independent, identically distributed random variables does not characterize the normal distribution (Fox [87]).

5. Approximations

The most common use of the normal distribution is as an approximation, either normality is ascribed to a distribution in the construction of a model, or a known distribution is replaced by a normal distribution with the same expected value and standard deviation. Examples of such replacement are the Fisher and Wilson-Hilferty approximations to the χ^2-distribution (Chapter 17), the normal approximation to the (central) t-distribution (Chapter 27) and the use of normal distribution to approximate the distribution of the arithmetic mean of a number (often not very large — say 8 or more) of independent and identically distributed random variables. Now, however, we are concerned with approximations *to* the normal distribution. It would be possible to regard the distributions which are approximated *by* the normal distribution as being, themselves, approximations *to* normal distributions. However, they are usually more complex than the normal distribution, and we would like to study approximations which are simpler than the normal distribution.

From the point of view of replacement of a normal distribution by another distribution we note that:

(*a*) A lognormal distribution can give a good representation of a normal distribution which has a small absolute value (less than 0.25, say) of the coefficient of variation (Section 6.4);

(*b*) A particular form of logistic distribution is very close to a normal distribution (see Chapter 22).

(*c*) A form of the Weibull distribution with the shape parameter ≈ 3.25 is almost identical with the unit normal distribution (see Chapter 20);

(*d*) Raab and Green [257] have suggested that the distribution with probability density function

$$(2\pi)^{-1}(1 + \cos x) \qquad (-\pi < x < \pi) \tag{25}$$

can be used to replace a normal distribution. The correspondence is not very precise (see Table 2 below comparing standardized percentile deviates of the two distributions) but will sometimes give useful analytical results. The replacement would only be used if substantial simplification in analysis were effected thereby.

The expected value and standard deviation of a random variable with distribution (25) are zero and $(\frac{1}{3}\pi^2 - 2)^{\frac{1}{2}} = 1.14$. The standardized range of the distribution $(-\pi,\pi)$ is thus from -2.77 to $+2.77$ standard deviations, and obviously the replacement gives a poor fit in the tails.

TABLE 2

Standardized Percentile Points of Distribution (25) and the Normal Distribution

Cumulative Probability	Standardized Value Normal	Standardized Value Distribution (25)
0.5	0.000	0.000
0.6	0.253	0.279
0.75	0.674	0.732
0.9	1.282	1.334
0.95	1.645	1.649
0.975	1.960	1.888
0.99	2.326	2.124
β_2	3.000	2.406

(*e*) Bell [7] has described even simpler approximations, using triangular distributions (Chapter 25). He pointed out that such approximations can be regarded as the second stage in a series of approximations by distributions of means of increasing numbers of independent rectangularly distributed variables (*cf.* the method of construction of 'random normal deviates' used by Buslenko *et al.* [26] and described in Section 1).

Chew [35] includes (*b*), (*d*) and (*e*) in a list of five possible replacements for normal distributions. The two other distributions he suggests are uniform (Chapter 25) and Laplace (Chapter 23). These also will be very crude approximations.

(*f*) Hoyt [146] has suggested using the distribution of the sum of three mutually independent random variables each uniformly distributed over the interval -1 to $+1$ as an approximation to the unit normal distribution. The density function is

$$\begin{array}{lll} \frac{1}{8}(3 - x^2) & \text{for} & |x| \leq 1 \\ \frac{1}{16}(3 - |x|)^2 & \text{for} & 1 \leq |x| \leq 3 \end{array}$$

This gives an error not exceeding 0.01 in the cumulative distribution function.

(*g*) Steffensen [288] has suggested the use of the distribution of a multiple of a chi random variable (i.e. $c\chi_\nu$), with ν sufficiently large. He called this a 'semi-normal' distribution.

(*h*) A different kind of approximation has been developed in connection with calculation of the functions $\Phi(\cdot)$, $Z(\cdot)$ in digital computers. These approximations usually employ polynomial expressions. They give quite high accuracy, sometimes only within definite limits on the values of the variable. Outside these limits they may give quite poor approximation.

Zelen and Severo [327] quote (*inter alia*) the following formulas, based on formulas given by Hastings [136]:

$$\Phi(x) \doteqdot 1 - (a_1 t + a_2 t^2 + a_3 t^3) Z(x) \tag{26}$$

with $t = (1 + 0.33267x)^{-1}$; $a_1 = 0.4361836$; $a_2 = -0.1201676$; $a_3 = 0.9372980$.

The error in $\Phi(x)$, for $x \geq 0$, is less than 1×10^{-5}.

$$\Phi(x) \doteqdot 1 - \tfrac{1}{2}(1 + a_1 x + a_2 x^2 + a_3 x^3 + a_4 x^4)^{-4} \tag{27}$$

with $a_1 = 0.196854$; $a_2 = 0.115194$; $a_3 = 0.000344$; $a_4 = 0.019527$.

The error in $\Phi(x)$, for $x \geq 0$, is less than 2.5×10^{-4}.

$$Z(x) \doteqdot (a_0 + a_2 x^2 + a_4 x^4 + a_6 x^6)^{-1} \tag{28}$$

with $a_0 = 2.490895$; $a_2 = 1.466003$; $a_4 = -0.024393$; $a_6 = 0.178257$.

The error in $Z(x)$ is less than 2.7×10^{-3}.

Very accurate results can be obtained with the formula (Hart [126])

$$(28)' \quad 1 - \Phi(x) \doteq (x\sqrt{2\pi})^{-1}[\exp(-x^2/2)]$$
$$\times \left[1 - \frac{(1 + bx^2)^{\frac{1}{2}}(1 + ax^2)^{-1}}{x\sqrt{\pi/2} + [\frac{1}{2}\pi x^2 + (1 + bx^2)^{\frac{1}{2}}(1 + ax^2)^{-1}\exp(-x^2/2)]^{\frac{1}{2}}}\right]$$

with

$$a = \frac{1}{2\pi}[1 + (1 + 6\pi - 2\pi^2)^{\frac{1}{2}}]$$
$$b = \frac{1}{2\pi}[1 + (1 + 6\pi - 2\pi^2)^{\frac{1}{2}}]^2$$

For $x > 2$, Schucany and Gray [275] have constructed the simpler formula

$$(28)'' \quad 1 - \Phi(x) \doteq ((x^2 + 2)\sqrt{2\pi})^{-1}x[\exp(-x^2/2)]\frac{x^6 + 6x^4 + 14x^2 - 28}{x^6 + 5x^4 - 20x^2 - 4}$$

which is even better than (28)′ for $x > 3$. (The proportionate error of (28)′ for $5 \leq x \leq 10$ is about 0.5×10^{-5}; that of (28)″ decreases from 0.39×10^{-5} for $x = 5$ to 0.42×10^{-7} for $x = 10$.)

By use of rather elaborate formulas, quite remarkable accuracy can be attained. Strecock [289] gives formulas which give values of erf(x) (see (5.1)) and of the inverse function (inverf(y) where erf(inverf(y)) = y), correct to 22 decimal places for $|x|$ (or $|\text{inverf}(y)|$) less than 7.85.

(*i*) Burr [25] has considered approximations to $\Phi(x)$ of form

$$G(x) = 1 - [1 + (\alpha + \beta x)^c]^{-k}.$$

He suggests taking $\alpha = 0.644693$, $\beta = 0.161984$, $c = 4.874$, $k = -6.158$. An even better approximation is obtained by using

$$H(x) = \tfrac{1}{2}[G(x) + 1 - G(-x)]$$

which is a symmetrical function of x. The discrepancy $|H(x) - \Phi(x)|$ reaches its maximum value of about 0.00046 when $x \doteq \pm 0.6$.

(*j*) McGillivray and Kaller [202] have considered the discrepancy between $\Phi(x)$ and $(\Phi(x) + a_{2r}Z(x)H_{2r-1}(x))$, where $H_{2r-1}(x)$ is the Hermite polynomial of order $(2r - 1)$ and a_{2r} is a constant chosen so that $1 + a_{2r}H_{2r}(x)$ cannot be negative. (This means that a_{2r} must be between zero and

$$A_{2r} = \left|\inf_x H_{2r}(x)\right|^{-1}.)$$

The second function $(\Phi(x) + a_{2r}Z(x)H_{2r-1}(x))$, is the cumulative distribution function of a symmetrical distribution having the first r even central moments (and, of course, all odd moments) the same as those for a unit normal distribution. The discrepancy cannot exceed

$$A_{2r} \sup_x \{Z(x)|H_{2r-1}(x)|\}.$$

The values of this quantity, for $r = 2, 3, 4$, are 0.10, 0.03, 0.005 respectively.

(Of course other distributions with the same (zero) odd central moments and first r even central moments *might* have greater discrepancies, but these results do give a useful idea of the accuracy obtained by equating moments.)

(k) Riffenburgh [263] has suggested that a symmetrical truncated unit normal distribution be approximated by the density function

$$\tfrac{1}{2}(Z(x) - Z(c))/(\Phi(c) - \tfrac{1}{2} - cZ(c)) \qquad (-c \leq x \leq c)$$

where $-c$, c are the points of truncation. Use of this approximation is recommended only when c exceeds 1 (preferably $c \geq 1.5$). Tables of the variance (to 3 decimal places) of the approximate distribution are given in [263] for $c = 0.8(0.1)1.2(0.05)4.00$, and also of $\Pr[X \leq x] - \tfrac{1}{2}$ (to 4 decimal places) for $c = 1.2(0.1)3.0$ and x at intervals of 0.05. (Riffenburgh has also developed test procedures based on this distribution.)

We now discuss some *bounds* on the value of $\Phi(x)$. Various inequalities for Mills' ratio (see Chapter 33) can be interpreted as bounds for $\Phi(x)$ or $Z(x)$. Using a simple geometrical argument (based on the joint distribution of two independent unit normal variables) it can be shown that

$$\tfrac{1}{2}[1 + (1 - e^{-x^2/2})^{\frac{1}{2}}] \leq \Phi(x) \leq \tfrac{1}{2}[1 + (1 - e^{-x^2})^{\frac{1}{2}}] \tag{29}$$

(see, for example, D'Ortenzio [69]). In [69] it is further shown, by a refinement of the argument, that, on the left hand side of (29), $(1 - e^{-x^2/2})$ can be replaced by

$$1 - e^{-x^2/2} + (2\pi^{-1} - \tfrac{1}{2})^2 e^{-x^2}$$

and, on the right hand side, $(1 - e^{-x^2})$ can be replaced by

$$1 - e^{-x^2} - (1 - 2\pi^{-1})^2 e^{-x^2}.$$

The approximation

$$\Phi(x) \doteqdot \tfrac{1}{2}[1 + \{1 - \exp(-2x^2/\pi)\}^{\frac{1}{2}}] \tag{30}$$

was obtained by Pólya [253]. This has a maximum error of 0.003, when $x = 1.6$. Cadwell [29] modified (30) to

$$\Phi(x) \doteqdot \tfrac{1}{2}[1 + \{1 - \exp(-2\pi^{-1}x^2 - \tfrac{2}{3}\pi^{-2}(\pi - 3)x^4)\}^{\frac{1}{2}}]. \tag{31}$$

Over the range $0 < x < 3.5$, the maximum error of (31) is 0.0007, when $x = 2.5$. Formula (31) should not be used for large values of x. Cadwell suggested, on empirical grounds, the addition of the terms

$$-0.0005x^6 + 0.00002x^8$$

to the exponent in (31). This reduces the maximum error to 0.00005.

A mechanical method of drawing a normal probability density curve has been described by Edwards [71].

Normal probability paper is graph paper with a natural scale in the horizontal (abscissa) direction, while the distances on the vertical (ordinate) scale are proportional to the corresponding normal deviates. The vertical scale is usually marked in percentages. Thus 50% correspond to the horizontal axis, 25% and 75% are at distances 0.6745 below and above this line, 5% and 95% are at distances 1.9600 below and above this line, and so on. (See Figure 2.)

If X has the distribution (1) and $\Pr[X \leq x]$ is plotted (as ordinate) against x (as abscissa), then a straight line is obtained. The slope of this line is σ^{-1} and its intercept on the horizontal axis is at $x = \xi$. If observed frequencies of the events $(X \leq x)$ are used in place of the actual probabilities an approximately straight line plot may be expected. A straight line fitted to these observed points gives estimates of σ and ξ. Such graphical methods of estimation can give good practical accuracy.

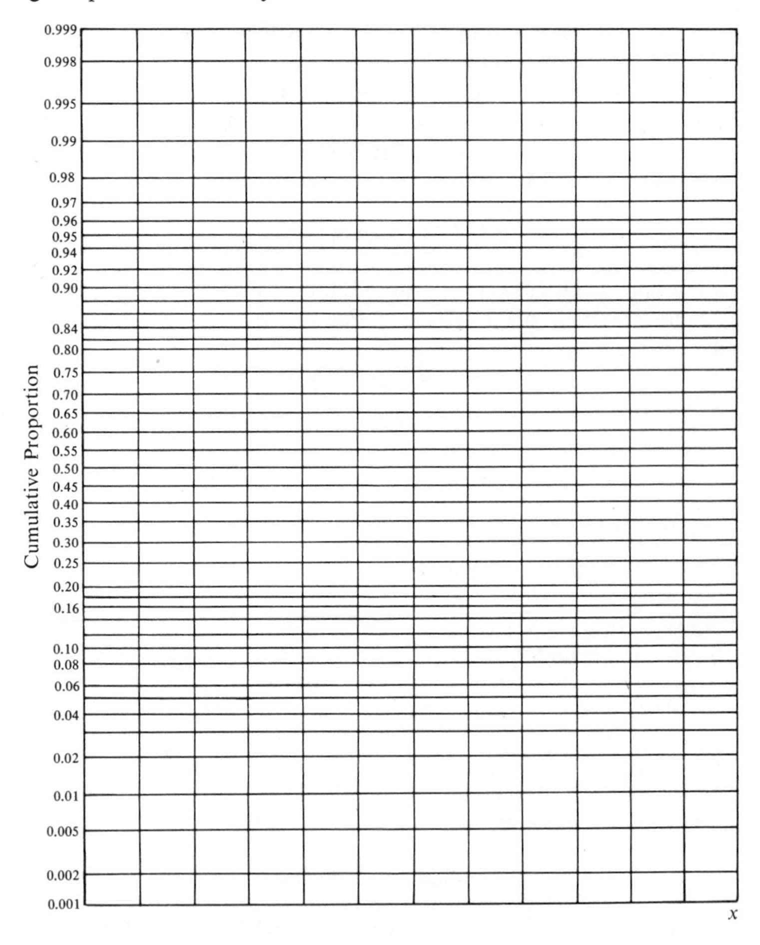

FIGURE 2

Normal Probability Paper

If the horizontal scale is logarithmic we have *lognormal probability* paper (see Chapter 14).

Half-normal probability paper is simply normal probability paper with negative abscissas omitted. It is used in connection with analysis of variance techniques developed by Daniel [55].

6. Estimation

The theory of estimation of ξ and σ has been fully worked out. To facilitate comprehension, this section is divided into four subsections. Subsections 6.1–6.3 describe techniques primarily appropriate to a complete sample — (though some apply also to censored data) — corresponding to values of n independent random variables each having distribution (1). Subsection 6.4 describes techniques suitable when the data have been censored by omission of certain order statistics. (*Truncated* normal distributions will be discussed in Section 7.)

The variety of applicable methods can be bewildering. In order to judge rapidly between them it is necessary to bear in mind accuracy, sensitivity to variations from normality and ease of calculation. The relative importance of these factors varies with circumstance, but they should always be taken into account.

6.1 *Estimation of* ξ

The arithmetic mean $\overline{X} = n^{-1} \sum_{j=1}^{n} X_j$ and the mean square deviation $S^2 = n^{-1} \sum_{j=1}^{n} (X_j - \overline{X})^2$ are jointly sufficient for ξ and σ, and $\overline{X}$ is sufficient for ξ alone. For most practical purposes, $\overline{X}$ is the best estimator for ξ, whether σ is known or not. It is the maximum likelihood estimator and is unbiased.

The only circumstances in which this estimator would not be used are

(*a*) when not all observations are available (this situation will be discussed more fully in Section 6.4), or

(*b*) when the accuracy of some values (e.g. outlying values) is doubtful.

In either case, estimation may be based on a central block of order statistics. As an extreme case (when n is odd), a single order statistic, the median, may be used to estimate ξ. This is an unbiased estimator of ξ, and has standard deviation approximately equal to

$$\tfrac{1}{2}\pi\sigma^2 n^{-1} = 1.5708\sigma^2 n^{-1}$$

compared with

$$\operatorname{var}(\overline{X}) = \sigma^2 n^{-1}.$$

The efficiency of the median, relative to $\overline{X}$, is thus approximately

$$100 \times (1.5708)^{-1}\% = 63.7\%.$$

Alternatively, the *j-th Winsorized mean*

$$(32) \qquad \xi_{(j)} = n^{-1}\left[jX'_{j+1} + \sum_{i=j+1}^{n-j} X'_i + jX'_{n-j}\right]$$

may be used ($j < [(n - 1)/2]$). It can be seen that $\xi_{(j)}$ is obtained by replacing each of $X'_1, X'_2, \ldots, X'_j$ by X'_{j+1}, and $X'_{n-j+1}, \ldots, X'_n$ by X'_{n-j}. This is also an unbiased estimator of ξ. It is interesting to note that Chernoff *et al.* [34] obtain a formula of the type (32), if only $X'_{j+1}, \ldots, X'_{n-j}$ are available, with the multipliers (j) of X'_{j+1} and X'_{n-j} replaced by

$$a = \frac{[Z(U_\epsilon)]^2\epsilon^{-1} + U_\epsilon Z(U_\epsilon)}{1 - 2\epsilon + 2U_\epsilon Z(U_\epsilon) + 2[Z(U_\epsilon)]^2\epsilon^{-1}}$$

where $\epsilon = j/(n + 1)$, and n^{-1} replaced by $(n - 2j + 2a)^{-1}$. In fact a is slightly less than j.

Instead of replacing the values of extreme observations by more central values, they may simply be omitted. The resulting unbiased estimator of ξ is the *j-th trimmed mean*

$$(33) \qquad \xi'_{(j)} = (n - 2j)^{-1} \sum_{i=j+1}^{n-j} X'_i.$$

Some relative efficiencies of $\xi'_{(j)}$, compared with $\overline{X}$, are shown in Table 3. (For efficiencies of $\xi_{(j)}$, see Table 9 in Section 6.4.)

TABLE 3

Efficiency of Trimmed Means, Relative to $\overline{X}$

n	j	Relative Efficiency of $\xi'_{(j)}$ (%)
5	2	69
10	2	89
10	3	81
15	2	92
15	4	83
15	6	73

It is apparent that the Winsorized mean $\xi_{(j)}$ is more efficient than $\xi'_{(j)}$. In fact, its efficiency compared with the best linear unbiased estimator *using the same order statistics*, never falls below 99.9% for $n \leq 20$ (Sarhan and Greenberg [270]).

Knowledge of σ is of no help in calculating point estimators of ξ. It is, however, of importance in calculating the standard deviations of such estimators, and in constructing confidence intervals for ξ.

If σ be known, $100(1 - \alpha)\%$ confidence limits for ξ are

$$\bar{X} \pm U_{1-\alpha/2}\sigma/\sqrt{n}. \tag{34}$$

Although the similar formulas

$$\begin{aligned} &\xi_{(j)} \pm U_{1-\alpha/2}\sqrt{\text{var}(\xi_{(j)})}; \\ &\xi'_{(j)} \pm U_{1-\alpha/2}\sqrt{\text{var}(\xi'_{(j)})} \end{aligned} \tag{35}$$

do not give *exact* limits (since $\xi_{(j)}, \xi'_{(j)}$ do not have normal distributions), they give limits which are useful for practical purposes, provided n is not too small (say $n \geq 15$).

If σ is not known, the above formulas cannot be used. It is natural to replace σ by an estimator of σ. If the sample size is large and a good (efficient) estimator of σ is used, this can be done with little serious effect on the confidence coefficient. The 'estimator' of σ most usually employed is

$$(1 - n^{-1})^{-\frac{1}{2}}S = \left[(n-1)^{-1}\sum_{j=1}^{n}(X_j - \bar{X})^2\right]^{\frac{1}{2}}$$

although this is not an unbiased estimator of σ. If this estimator is used, then $U_{1-\alpha/2}$ in (34) should be replaced by $t_{n-1,1-\alpha/2}$, the upper $50\alpha\%$ point of the t distribution with $(n - 1)$ degrees of freedom (see Chapter 27). The effect of replacement of σ by $(1 - n^{-1})^{-\frac{1}{2}}S$ in formulas (35) is not so clear; but there should be a comparable increase in the multiplying factor $U_{1-\alpha/2}$.

It can be shown (see Section 3) that $\bar{X}$ and *any* function of the deviations $\{X_j - \bar{X}\}$, only, are mutually independent. This facilitates computation of percentage points of distributions of statistics of form

$$\frac{\sqrt{n}(\bar{X} - \xi)/\sigma}{f(X_1 - \bar{X}, \ldots, X_n - \bar{X})} = T_{\{f\}},$$

say with various functions $f(X_1 - \bar{X}, \ldots, X_n - \bar{X})$ in the denominator, (subject to the restrictions that $f(\cdot)$ is positive with probability one, and

$$f(ay_1, ay_2, \ldots, ay_n) = af(y_1, y_2, \ldots, y_n)$$

for any $a \geq 0$). For example we might have

$$\begin{aligned} f(X_1 - \bar{X}, \ldots, X_n - \bar{X}) & \\ &= \text{range} \\ &= \max(X_1 - \bar{X}, \ldots, X_n - \bar{X}) - \min(X_1 - \bar{X}, \ldots, X_n - \bar{X}) \\ &= W \end{aligned}$$

or

$$\begin{aligned} f(X_1 - \bar{X}, \ldots, X_n - \bar{X}) &= \text{mean deviation} \\ &= n^{-1}\sum_{j=1}^{n}|X_j - \bar{X}| \\ &= M. \end{aligned}$$

Indeed, any of the estimators of σ to be described in Section 6.2 might be used as $f(\cdot)$.

Under the conditions stated, the distribution of $\sigma^{-1}f(X_1 - \overline{X},\ldots,X_n - \overline{X})$ does not depend on σ. The distribution of $T_{\{f\}}$, therefore, does not depend on ξ or σ, and so it is possible, at any rate in principle, to construct tables of percentage points $T_{\{f\},\alpha}$ of this distribution, defined by $f(\cdot)$ and α alone. The relation

$$\Pr\left[T_{\{f\},\alpha_1} < \frac{\sqrt{n}\,(\overline{X} - \xi)}{f(X_1 - \overline{X},\ldots,X_n - \overline{X})} < T_{\{f\},1-\alpha_2} \mid \xi,\sigma\right] = 1 - (\alpha_1 + \alpha_2)$$

can be rearranged to show that the limits

$$\begin{aligned} \overline{X} - T_{\{f\},\alpha_1} f(X_1 - \overline{X},\ldots,X_n - \overline{X})/\sqrt{n} \\ \overline{X} - T_{\{f\},1-\alpha_2} f(X_1 - \overline{X},\ldots,X_n - \overline{X})/\sqrt{n} \end{aligned} \tag{36}$$

form a $100(1 - \alpha_1 - \alpha_2)\%$ confidence interval for ξ.

In order to calculate such limits it is necessary to have tables of percentage points of the distribution of $T_{\{f\}}$. Such tables are available:

for $f(\cdot) = (1 - n^{-1})^{-\frac{1}{2}}S$, as described in Chapter 27 (tables of t distribution);
for $f(\cdot) = W$ (in Lord [189]); and
for $f(\cdot) = M$ (in Herrey [143]).

If tables are not available, approximations may be used, such as, for example, the approximations to the distributions of M and W to be described in Section 6.2.

6.2 *Estimation of σ*

The maximum likelihood estimator of σ (ξ not being known) is

$$S = \left[n^{-1}\sum_{j=1}^{n}(X_j - \overline{X})^2\right]^{\frac{1}{2}} \tag{37}$$

If ξ is known, the maximum likelihood estimator is

$$\left[n^{-1}\sum_{j=1}^{n}(X_j - \xi)^2\right]^{\frac{1}{2}} = [S^2 + (\overline{X} - \xi)^2]^{\frac{1}{2}}. \tag{38}$$

It is, however, very unusual to know ξ exactly, and we will not discuss this estimator further, except to note that neither (36) nor (37) is an unbiased estimator of σ. In fact

$$\begin{aligned} E[S] &= \sigma(2/n)^{\frac{1}{2}}\Gamma(\tfrac{1}{2}n)/\Gamma(\tfrac{1}{2}n - \tfrac{1}{2}) \\ &= \sigma/a_n \end{aligned} \tag{39}$$

and

$$E[\{S^2 + (\overline{X} - \xi)^2\}^{\frac{1}{2}}] = \sigma(2/n)^{\frac{1}{2}}\Gamma(\tfrac{1}{2}n + \tfrac{1}{2})/\Gamma(\tfrac{1}{2}n).$$

In order to obtain an unbiased estimator of σ, S must be multiplied by a_n. A few values of a_n are shown in Table 4. Values of $a_n' = a_n\sqrt{1 - n^{-1}}$, such that $a_n'E[\{(n - 1)^{-1}\sum(X_j - \overline{X})^2\}^{\frac{1}{2}}] = \sigma$, are also shown.

TABLE 4

Multipliers a_n, a'_n, such that
$E[a_n S] = \sigma = E[a'_n \sqrt{V}]$

n	a_n	a'_n
2	1.77245	1.25331
3	1.38198	1.12838
4	1.25331	1.08540
5	1.18942	1.06385
6	1.15124	1.05094
7	1.12587	1.04235
8	1.10778	1.03624
9	1.09424	1.03166
10	1.08372	1.02811

For n greater than 10 the formulas

$$a_n \doteqdot 1 + \tfrac{3}{4}(n-1)^{-1}; \qquad a'_n = 1 + \tfrac{1}{4}(n-1)^{-1}$$

give quite good approximations.

$$V = n(n-1)^{-1}S^2 = (n-1)^{-1}\sum_{j=1}^{n}(X_j - \overline{X})^2 \tag{40}$$

is an unbiased estimator of σ^2.

Jarrett [151] has given an interesting historical account of tables of these multiplying factors. (See also Cureton [53] and Bolch [17].)

(Note that the value of a minimizing the *mean square error* of aS^2 as an estimator of σ^2 is $(n^{-1}+1)^{-1}$. The value of b minimizing the mean square error of bS as an estimator of σ is a'_{n+1}. (Markowitz [200].))

The variance of the unbiased estimator, $a_n S$, of σ is

$$\text{var}(a_n S) = [a_n^2(1 - n^{-1}) - 1]\sigma^2. \tag{41}$$

The variance of V is

$$\text{var}(V) = 2(n-1)^{-1}\sigma^4. \tag{42}$$

Unbiased estimators of σ can also be obtained by multiplying the mean deviation (M) and the range (W) by appropriate factors (depending, of course, on n). The resulting unbiased estimators $b_n M$, $c_n W$ are identical with $a_n S$ for $n = 2$, and have greater variances than $a_n S$ for $n > 2$. Values of b_n can be calculated from the simple formula $b_n = \sqrt{\frac{\pi}{2} n(n-1)^{-1}}$. Values of c_n for $n = 2(1)20$ are given in Pearson and Hartley [240].

Relative efficiencies (inverse ratio of variance) of $b_n M$, $c_n W$ and other unbiased estimators of σ compared with $a_n S$, are shown in Table 7 page 70. From this table it can be seen that the estimator based on range is (slightly) more efficient than that based on mean deviation for $n = 4, 5$, but less efficient for $n \geq 6$. (For $n = 2, 3$ the two estimators give identical estimators of σ.)

The formula for b_n, quoted above, follows from

$$E[M] = \sigma\sqrt{(2/\pi)(1 - n^{-1})}. \tag{43}$$

The variance of M is

$$\text{var}(M) = \frac{2\sigma^2}{n\pi}\left(1 - \frac{1}{n}\right)\left\{\tfrac{1}{2}\pi + \sqrt{n(n-2)} - n + \sin^{-1}\frac{1}{n-1}\right\}. \tag{44}$$

For $n \geq 5$, a very good approximation (error no more than about 0.00001) is

$$\sigma^{-2}\,\text{var}(M) \doteqdot n^{-1}(1 - 2\pi^{-1})(1 - 0.12n^{-1}). \tag{45}$$

Approximate formulas for the moment ratios of M are

$$\beta_1(M) \doteqdot 1.05n^{-1}; \qquad \beta_2(M) = 3 + 0.91n^{-1}.$$

Godwin and Hartley [106] calculated a table of the probability integral of the distribution of M, giving $\Pr[M \leq m\sigma]$ to 5 decimal places for $m = 0.00(0.02)3.00$ and $n = 2(1)10$; and also multipliers M_α for percentile points such that

$$\Pr[M \leq M_\alpha\sigma] = \alpha,$$

to 3 decimal places, for $n = 2(1)10$ and $\alpha = 0.001, 0.005, 0.01, 0.025, 0.05, 0.1, 0.9, 0.95, 0.975, 0.99, 0.995$, and 0.999. For $n = 10$, there are also given approximate values calculated from the formula

$$M_\alpha \doteqdot E[M/\sigma] + U_\alpha(\sqrt{\text{var}(M)}/\sigma). \tag{46}$$

Although the upper and lower 2.5%, 5% and 10% values are not too inaccurate, approximation is poor for the more extreme values. A better approximation was obtained by Cadwell [30], by regarding $(M/\sigma)^{1.8}$ as having (approximately) a $c\chi^2_\nu$ distribution with

$$\begin{cases} \log c = -\log 2 - 1.8[\log(\tfrac{5}{9} + \tfrac{1}{2}\nu) - \log\Gamma(\tfrac{1}{2}\nu) \\ \qquad\qquad - \tfrac{1}{2}\log\{(2/\pi)(1 - n^{-1})\}] \\ \nu = \nu_0 + 0.196 - 0.159\nu_0^{-1} \\ \nu_0 = 0.617[E[M]]^2/\text{var}(M). \end{cases} \tag{47}$$

The approximation was obtained by first finding values of λ, c, and ν to make $(M/\sigma)^\lambda$ and $c\chi^2_\nu$ have the same first three moments. The values depend, of course, on n, and are shown in Table 5 (based on [30]). This table also shows the results of similar calculations for the range, W. It can be seen that, for M, the values of λ do not vary much, once n exceeds 5, say. An 'average' value of 1.8 was chosen; the values of ν and c were then adjusted to make the first *two* moments of $(M/\sigma)^{1.8}$ and $c\chi^2_\nu$ agree. It might be thought that $\lambda = 1.7$ would have been a better choice, but the chosen value of λ (1.8) does give closer approximation for smaller values of n, without affecting the accuracy too severely for larger values of n.

TABLE 5

Values of ν, λ and $\log c$ such that First Three Moments of $(T/\sigma)^\lambda$ and $c\chi^2_\nu$ Agree

	T = Range			T = Mean Deviation		
n	ν	λ	$\log_{10} c$	ν	λ	$\log_{10} c$
2	1.00	2	0.3010	1.00	2	$\bar{1}.6990$
3	2.05	1.96	0.2368	2.05	1.96	$\bar{1}.4234$
4	3.20	1.90	0.1547	3.35	1.84	$\bar{1}.2370$
5	4.5	1.83	0.0607	4.6	1.80	$\bar{1}.1149$
6	6.0	1.75	$\bar{1}.9588$	5.9	1.77	$\bar{2}.0177$
7	7.7	1.67	$\bar{1}.8574$	7.2	1.74	$\bar{2}.9388$
8	9.5	1.60	$\bar{1}.7685$	8.4	1.74	$\bar{2}.8765$
9	12	1.51	$\bar{1}.6457$	9.6	1.74	$\bar{2}.8222$
10	14	1.46	$\bar{1}.5785$	11	1.72	$\bar{2}.7674$
12	19	1.36	$\bar{1}.4252$	13	1.74	$\bar{2}.6964$
14	26	1.24	$\bar{1}.2479$	16	1.70	$\bar{2}.6126$
16	34	1.14	$\bar{1}.0964$	18	1.72	$\bar{2}.5617$
18	46	1.03	$\bar{2}.9142$	21	1.70	$\bar{2}.4988$
20	60	0.94	$\bar{2}.7569$	23	1.71	$\bar{2}.5491$

In the same paper [30], Cadwell discussed approximations to the distributions of the arithmetic mean of a number (k) of (independent) (M/σ)'s or (W/σ)'s. He also considered the distributions of the ratios

$$\max_j (M_j)/\min_j (M_j) \qquad \text{and} \qquad \max_j (W_j)/\min_j (W_j),$$

analogous to the ratios of maximum to minimum of a number of independent mean square deviations ($S_1^2, S_2^2, \ldots, S_k^2$) each based on the same number of observed values. Cadwell gives tables of approximate upper 5% and 1% percentage points of these statistics for $n = 3(1)10$ and $k = 2(1)12$ (for the 5% points, $n = 12, 15, 20, 30, 60$ are also included for the mean deviation ratio, and $n = 12, 15, 20$ for the range ratio). Similar tables for the statistics

$$\max_j (S_j^2)/\min_j (S_j^2)$$

are given by Pearson and Hartley [240]. In [240], also, are some useful tables of values connected with the distribution of (M/σ), including the expected value, variance, β_1, and β_2 for $n = 2(1)20, 30, 60$, and the upper and lower 0.1, 0.5, 2.5, 5 and 10 per cent points for $n = 2(1)10$.

Among the few simple exact results concerning the distribution of range, we note:

for $n = 2$; $E[W] = 2\sigma/\sqrt{\pi}$; $\text{var}(W) = 2\sigma^2(1 - 2\pi^{-1})$;
for $n = 3$; $E[W] = 3\sigma/\sqrt{\pi}$; $\text{var}(W) = \sigma^2[2 - (9 - 3\sqrt{3})\pi^{-1}]$;
and for $n = 4$; $E[W] = (3\sigma/\sqrt{\pi})(1 + 2\pi^{-1}\sin^{-1}\frac{1}{3})$.

Godwin [105] gives a number of other exact values of first and second moments.

However, quite extensive tables of the distribution and moments of (W/σ) are available. A historical survey (up to 1960) of these tables has been given by Harter [127], who also provided tables of percentage point multipliers, W_α, to 6 decimal places for $n = 2(1)20(2)40(10)100$ and $\alpha = 0.0001$, 0.0005, 0.001, 0.005, 0.01, 0.025, 0.05, 0.1(0.1)0.9, 0.95, 0.975, 0.99, 0.995, 0.999, 0.9995, and 0.9999. There are also, in [127], tables of the expected value, variance, β_1 and β_2 of (W/σ) to 8 (or more) significant figures for $n = 2(1)100$. Pearson and Hartley [240] give tables of $\Pr[W \leq w\sigma]$ to 4 decimal places for $n = 2(1)20$ and $w = 0.00(0.05)7.25$. They also give the upper and lower 0.1, 0.5, 1, 2.5, 5 and 10 percent points of the distribution of (W/σ) to 2 decimal places, and expected value and variance (5 decimal places), β_1 (4 decimal places) and β_2 (3 decimal places) for $n = 2(1)20$, 30 and 60.

From Table 5, it can be seen that it is not likely that a single value of λ can be found such that $(W/\sigma)^\lambda$ is well approximated by a distribution of $c\chi_\nu^2$ (for suitably chosen c and ν) for a range of values of $n \leq 20$.

Pearson [236] and Cox [48] have investigated this kind of approximation in some detail. From their investigations it appears that for smaller values of n (say $n \leq 8$) an approximation of form $c\chi_\nu$ is preferable (indeed, it is exact for $n = 2$); an approximation of form $c\chi_\nu^2$ is better for larger values of n.

Using tables of percentage points of (M/σ), (W/σ) and χ^2 (see Chapter 17) it is possible to construct confidence intervals for σ by rewriting the equation

$$\Pr[T_{\alpha_1}\sigma < T < T_{1-\alpha_2}\sigma \mid \sigma] = 1 - \alpha_1 - \alpha_2$$

in the form

$$\Pr[T/T_{1-\alpha_2} < \sigma < T/T_{\alpha_1} \mid \sigma] = 1 - \alpha_1 - \alpha_2 \tag{48}$$

showing that $(T/T_{1-\alpha_2}, T/T_{\alpha_1})$ is a $100(1 - \alpha_1 - \alpha_2)\%$ confidence interval for σ. Here T can be replaced by M, W or $\sqrt{n}\,S$, and T_α by $(M/\sigma)_\alpha$, $(W/\sigma)_\alpha$ or $\sqrt{n}\,\chi_{n-1,\alpha}$ respectively.

The maximum likelihood estimator, S, of σ (like its unbiased counterpart a_nS) is not a linear function of the observed values of $X_1, X_2, \ldots, X_n$. It is, however, possible to construct a *best linear unbiased estimator* of σ, using the order statistics $X_1', X_2', \ldots, X_n'$. Such estimators (using all the sample values) are of form

$$D = \sum_{j=1}^{[n/2]} \alpha_j(X_{n-j+1}' - X_j'). \tag{49}$$

Values of α_j have been calculated for $n = 2(1)20$, and have been published in Sarhan and Greenberg [270]. The efficiency of D relative to a_nS is always greater than 98%. Although this is very satisfactory, these estimators are not used often, because if *all* values are to be used, it is just as easy to calculate $a_n S$, and it does not require such extensive auxiliary tables. If a linear estimator is desired (e.g. to reduce effects of inaccurate outlying observations) there are other linear estimators, nearly as efficient as D, but with simpler formulas for the coefficients.

We take especial note of *Gini's mean difference*

$$G = \binom{n}{2}^{-1} \sum_{i<j}^{n} \sum^{n} |X_i - X_j| \tag{50}$$

$$= \frac{4}{n(n-1)} \sum_{j=1}^{[n/2]} \{\tfrac{1}{2}(n+1) - j\}(X'_{n-j+1} - X'_j).$$

We have

$$E[G] = (2/\sqrt{\pi})\sigma$$

$$\text{var}[G] = 4[n(n-1)]^{-1}[\tfrac{1}{3}(n+1) + 2\pi^{-1}(n-2)\sqrt{3} - 2\pi^{-1}(2n-3)]\sigma^2$$

The statistic $\frac{1}{2}\sqrt{\pi}\, G$ is an unbiased estimator of σ. The first three lines of Table 7 (taken from Nair [215]) show the efficiencies of $M\sqrt{\pi/2}$, $\frac{1}{2}\sqrt{\pi}\, G$ and D, the best linear unbiased estimator of σ, for $n = 2(1)10$. It can be seen that $\frac{1}{2}\sqrt{\pi}\, G$ is very nearly as efficient as D. As n tends to infinity, the efficiency of $\frac{1}{2}\sqrt{\pi}\, G$, relative to $a_n S$, tends to

$$(\tfrac{2}{3} + 4\sqrt{3} - 8)^{-1} = 97.8\%.$$

The asymptotically efficient estimator obtained by the method of Chernoff, *et al.* [34] (see Section 2 of Chapter 12) is obtained by putting α_j in (49), equal to $U_{j/(n+1)}$, for all j.

We also mention an estimator of σ, suggested by Gupta [114] as an approximation to D. This is obtained by replacing the coefficients α_j in (49) by

$$\alpha'_j = E[U'_{n-j+1}] \Big/ \sum_{j=1}^{n} \{E[U'_j]\}^2 \tag{51}$$

where $U'_1 \le U'_2 \le \cdots \le U'_n$ are order statistics corresponding to n independent unit normal variables, so that the estimator is

$$D' = \left\{\sum_{j=1}^{n} U'_j E[U'_j]\right\} \Big/ \sum_{j=1}^{n} \{E[U'_j]\}^2. \tag{52}$$

For large n this estimator is very nearly the same as the asymptotically efficient estimator just described. Shapiro and Wilk [278] described estimators similar to D', except that the ratio of α'_1 to the remaining α''s is modified.

In view of the accuracy attainable with $\frac{1}{2}\sqrt{\pi}\, G$ it does not seem necessary to consider the use of D'.

However, we note an estimator proposed by Mead [206] which is based on the same general idea as Gupta's estimator and may be convenient to apply in special circumstances (e.g. using certain kinds of rapid measuring apparatus).

Suppose that the number of observations (n) is a multiple of m, say km. The data are then sorted into m groups, so that the m least values are in the first group, the next m least values is the second group, and so on, the last group consisting of the m greatest values.

If the unit normal distribution is truncated between u_{i-1} and u_i ($> u_{i-1}$) then (see Section 7.1) the expected value of the truncated distribution is

$$\lambda_{(i)} = -[Z(u_i) - Z(u_{i-1})]/[\Phi(u_i) - \Phi(u_{i-1})].$$

Mead's estimator is then

$$\sum_{i=1}^{k} \lambda_{(i)} \text{ [mean of } i\text{th group]} \Big/ \sum_{i=1}^{k} \lambda_{(i)}^2 \tag{53}$$

with

$$u_0 = -\infty; \qquad u_i = U_{ik^{-1}}; \qquad u_k = \infty; \qquad (i = 1,\ldots,(k-1)$$

(note that $\Phi(u_i) - \Phi(u_{i-1}) = k^{-1}$ for all i). Denoting the mean of the ith group by $\bar{Y}_i$, the estimators are: for $k = 2$, $0.62666\,(\bar{Y}_2 - \bar{Y}_1)$; for $k = 3$, $0.45838\,(\bar{Y}_3 - \bar{Y}_1)$; for $k = 4$, $0.36927\,(\bar{Y}_4 - \bar{Y}_1) + 0.09431\,(\bar{Y}_3 - \bar{Y}_2)$; for $k = 5$, $0.31213\,(\bar{Y}_5 - \bar{Y}_1) + 0.11860\,(\bar{Y}_4 - \bar{Y}_2)$; for $k = 6$, $0.27177\,(\bar{Y}_6 - \bar{Y}_1) + 0.12373\,(\bar{Y}_5 - \bar{Y}_2) + 0.03844\,(\bar{Y}_4 - \bar{Y}_3)$. Mead obtained the following values for the asymptotic efficiency of this estimator (n large) relative to $a_n S$.

k	2	3	4	5	6
Asymptotic Efficiency (%)	87.6	93.3	95.6	96.8	97.8

Yet another class of linear estimators has been studied by Oderfeld and Pleszczyńska [222], [251]. These are linear functions of the order statistics $Y_1', \ldots, Y_n'$ corresponding to the random variables $Y_j = |X_j - \bar{X}|$. These authors estimated the values of the coefficients in

$$\sum_{j=1}^{n} \alpha_j Y_j'$$

on an empirical basis, using results of sampling experiments. In [222], the only value used for n was 5, and the suggested estimator was

$$-0.065\,Y_1' + 0.150\,Y_2' + 0.175\,Y_3' + 0.312\,Y_4' + 0.405\,Y_5'.$$

In [251] the smallest absolute deviate was omitted (i.e. α_1 taken equal to zero). Coefficients (α_j) were estimated for $n = 3(1)10$.

In all cases the largest coefficient was α_n, indicating the relative importance of extreme observations in estimating σ.

These estimators appear (from empirical evidence) to have quite high (over 95%) efficiencies relative to $a_n S$, but no higher than, for example, estimators based on Gini's mean difference.

For $n \leq 10$ (at least) there are even simple unbiased estimators of σ which are not much less efficient than $\frac{1}{2}\sqrt{\pi}\, G$ or D. These are based on the *thickened ranges*

$$J_{(r)} = \sum_{j=1}^{r} (X'_{n-j+1} - X'_j) \tag{54}$$

(Jones [153]).

Values of the multiplying factor to be applied to $J_{(2)}$ to make it an unbiased estimator of σ are shown in Table 6 (taken from [153]).

TABLE 6

Multiplying Factor for $J_{(2)}$

n	Factor
4	0.37696
5	0.30157
6	0.26192
7	0.23702
8	0.21970
9	0.20684
10	0.19684

Relative efficiencies of these estimators (compared with $a_n S$) are also shown in Table 7. It can be seen that if $J_{(1)}$ ($\equiv W$) be used for $n \leq 5$ and $J_{(2)}$ for $6 \leq n \leq 10$, the relative efficiency never falls below 98%.

For large n, Prescott [255] has given the approximate formulas

$$E[J_{(r)}] = 2r\sigma Z(U_p)/p$$

$$\text{var}(J_{(r)}) = \frac{2r^2\sigma^2}{np^2}[p + \{pU_p - Z(U_p)\}\{(1 - 2p)U_p + 2Z(U_p)\}]$$

where $p = r/n$ is not too small.

For n large, maximum efficiency (96.65%) is attained with $p = 0.225$. Prescott suggests using $p = \frac{1}{6}$ since the efficiency is still over 90% and the easily remembered quantity

$$\frac{1}{3r} J_{(r)}$$

is very nearly an unbiased estimator of σ.

TABLE 7

Relative Efficiencies of Unbiased Estimators of σ

Estimator based on:		n = 2	3	4	5	6	7	8	9	10
Mean deviation		100.0	99.19	96.39	94.60	93.39	92.54	91.90	91.4	91.0
Gini mean difference		100.0	99.19	98.75	98.50	98.34	98.24	98.16	98.1	98.1
Best linear		100.0	99.19	98.92	98.84	98.83	98.86	98.90	98.9	99.0
Range ($J_{(1)} \equiv W$)		100.0	99.19	97.52	95.48	93.30	91.12	89.00	86.9	85.0
Thickened range	$J_{(2)}$	—	—	91.25	93.84	95.71	96.67	96.97	96.8	96.4
	$J_{(3)}$	—	—	—	—	90.25	91.78	93.56	95.0	95.9
	$J_{(4)}$	—	—	—	—	—	—	89.76	90.7	92.2
	$J_{(5)}$	—	—	—	—	—	—	—	—	89.4

Dixon [65] has considered estimators of form:

$$k' \sum_j W_{(j)} = k' \sum_j (X'_{n-j+1} - X'_j) \tag{55}$$

where the summation $\sum_j$ is over some set of values of j. The statistic $W_{(j)}$ is sometimes called the jth *quasi-range*, occasionally the jth *subrange*. Evidently

$$J_{(r)} = \sum_{j=1}^{r} W_{(j)} \qquad (\text{and } W_{(1)} \equiv W). \tag{56}$$

Dixon found that, for $n \leq 10$, the most efficient estimators of form (55) are those based on the range ($W \equiv W_{(1)}$), or thickened range $J_{(2)}$ just described. For $n = 11 - 20$ he obtained the most efficient unbiased estimators (in the class (55)) given in Table 8.

The efficiencies compare quite well with those of D, though they are not generally as high as those of Gini's mean difference.

Note that those unbiased linear estimators which are optimal (in various restricted classes) give large relative weight to the extreme observations. (Even S, which appears to be symmetrical, can be regarded (very roughly) as using weights proportional to the absolute magnitude of deviation from the sample mean.) Although we have obtained high efficiencies with these estimators, the calculations are all based on the complete validity of the normal distribution as applied to the data. Distributions of extreme order statistics are likely to be especially sensitive to departures from normality, and it is sometimes more

TABLE 8

Unbiased Estimators of σ

n	Estimator	Efficiency (%) relative to $a_n S$
11	$0.1608(W + W_{(2)} + W_{(4)})$	96.7
12	$0.1524(W + W_{(2)} + W_{(4)})$	97.2
13	$0.1456(W + W_{(2)} + W_{(4)})$	97.5
14	$0.1399(W + W_{(2)} + W_{(4)})$	97.7
15	$0.1352(W + W_{(2)} + W_{(4)})$	97.7
16	$0.1311(W + W_{(2)} + W_{(4)})$	97.5
17	$0.1050(W + W_{(2)} + W_{(3)} + W_{(5)})$	97.8
18	$0.1020(W + W_{(2)} + W_{(3)} + W_{(5)})$	97.8
19	$0.09939(W + W_{(2)} + W_{(3)} + W_{(5)})$	97.9
20	$0.10446(W + W_{(2)} + W_{(4)} + W_{(6)})$	98.0

important to guard against this possibility than to squeeze the last drop of formal 'efficiency' from the data.

The simplest method of this kind uses only a single pair of symmetrically placed order statistics, in the form of a quasi-range, giving an estimator

$$\beta_r W_{(r)} = \beta_r(X'_{n-r+1} - X'_r).$$

($W_{(r)}$ is called a *quasi-range* — see Equation (77).)

For n large it is best to take $r \doteqdot 0.069n$. The efficiency (relative to $a_n S$) of the corresponding unbiased estimator of σ is about 65% (Pearson [244]). (For estimating ξ by a statistic of form

$$\beta_{\frac{1}{2}r}(X'_{n-r+1} + X'_r)$$

the best choice is $r \doteqdot 0.270n$, and the efficiency is about 67%.)

Kulldorff [174] has studied the construction of estimators of this kind in some detail. He found that if estimators of form

$$\sum_{i=1}^{k} \beta_{r_i} W_{(r_i)} \tag{57}$$

were considered, quite good results could be obtained by taking β_{r_i} proportional to i giving estimators of form

$$\gamma \sum_{i=1}^{k} i W_{(r_i)}. \tag{58}$$

For $k = 2$, Kulldorff found that the best values of r_1 and r_2 (subject to an estimator of form (58) being used) to take are $0.0235n$ and $0.1279n$ respectively, with $\gamma = 0.1174$; for $k = 3$, optimal values are $r_1 = 0.0115n$, $r_2 = 0.0567n$, $r_3 = 0.1704n$, with $\gamma = 0.0603$. The corresponding relative efficiencies are approximately 82% for $k = 2$, and 89% for $k = 3$. Note that these results apply to 'large' samples (large values of n), and cannot be expected to apply when $n \leq 20$, as in the discussion of the other estimators.

Using the large-sample approximations, Eisenberger and Posner [73] have constructed 'best linear unbiased estimators' of mean and standard deviation using only a fixed number (k) of quantiles, and excluding quantiles more extreme than 1 and 99%, or 2.5 and 97.5%, for $k = 2(2)20$. They also give (for the same values of k) pairs of linear estimators minimizing

$$\text{(variance of estimator of mean)} + \lambda \cdot \text{(variance of estimator of standard deviation)}$$

for $\lambda = 1, 2$ and 3.

Formulas appropriate for censored samples (described in Section 6.4) can also be used for complete samples, if it is felt wise to ignore certain observed values.

For the linear estimators based on order statistics (apart from those based on W alone) there are no easily available tables of percentage points. Such tables would be needed to construct confidence intervals for σ by rearranging the relation

$$\Pr[T_{\alpha_1}\sigma < T < T_{1-\alpha_2}\sigma \mid \sigma] = 1 - \alpha_1 - \alpha_2$$

in the form

$$\Pr[T/T_{1-\alpha_2} < \sigma < T/T_{\alpha_1} \mid \sigma] = 1 - \alpha_1 - \alpha_2$$

(see (48)). Even for those cases where such tables are available (M and W) only symmetrical intervals (with $\alpha_1 = \alpha_2$) are practically useful. Except in connection with intervals based on S, no attempt has been made to construct shortest confidence intervals for σ. Nair [214] suggested calculation of upper and lower 1% and 5% points of the distribution of the second thickened range $J_{(2)}$, but such tables have not been published.

In many cases an approximation using a distribution of the form of that of a multiple of a chi random variable ($c\chi_\nu$) (Chapter 17) may give results which are not seriously inaccurate for practical purposes.

6.3 *Estimation of functions of ξ and σ*

Certain functions of both ξ and σ are sometimes the primary target of statistical estimation. Among these we particularly note the $100\alpha\%$ percentile point ($\xi + U_\alpha\sigma$) and the proportion (of population) less than a fixed number x,

$$P_x = (\sqrt{2\pi}\,\sigma)^{-1}\int_{-\infty}^{(x-\xi)/\sigma} e^{-\frac{1}{2}u^2}\,du. \tag{59}$$

These quantities may be estimated by general methods, not using the special form of distribution. However when the validity of the assumption of normality has been clearly established, it is to be expected that more accurate estimates can be obtained by using this knowledge of the form of distribution.

Evidently, if $\hat{\xi}'$, $\hat{\sigma}'$ be any unbiased estimators of ξ, σ respectively, then

$(\hat{\xi}' + U_\alpha \hat{\sigma}')$ is an unbiased estimator of $\xi + U_\alpha \sigma$. If $\bar{X}$ be used as the estimator of ξ, and $\hat{\sigma}'$ is any of the unbiased estimators of σ described in Section 6.2, then, since $\bar{X}$ and $\hat{\sigma}'$ are independent

$$\text{var}(\bar{X} + U_\alpha \hat{\sigma}') = \sigma^2 n^{-1} + U_\alpha^2 \,\text{var}(\hat{\sigma}'). \tag{60}$$

Combination of $\bar{X}$ with the best linear unbiased estimator of σ (or one of the other, nearly as efficient, linear estimators of σ) will give a good linear estimator of $\xi + U_\alpha \sigma$.

If $a_n S$ be used as estimator of σ the distribution of the estimator $(\bar{X} + U_\alpha a_n S)$ may be evaluated in the following way.

$$\begin{aligned} (61) \quad & \Pr[\bar{X} + U_\alpha a_n S \le K] \\ &= \Pr[(\xi + U\sigma/\sqrt{n}) + (U_\alpha a_n \sigma/\sqrt{n})\chi_{n-1} \le K] \\ &= \Pr[\{U + \sqrt{n}\,(\xi - K)/\sigma\}(\chi_{n-1}/\sqrt{n-1})^{-1} \le -U_\alpha a_n \sqrt{n-1}] \\ &= \Pr[t'_{n-1}(\sqrt{n}\,(\xi - K)/\sigma) \le -U_\alpha a_n \sqrt{n-1}] \end{aligned}$$

where $t'_{n-1}(\lambda)$ denotes a noncentral t variable (see Chapter 31) with $(n-1)$ degrees of freedom and noncentrality parameter.

If other estimators of σ are used, approximate results, of similar form, can be obtained by approximating the distribution of $\hat{\sigma}'$ by that of $c\chi_\nu$, with suitable values of c and ν. It will usually be troublesome to assess the accuracy of these approximations. If effects of unreliable outlying observations are to be specially avoided, then estimators of ξ and σ not using such observations may be used. However, if the reason for this precaution is that possible lack of normality is suspected, it is doubtful whether $\xi + U_\alpha \sigma$ should be estimated at all.

Coming now to the estimation of quantities like

$$\Pr[X \le x] = (\sqrt{2\pi})^{-1} \int_{-\infty}^{(x-\xi)/\sigma} e^{-\frac{1}{2}u^2}\, du$$

it is clear that the maximum likelihood estimator is obtained by replacing ξ by $\bar{X}$, and σ by S'. The resulting estimator is, in general, biased. (It is unbiased if it so happens that $x = \xi$.)

To obtain the minimum variance unbiased estimator, the Blackwell-Rao theorem may be used. The estimator

$$T = \begin{cases} 1 \text{ if } X_1 \le x \\ 0 \text{ if } X_1 > x \end{cases} \tag{62}$$

is an unbiased estimator of $\Pr[X \le x]$ and $\bar{X}$ and S are jointly complete sufficient statistics for ξ and σ. Hence the minimum variance unbiased estimator of $\Pr[X \le x]$ is

$$\begin{aligned} E[T \mid \bar{X},S] &= \Pr[X_1 \le x \mid \bar{X},S] \\ &= \Pr\left[\frac{X_1 - \bar{X}}{S} \le \frac{x - \bar{X}}{S} \,\middle|\, \bar{X},S\right]. \end{aligned}$$

Since the conditional distribution of $(X_1 - \overline{X})/S$ is independent of both $\overline{X}$ and S, it is the same as the unconditional distribution of $(X_1 - \overline{X})/S$. Making an orthogonal transformation with one new variable and one equal to $\sqrt{n}\,\overline{X}$, equal to $\sqrt{n/(n-1)}\,(X_1 - \overline{X})$ it can be seen that $(X_i - \overline{X})/S$ is distributed symmetrically about zero as the signed square root of $(n-1)$ times a beta variable with parameters $\frac{1}{2}$, $\frac{1}{2}n - 1$ (see Chapter 24). Hence the minimum variance unbiased estimator of $\Pr[X \leq x]$ is

$$(63)\qquad \begin{cases} [B(\tfrac{1}{2},\tfrac{1}{2}n-1)]^{-1}\displaystyle\int_{-1}^{(x-\overline{X})/(S\sqrt{n-1})} (1-v^2)^{\frac{1}{2}(n-4)}\,dv & \text{for } |x-\overline{X}| \leq S\sqrt{n-1} \\ 0 & \text{for } x < \overline{X} - S\sqrt{n-1} \\ 1 & \text{for } x > \overline{X} + S\sqrt{n-1}. \end{cases}$$

(Numerical evaluation can be effected using tables of the incomplete beta function, as described in Chapter 24.)

At this point we note that, if X_{n+1} is independent of, and has the same distribution as each X_j then $(X_{n+1} - \overline{X})/S$ is distributed as $[(n+1)/(n-1)]^{\frac{1}{2}}$ times t with $(n-1)$ degrees of freedom. Hence the interval

$$(\overline{X} + t_{n-1,\alpha_1}[(n+1)/(n-1)]^{\frac{1}{2}}S,\quad \overline{X} + t_{n-1,1-\alpha_2}[(n+1)/(n-1)]^{\frac{1}{2}}S)$$

contains on average a proportion $(1 - \alpha_1 - \alpha_2)$ of the population values. It is thus a form of *tolerance interval* for the normal distribution. Unlike the tolerance intervals described in Chapter 12, the construction of this interval makes use of knowledge of the form of population distribution. It cannot be used for other populations without the possibility of introducing bias.

Wald and Wolfowitz [311] have shown that a good approximation to *tolerance limits*, such that there is a probability equal to $(1 - \alpha)$ that the limits include *at least* a specified proportion γ of the population is

$$(64)\qquad \overline{X} \pm \lambda_\gamma \sqrt{n}\, S/\chi_{n-1,\alpha}$$

where λ_γ satisfies the equation

$$\Phi(n^{-\frac{1}{2}} + \lambda_\gamma) - \Phi(n^{-\frac{1}{2}} - \lambda_\gamma) = \gamma.$$

The construction of exact *one-sided* tolerance limits can be simply effected, using the noncentral t distribution (see Chapter 31). We note that the population proportion less than $(\overline{X} + kS)$ is

$$\Phi(\{\overline{X} + kS - \xi\}/\sigma)$$

and this is at least γ if

$$(\overline{X} + kS - \xi)/\sigma \geq U_\gamma.$$

This inequality can be rearranged in the form

(65)
$$\frac{\sqrt{n}\,\{(\bar{X}-\xi)/\sigma\}-\sqrt{n}\,U_\gamma}{\sqrt{n}\,(S/\sigma)/\sqrt{n-1}} \geq -k\sqrt{n-1}\cdot$$

The statistic on the left-hand side of (65) has a noncentral t distribution with $(n-1)$ degrees of freedom and noncentrality parameter $(-\sqrt{n}\,U_\gamma)$. In order that the probability that at least a proportion γ of the population is less than $(\bar{X}+kS)$, should be equal to $(1-\alpha)$, we make $(-k\sqrt{n-1})$ equal to the lower $100\alpha\%$ point of the noncentral t distribution, i.e.

(66)
$$k = -t'_{n-1,\alpha}(-\sqrt{n}\,U_\gamma)/\sqrt{n-1}.$$

Sometimes it is desired to estimate the *mean square error* $(\xi-\xi_0)^2+\sigma^2$, where ξ_0 is a specified number. The mean square

$$n^{-1}\sum_{j=1}^{n}(X_j-\xi_0)^2$$

is an unbiased estimator of this quantity. It is distributed as

$$n^{-1}\sigma^2 \times (\text{noncentral } \chi^2 \text{ with } n \text{ degrees of freedom and noncentrality parameter } n(\xi-\xi_0)^2/\sigma^2)$$

(see Chapter 28), and has variance

$$2n^{-2}\sigma^4[n+2n(\xi-\xi_0)^2/\sigma^2] = 2n^{-1}\sigma^4[1+2(\xi-\xi_0^2)/\sigma^2].$$

A natural estimate of the coefficient of variation (σ/ξ) is the ratio $a_nS/\bar{X}$, or more generally $a'_nS/\bar{X}$, a'_n being suitably chosen. Since the expected value of $S/\bar{X}$ is infinite it is not possible to obtain an unbiased estimator of this form. We can, however, construct an *approximate* confidence interval for σ/ξ. We will suppose that $\Pr[\bar{X}<0]$ can be neglected (i.e. σ/ξ sufficiently small — less than $\frac{1}{4}$, say). Then, since $\bar{X}/S$ is distributed as $(n-1)^{-\frac{1}{2}}$ times noncentral t with $(n-1)$ degrees of freedom and noncentrality parameter $\sqrt{n}\,\xi/\sigma$, it follows that (in the notation of Chapter 31)

$$\Pr[t'_{n-1,\alpha_1}(\sqrt{n}\,\xi/\sigma) \leq \sqrt{n-1}\,\bar{X}/S \leq t'_{n-1,1-\alpha_2}(\sqrt{n}\,\xi/\sigma)] = 1-\alpha_1-\alpha_2$$

or

(67)
$$\Pr[n^{-\frac{1}{2}}g_{1-\alpha_2}(\sqrt{n-1}\,\bar{X}/S) \leq \xi/\sigma \leq n^{-\frac{1}{2}}g_{\alpha_1}(\sqrt{n-1}\,\bar{X}/S)] \doteqdot 1-\alpha_1-\alpha_2$$

where $g_\alpha(z)$ is the solution (for g) of the equation

(68)
$$t'_{n-1,\alpha}(g) = z$$

(assuming $\bar{X}$ not too small). Assuming now that $\xi>0$, (67) can be rewritten

$$\Pr[\sqrt{n}/g_{\alpha_1}(\sqrt{n-1}\,\bar{X}/S) < \sigma/\xi < \sqrt{n}/g_{1-\alpha_2}(\sqrt{n-1}\,\bar{X}/S)] \doteqdot 1-\alpha_1-\alpha_2.$$

It is necessary to use tables of the noncentral t distribution (see Chapter 31, Section 7) to calculate even these approximate limits.

More easily calculable, but rather rough, approximate limits are obtained from the formula

$$\text{(69)} \quad \begin{aligned} \text{Lower limit} &= V[1 - n^{-\frac{1}{2}} U_{\alpha_2} \sqrt{\tfrac{1}{2} + V^2}]^{-1} \\ \text{Upper limit} &= V[1 - n^{-\frac{1}{2}} U_{1-\alpha_1} \sqrt{\tfrac{1}{2} + V^2}]^{-1} \end{aligned}$$

where $V = S/\overline{X}$. These are based on the assumption that $(S - k\overline{X})$ is approximately normally distributed with expected value $(\sigma - k\xi)$ and variance $n^{-1}\sigma^2(1 + \frac{1}{2}k^2)$ so that (remembering $\xi \gg \sigma$)

$$\Pr[S/\overline{X} < k] \doteqdot \Phi(\sqrt{n/(1 + \tfrac{1}{2}k^2)}\,\{1 - k\xi/\sigma\})$$

i.e. $\sqrt{n/(1 + \frac{1}{2}V^2)}\,(1 - V\xi/\sigma)$ has approximately a unit normal distribution.

A similar argument indicates that if X_1, X_2 are independent normal random variables and $E[X_j] = \xi_j$, $\text{var}(X_j) = \sigma_j^2$ $(j = 1,2)$ with $\xi_2 \gg \sigma_2$ then putting $X_1/X_2 = R$, the distribution of

$$(R\xi_2 - \xi_1)/(R^2\sigma_2^2 + \sigma_1^2)^{\frac{1}{2}}$$

is approximately unit normal.

Koopmans *et al.* [169] have pointed out that if the distribution of each of the independent variables is *lognormal* (see Chapter 14) then construction of *exact* confidence intervals for the coefficient of variation is straightforward. Since it is possible to approximate a normal distribution quite closely by a lognormal distribution (see Chapter 14, Section 3) it is likely that the same formulas will give good results for normal variables (though they will not, of course, give *exactly* specified values for confidence coefficients). The (approximate) confidence limits, in terms of the original variables $X_1, \ldots, X_n$, obtained by this method are

$$\text{(70)} \quad \left[\exp\left\{\sum_{j=1}^{n} (\log X_j - \overline{\log X})^2/\chi^2_{n-1,1-\alpha_1}\right\} - 1\right]^{\frac{1}{2}}$$

and

$$\left[\exp\left\{\sum_{j=1}^{n} (\log X_j - \overline{\log X})^2/\chi^2_{n-1,\alpha_2}\right\} - 1\right]^{\frac{1}{2}}$$

where $\overline{\log X} = n^{-1} \sum_{j=1}^{n} \log X_j$.

The cumulative distribution function of the rth quasi-range for random samples from a unit normal distribution is (Jones *et al.* [154]).

$$F_{W_{(r)}}(w) = \sum_{i=0}^{r} \frac{n^{(2r-i+1)}}{r!(r-i)!} \sum_{j=0}^{r-i} \sum_{k=0}^{n-2r+i-1} (-1)^{n-2r+i-1-j+k} P \qquad (w > 0)$$

where

$$P = \binom{r-i}{j}\binom{n-2r+i-1}{k} \Pr\left[\bigcap_{l=1}^{n-r+i+j-1} (Y_l \leq \delta_{j+h-l} w/\sqrt{2})\right]$$

and $\delta_h = 0, 1$ for $h < 0, \geq 0$ respectively; the Y's are standardized multinormal variables (Chapter 35) with all correlations equal to $\frac{1}{2}$.

6.4 *Estimation from Censored Data*

We will consider situations in which the r_1 least, and r_2 greatest, observations are censored (i.e. not recorded) leaving only $X'_{r_1+1}, \ldots, X'_{n-r_2}$. Best linear unbiased estimators, based on these order statistics, are particularly useful in these circumstances, as the maximum likelihood estimators of ξ and σ are much more difficult to calculate than they are for complete samples. We will first, however, discuss maximum likelihood estimators, and possible approximations thereto.

The joint probability density function of $X'_{r_1+1}, \ldots, X'_{n-r_2}$ is

$$(71)\quad \begin{aligned} &p(x_{r_1+1}, \ldots, x_{n-r_2+1}) \\ &= \frac{n!}{k_1!k_2!}[\Phi((x_{r_1+1} - \xi)/\sigma)]^{k_1}[1 - \Phi((x_{n-r_2} - \xi)/\sigma)]^{k_2} \\ &\quad \times \sigma^{-(n-r_1-r_2)} \prod_{j=r_1+1}^{n-r_2} Z((x_j - \xi)/\sigma). \end{aligned}$$

The maximum likelihood estimators $\hat{\xi}$, $\hat{\sigma}$ of ξ, σ respectively satisfy the following equations (using the notation $\hat{U}'_j = (X'_j - \hat{\xi})/\hat{\sigma}$):

$$(72.1)\quad \begin{aligned} &r_1[-Z(\hat{U}'_{r_1+1})/\Phi(\hat{U}'_{r_1+1})] + \sum_{j=r_1+1}^{n-r_2} \hat{U}'_j \\ &\qquad + r_2[Z(\hat{U}'_{n-r_2})/\{1 - \Phi(\hat{U}'_{n-r_2})\}] = 0 \end{aligned}$$

$$(72.2)\quad \begin{aligned} &r_1[1 - \hat{U}'_{r_1+1}Z(\hat{U}'_{r_1+1})/\Phi(\hat{U}'_{r_1+1})] + \sum_{j=r_1+1}^{n-r_2} \hat{U}'^2_j \\ &\qquad + r_2[1 + \hat{U}'_{n-r_2}Z(\hat{U}'_{n-r_2})/\{1 - \Phi(\hat{U}'_{n-r_2})\}] = n. \end{aligned}$$

From equations (73) below it can be seen that, in effect, in (72.1) the censored "observations" are replaced by the expected value of the appropriate tail of the normal distribution truncated at X'_{r_1+1} or X'_{n-r_2}, as the case may be, and in (72.2) the squared standardized deviates of these "observations" are replaced by the corresponding expected values for the tails.

Bearing this in mind, we obtain approximate solutions of Equations (72) by replacing $(X'_{r_1+1} - \hat{\xi})/\hat{\sigma}$ and $(X'_{n-r_2} - \hat{\xi})/\hat{\sigma}$ by U_{α_1} and $U_{1-\alpha_2}$ respectively (except in the summations) with $\alpha_j = (r_j + 1)/(n + 1)$ for $j = 1, 2$. Introducing the following notation for the moments of the tails (of singly truncated normal distributions — see Section 7):

$$(73)\quad \begin{aligned} -Z(U_\alpha)/\Phi(U_\alpha) &= \mu'_{1,(\alpha-)};\ Z(U_\alpha)/\{1 - \Phi(U_\alpha)\} = \mu'_{1,(\alpha+)} \\ 1 - U_\alpha Z(U_\alpha)/\Phi(U_\alpha) &= \mu'_{2,(\alpha-)};\ 1 + U_\alpha Z(U_\alpha)/\{1 - \Phi(U_\alpha)\} = \mu'_{2,(\alpha+)} \end{aligned}$$

we obtain the approximate equations

$$(74.1)\qquad \overline{X}' + (n - r_1 - r_2)^{-1}(r_1\mu'_{1,(\alpha_1-)} + r_2\mu'_{1,(1-\alpha_2+)})\hat{\sigma} \doteqdot \hat{\xi}$$

$$(74.2)\qquad \left[\sum_{j=r_1+1}^{n-r_2} (X'_j - \overline{X}')^2\right][n - r_1\mu'_{2,(\alpha_1-)} - r_2\mu'_{2,(1-\alpha_2+)} - (n - r_1 - r_2)^{-1}(r_1\mu_{1,(\alpha_1-)} + r_2\mu_{1,(1-\alpha_2+)})^2]^{-1} = \hat{\sigma}^2$$

with $\overline{X}' = (n - r_1 - r_2)^{-1} \sum_{j=r_1+1}^{n-r_2} X'_j$.

Values of $\mu'_{1,\alpha\pm}$ and $\mu'_{2,\alpha\pm}$ can be obtained from tables mentioned in Section 1 (also in Harter and Moore [131]).

Having obtained first approximations to $\hat{\xi}$ and $\hat{\sigma}$ from (74.1) and (74.2), these can be used to calculate new values of $\mu'_{j,\pm\alpha}$, now using $(X'_{r_1+1} - \hat{\xi})/\hat{\sigma}$ and $(X'_{n-r_2} - \hat{\xi})/\hat{\sigma}$ in place of U_{α_1} and $U_{1-\alpha_2}$ respectively. Then (72.1) and (72.2), in turn, give new approximations to $\hat{\xi}$, $\hat{\sigma}$ and so on.

Cohen [43] has given a chart which may be used in the solution of (72.1) and (72.2).

Approximations (for large values of n) to the variances and covariance of $\hat{\xi}$ and $\hat{\sigma}$ are the same as for the corresponding truncated distributions, with truncation points corresponding to $\alpha_j = r_j/n$ ($j = 1,2$). Some numerical values are given in Section 7.

Although solution of the maximum likelihood equations is practicable if the techniques described above or variants thereof are used, it is often convenient to use simpler estimating formulas.

Making the further approximation $r_j/n \doteqdot \Phi(U_{\alpha_j})$ ($j = 1,2$), Equations (74.1) and (74.2) can be written

$$(75.1)\qquad \overline{X}' + \hat{\sigma}[Z(U_{1-\alpha_2}) - Z(U_{\alpha_1})][1 - \alpha_1 - \alpha_2]^{-1} = \hat{\xi}$$

$$(75.2)\qquad \left[n^{-1} \sum_{j=r_1+1}^{n-r_2} (X'_j - \overline{X}')^2\right] \times [1 - \alpha_1 - \alpha_2 + U_{\alpha_1}Z(U_{\alpha_1}) - U_{1-\alpha_2}Z(U_{1-\alpha_2}) - \{Z(U_{\alpha_1}) - Z(U_{1-\alpha_2})\}^2(1 - \alpha_1 - \alpha_2)^{-1}]^{-1} = \hat{\sigma}^2$$

Tiku [303] has suggested that the approximate formulas

$$Z(x)/\Phi(x) = a_1 + b_1x, \qquad Z(x)/[1 - \Phi(x)] = a_2 + b_2x$$

be used to simplify equations (72.1) and (72.2). (The values of a_1, b_1, a_2, b_2 are chosen to give good fits over the range $U_{\alpha_1} \leq x \leq U_{1-\alpha_2}$.)

This leads to the following equations for the estimators of ξ and σ (denoted here by $\hat{\xi}'$, $\hat{\sigma}'$):

$$\hat{\xi}' = K + L\hat{\sigma}'$$

$$(1 - \alpha_1 - \alpha_2)\hat{\sigma}'^2 - \{\alpha_2a_2X'_{n-r_2} - \alpha_1a_1X'_{r_{1+1}} - (\alpha_2a_2 - \alpha_1a_1)K\}\hat{\sigma}'^2 - \left\{n^{-1} \sum_{j=r_1+1}^{n-r} X'^2_j + \alpha_2b_2X'^2_{n-r_2} - \alpha_1b_1X'^2_{r_{1+1}}\right\} - (1 - \alpha_1 - \alpha_2 + \alpha_2b_2 + \alpha_1b_1)K^2 = 0$$

with

$$K = \frac{n^{-1} \sum_{j=r_1+1}^{n-T_2} X'_j + \alpha_2 b_2 X'_{n-r_2} - \alpha_1 b_1 X'_{r_1+1}}{[1 - \alpha_1 - \alpha_2 + \alpha_2 b_2 - \alpha_1 b_1]}$$

and

$$L = (\alpha_2 a_2 - \alpha_1 a_1)/[1 - \alpha_1 - \alpha_2 + \alpha_2 b_2 - \alpha_1 b_1]$$

(the quadratic in $\hat{\sigma}'$ has only one positive root).

For symmetrical censoring ($\alpha_1 = \alpha_2 = \alpha$) we have $a_2 = a_1 = a$, say, and $b_2 = -b_1 = b$, leading to

$$\hat{\xi}' = \left[n^{-1} \sum_{j=r+1}^{n-r} X'_j + \alpha b(X'_{r+1} + X'_{n-r})\right] \Big/ (1 - 2\alpha + 2\alpha b)$$

and

$$(1 - 2\alpha)\hat{\sigma}'^2 - \alpha a(X'_{n-r} - X'_{r+1})\hat{\sigma}' - \left[n^{-1} \sum_{j=r+1}^{n-r} X'_j + \alpha b(X'^2_{n-r} + X'^2_{r+1}) - (1 - 2\alpha + 2\alpha b)K^2\right] = 0$$

(Note the $\hat{\xi}'$ is of similar form to the Winsorized mean (32).)

Tiku gives tables to assist in obtaining good values for a and b.

For symmetrical censoring, the Winsorized mean (32) described in Section 6.1 is a natural choice as estimator of ξ. For a moderate degree of asymmetry in censoring (i.e. $|r_1 - r_2|$ small) it may be worthwhile ignoring the excess observations to one side and using the Winsorized mean of the largest available symmetrically censored set of sample values. Table 9 (from Dixon [65]) gives the efficiency of the Winsorized mean

TABLE 9

Efficiency (%) of Winsorized Mean Based on $X'_{j+1}, \ldots, X'_{n-j}$ Relative to the Best Linear Unbiased Estimator of ξ Based on $X'_j, X'_{j+1}, \ldots, X'_{n-j+1}$

n \ j	1	2	3	4	5	6
3	100.0					
4	96.2					
5	96.4	100.0				
6	96.9	96.4				
7	97.3	96.3	100.0			
8	97.7	96.7	96.7			
9	98.0	97.1	96.5	100.0		
10	98.2	97.4	96.8	97.1		
12	98.6	97.9	97.4	97.0	97.4	
14	98.8	98.3	97.9	97.5	97.1	97.6
16	99.0	98.6	98.2	97.9	97.5	97.3
18	99.1	98.8	98.5	98.2	97.9	97.6
20	99.2	98.9	98.7	98.4	98.2	98.0

$$\xi_{(j)} = n^{-1}[(j+1)X'_{j+1} + X'_{j+2} + \cdots + X'_{n-j+1} + (j+1)X'_{n-j}]$$

relative to the best linear unbiased estimator of ξ based on $X'_j, X'_{j+1}, \ldots, X'_{n-j}$. It can be seen that the loss in efficiency from ignoring the value X'_j is trifling.

Even if one of r_1 and r_2 is zero, the Winsorized mean can still be used with little loss of efficiency. Even this can be reduced by using a modified Winsorized mean of form (for $r_1 = 0, r_2 = r$)

$$(n + a - 1)^{-1}[aX'_1 + X'_2 + \cdots + (r+1)X'_{n-r}] \tag{76}$$

with a chosen to make this an unbiased estimator of ξ. Minimum values of the efficiency of this estimator (relative to the best linear unbiased estimator) are shown below

r	Minimum Relative Efficiency
1	99.8%
2	99.2%
3	98.5%
4	97.7%
5	96.9%
6	96.0%

Values of a are given in Dixon [65] (Table II), or may easily be calculated from tables of expected values of ordered normal variables (see Section 1).

Estimation of σ is much more seriously, and adversely, affected by omission of extreme values, than is estimation of ξ. However, there is the compensating feature that the estimates are generally less sensitive to departures from exact normality of distribution (as already noted in Section 6.2). The simplest estimators of σ which may be used when analyzing censored samples are those based on the *quasi-ranges*

$$W_{(j)} = X'_{n-j+1} - X'_j. \tag{77}$$

Dixon found that for symmetrical censoring with $r_1 = r_2 = r \leq 6$, and $n \leq 20$, estimators based on

(i) $W_{(r+1)} + W_{(r+2)}$ for $r = 1, 11 \leq n \leq 15$ and $r = 2, 16 \leq n \leq 19$

(ii) $W_{(r+1)} + W_{(r+3)}$ for $r = 1, 16 \leq n \leq 20$ and $r = 2, n = 20$

(iii) $W_{(r+1)}$ for all other $r \leq 6, n \leq 20$

have efficiencies, relative to the best linear unbiased estimators of σ, of at least 96.5% A similar situation exists when $|r_1 - r_2| = 1$. Dixon also gives simple formulas for linear unbiased estimators for *single* censoring with $r_1 = 0$,

$r_2 \leq 6, n \leq 20$, which have a minimum efficiency of 93.7% relative to the best linear unbiased estimators.

7. Related Distributions

Section 6 contains references to the distributions of arithmetic mean, median, variance, range, mean deviation, etc. in 'random samples from normal populations'. Many chapters of this volume discuss other distributions related to the normal distribution. In the present section we will confine ourselves to discussion of *truncated* normal distributions and *mixtures* of normal distributions, together with some brief references to distributions of other statistics based on 'random normal samples'.

7.1 *Truncated Normal Distributions*

A random variable X has a *doubly truncated* normal distribution if its probability density function is

$$(78) \qquad \frac{1}{\sqrt{2\pi}\,\sigma} e^{-\frac{1}{2\sigma^2}(x-\xi)^2} \left[\frac{1}{\sqrt{2\pi}\,\sigma} \int_A^B e^{-\frac{1}{2\sigma^2}(t-\xi)^2} dt \right]^{-1}$$

$$= \sigma^{-1} Z\left(\frac{x-\xi}{\sigma}\right) \left[\Phi\left(\frac{B-\xi}{\sigma}\right) - \Phi\left(\frac{A-\xi}{\sigma}\right) \right]^{-1} \qquad (A \leq X \leq B).$$

The *lower* and *upper truncation points* are A, B respectively; *the degrees of truncation* are $\Phi\left(\frac{A-\xi}{\sigma}\right)$ (from below) and $\left(1 - \Phi\left(\frac{B-\xi}{\sigma}\right)\right)$ (from above). If A is replaced by $-\infty$, or B by ∞, the distribution is *singly truncated* from *above*, or *below*, respectively. More elaborate forms of truncation (e.g. omission of a number of intervals of X) will not be considered here.

Some typical doubly and singly truncated normal probability density functions are shown in Figure 3. These are classified according to the degrees of truncation. It can be seen that when these are large, the distribution bears little resemblance to a normal distribution. It is, indeed, more like a rectangular or trapezoidal distribution (Chapter 25).

The particular case $A = \xi$, $B = \infty$ produces a *half-normal* distribution. This is, in fact, the distribution of $\xi + \sigma|U|$ where U is a unit normal variable.

We will discuss, in detail, only doubly truncated normal distributions. Treatment of singly truncated normal distributions follows similar lines.

The expected value of X (from (78)) is given by

$$(79) \qquad E[X] = \xi + \frac{Z\left(\frac{A-\xi}{\sigma}\right) - Z\left(\frac{B-\xi}{\sigma}\right)}{\Phi\left(\frac{B-\xi}{\sigma}\right) - \Phi\left(\frac{A-\xi}{\sigma}\right)} \sigma$$

and the variance of X by

FIGURE 3

Density Functions of Some Truncated Unit Normal Distributions

Points of Truncation, $A < B$

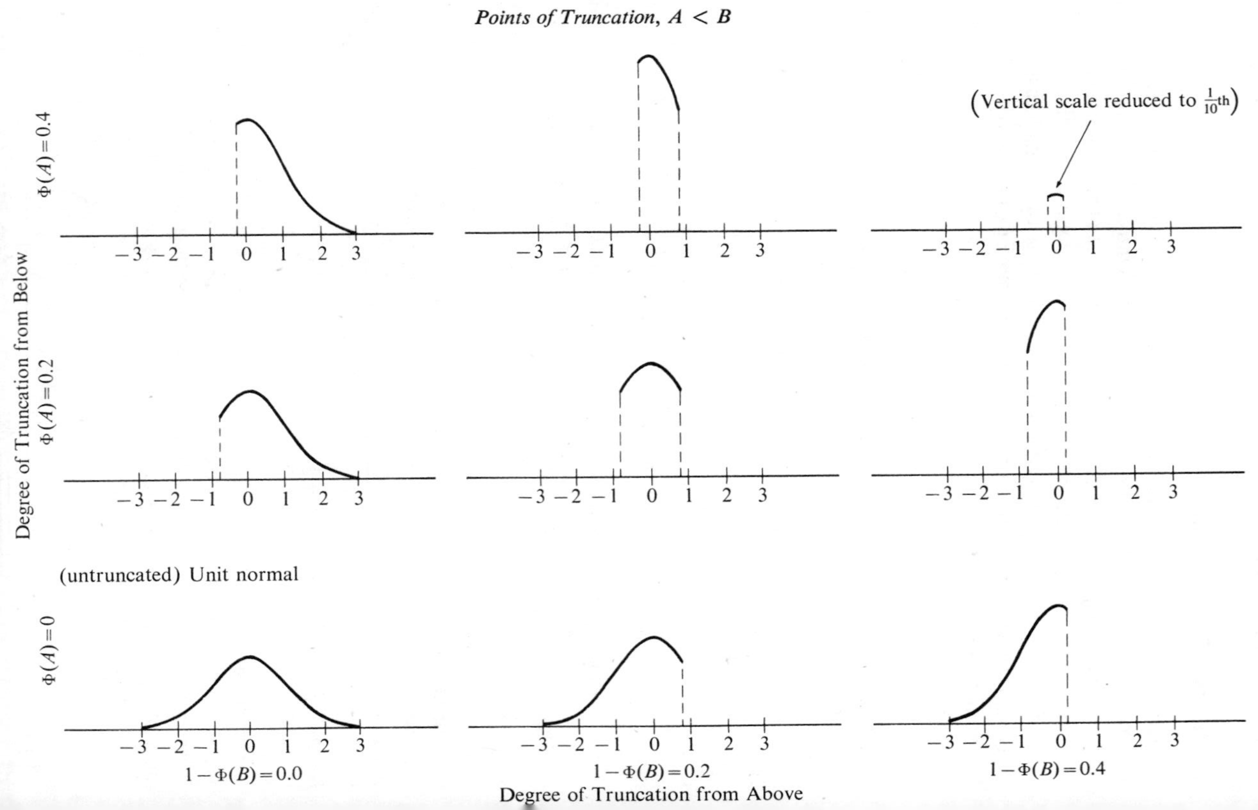

(80)
$$\operatorname{var}(X) = \left[1 + \frac{\left(\frac{A-\xi}{\sigma}\right) Z\left(\frac{A-\xi}{\sigma}\right) - \left(\frac{B-\xi}{\sigma}\right) Z\left(\frac{B-\xi}{\sigma}\right)}{\Phi\left(\frac{B-\xi}{\sigma}\right) - \Phi\left(\frac{A-\xi}{\sigma}\right)} - \left\{\frac{Z\left(\frac{A-\xi}{\sigma}\right) - Z\left(\frac{B-\xi}{\sigma}\right)}{\Phi\left(\frac{B-\xi}{\sigma}\right) - \Phi\left(\frac{A-\xi}{\sigma}\right)}\right\}^2\right]\sigma^2$$

Note that if $A - \xi = -(B - \xi) = -\delta$ then

(81)
$$\begin{cases} E[X] = \xi \\ \operatorname{var}(X) = \left[1 - \dfrac{2\,\delta Z(\delta)}{2\Phi(\delta) - 1}\right]\sigma^2 \end{cases}$$

The mean deviation of X is

(82)
$$2\left[Z(J) - Z\left(\frac{B-\xi}{\sigma}\right) - \left\{\Phi\left(\frac{B-\xi}{\sigma}\right) - \Phi(J)\right\} J\right] \times \left[\Phi\left(\frac{B-\xi}{\sigma}\right) - \Phi\left(\frac{A-\xi}{\sigma}\right)\right]^{-1}$$

where

$$J = \frac{Z\left(\frac{A-\xi}{\sigma}\right) - Z\left(\frac{B-\xi}{\sigma}\right)}{\Phi\left(\frac{B-\xi}{\sigma}\right) - \Phi\left(\frac{A-\xi}{\sigma}\right)}.$$

Some values of $E(X)$, $\sqrt{\operatorname{var}(X)}$ and the ratio (mean deviation)/(standard deviation) are given in Table 10.

The moments of the truncated distribution can also be expressed in terms of the Hh functions (Fisher [84]) defined by

(83)
$$Hh_n(y) = (n!)^{-1}\int_y^\infty (t - y)^n e^{-\frac{1}{2}t^2}\, dt.$$

(Note that $Hh_n(y) = \int_y^\infty Hh_{n-1}(t)\, dt$; $Hh_0(y) = \sqrt{2\pi}\,(1 - \Phi(y))$; $Hh_{-1}(y) = e^{-\frac{1}{2}y^2}$.)

The values of A and B are usually known, at least fairly closely. If A and B be known, maximum likelihood estimation of ξ and σ is equivalent to estimation by equating first and second sample and population moments. The equations satisfied by the estimators $\hat{\xi}$ and $\hat{\sigma}$ are similar in form to those for censored samples (72.1) (72.2), with X'_{r_1+1}, X'_{n-r_2} replaced by A, B respectively and with the multipliers $r_1 n^{-1}$, $r_2 n^{-1}$ replaced by the degrees of truncation $\Phi\left(\frac{A-\xi}{\sigma}\right)$, $1 - \Phi\left(\frac{B-\xi}{\sigma}\right)$ respectively.

TABLE 10

Expected Value, Standard Deviation and (Mean Deviation)/(Standard Deviation) for Truncated Normal Distributions

Degrees of Truncation				
$q_1 = \Phi\left(\frac{A-\xi}{\sigma}\right)$ (below)	$q_2 = 1 - \Phi\left(\frac{B-\xi}{\sigma}\right)$ (above)	[$-\xi$ + (Expected value)]$\times\sigma^{-1}$	(Standard deviation)$\times\sigma^{-1}$	m.d/s.d
0	0.1	−0.195	0.844	0.814
0	0.2	−0.350	0.764	0.812
0	0.3	−0.497	0.702	0.809
0	0.4	−0.644	0.650	0.805
0	0.5	−0.798	0.603	0.801
0	0.6	−0.966	0.558	0.795
0	0.7	−1.159	0.514	0.791
0	0.8	−1.400	0.468	0.785
0	0.9	−1.755	0.411	0.777
0.1	0.1	0.000	0.662	0.844
0.1	0.2	−0.149	0.566	0.850
0.1	0.3	−0.287	0.489	0.853
0.1	0.4	−0.422	0.420	0.854
0.1	0.5	−0.559	0.355	0.856
0.2	0.2	0.000	0.463	0.856
0.2	0.3	−0.135	0.382	0.859
0.2	0.4	−0.266	0.309	0.861
0.2	0.5	−0.397	0.239	0.863
0.3	0.3	0.000	0.297	0.862
0.3	0.4	−0.129	0.222	0.864
0.3	0.5	−0.256	0.151	0.865
0.4	0.4	0.000	0.146	0.865
0.4	0.5	−0.126	0.073	0.866

Harter and Moore [131] have given a table of asymptotic variances of, and correlation between, ξ and $\hat{\sigma}$ (for large n). Table 11 is based on their figures, and includes figures for the asymptotic variance of the maximum likelihood estimator of ξ when σ is known, and of σ when ξ is known.

If $\Phi\left(\frac{A-\xi}{\sigma}\right)$ is greater than $\left[1 - \Phi\left(\frac{B-\xi}{\sigma}\right)\right]$ the variable X can be replaced by $-X$, ξ by $-\xi$, A by $-B$ and B by $-A$, then, $\Phi\left(\frac{-B+\xi}{\sigma}\right)$ will be less than $\left[1 - \Phi\left(\frac{-A+\xi}{\sigma}\right)\right]$, and the tables may be applied.

The tables in [131] give 6 decimal places. Earlier tables by Gupta [114] give the first three columns to 5 decimal places for singly truncated distributions ($A = -\infty$ and so $\Phi\left(\frac{A-\xi}{\sigma}\right) = 0$), for $\Phi\left(\frac{B-\xi}{\sigma}\right) = 0.05(0.05)0.95(0.01)0.99$.

The values given in these tables also give the asymptotic variances for estimators of ξ and σ from singly and doubly censored sets of independent identically distributed normal variables. They are, however, only *asymptotic* values, and some care is needed in using them when the sample size n is not large (say, less than 100).

Sampling experiments (Harter and Moore [131]) with $n = 10$ and $n = 20$, have indicated that there is a negative bias in both $\hat{\xi}$ and $\hat{\sigma}$, which increases with the degree of truncation.

For a moderate degree of truncation (or censoring) knowledge of either parameter does not result in much reduction of variance of the estimator of the other parameter.

Occasionally, the points of truncation (A and B) are not known, and must be estimated from the data, as well as ξ and σ. With sufficiently large samples it will probably be good enough to take A equal to a value slightly less than the least observed value, and B slightly larger than the greatest observed value, but sometimes more elaborate methods may be needed.

Maximum likelihood estimation of the four parameters A, B, ξ, σ has been described by Cohen [40]. Estimation using the first four moments has been described by Shah and Jaiswal [277]. There are marked similarities between the equations obtained by the two methods. Cohen gives tables of auxiliary quantities useful in solving the maximum likelihood equations. These tables have been extended by Lifsey [182].

Also, Cohen [41] has described how the parameters ξ and σ can be estimated when there is single truncation (i.e. $A = -\infty$ or $B = \infty$), using the first three sample moments. As the point of truncation is supposed known, it is not really necessary to use the third moment. However, by introducing the third moment, simple explicit formulas are obtained. Supposing $B = \infty$ (i.e. left truncation only) the estimators of ξ and σ^2 are

$$(84)\qquad \begin{cases} A - (2m_1'^2 - m_2')^{-1}(2m_1'm_2' - m_3') \\ (2m_1'^2 - m_2')^{-1}(m_1'm_3' - m_2'^2) \end{cases}$$

respectively, where m_r' is the rth sample moment about the point of truncation, A.

Cohen found that the asymptotic efficiency of these estimators (relative to maximum likelihood estimators) is never less than 77% for ξ, 72% for σ^2. For very small or very large degrees of truncation the efficiencies are greater.

The use of the third moment may be expected to have introduced unnecessary inaccuracies. Pearson and Lee [248] gave formulas from which estimators which can be obtained, equating only first and second sample and population moments; the estimators are also, in fact, maximum likelihood estimators. The equations for the estimators $\hat{\xi}$, $\hat{\sigma}$ are (again assuming $B = \infty$)

$$(85.1)\qquad \hat{\sigma} = \left(\frac{Z(\hat{\delta})}{1 - \Phi(\hat{\delta})} - \hat{\delta}\right)^{-1} \overline{X}$$

TABLE 11

Asymptotic Values of $n \times (Variance)/\sigma^2$ for Maximum Likelihood Estimators

Proportions Censored		ξ and σ unknown			σ known	ξ known
		Variance of		Correlation	Variance	Variance
$\Phi\left(\frac{A-\xi}{\sigma}\right)$	$1-\Phi\left(\frac{B-\xi}{\sigma}\right)$	$\hat{\xi}$	$\hat{\sigma}$	$\hat{\xi}, \hat{\sigma}$	$\hat{\xi}$	$\hat{\sigma}$
0.0	0.0	1.00	0.50	0.000	1.00	0.50
0.0	0.1	1.02	0.59	0.053	1.02	0.58
0.0	0.2	1.06	0.69	0.125	1.05	0.68
0.0	0.3	1.14	0.82	0.214	1.09	0.78
0.0	0.4	1.27	0.99	0.320	1.14	0.89
0.0	0.5	1.52	1.24	0.441	1.22	1.00
0.0	0.6	1.99	1.62	0.572	1.34	1.09
0.0	0.7	3.02	2.25	0.700	1.53	1.14
0.0	0.8	5.78	3.54	0.822	1.87	1.15
0.0	0.9	17.79	7.51	0.918	2.78	1.18
0.1	0.1	1.04	0.70	0.000	1.04	0.70
0.1	0.2	1.07	0.85	0.075	1.06	0.84
0.1	0.3	1.14	1.04	0.172	1.11	1.01
0.1	0.4	1.27	1.32	0.293	1.17	1.20
0.1	0.5	1.54	1.74	0.417	1.25	1.41
0.1	0.6	2.13	2.46	0.597	1.37	1.58
0.1	0.7	3.67	3.95	0.757	1.57	1.69
0.1	0.8	9.77	8.66	0.896	1.94	1.71
0.2	0.2	1.10	1.05	0.000	1.10	1.05
0.2	0.3	1.15	1.34	0.103	1.14	1.32
0.2	0.4	1.28	1.78	0.239	1.20	1.68
0.2	0.5	1.56	2.54	0.413	1.29	2.10
0.2	0.6	2.30	4.09	0.618	1.42	2.53
0.2	0.7	5.18	8.93	0.827	1.64	2.82
0.3	0.3	1.19	1.79	0.000	1.19	1.80
0.3	0.4	1.29	2.57	0.150	1.26	2.51
0.3	0.5	1.57	4.16	0.368	1.35	3.59
0.3	0.6	2.69	9.04	0.665	1.50	5.04
0.4	0.4	1.33	4.17	0.000	1.33	4.17
0.4	0.5	1.57	9.09	0.286	1.44	8.35

$$(\hat{\sigma}/\overline{X})(\hat{\sigma}\overline{X}^{-1} - \hat{\delta}) = n^{-1} \sum_{j=1}^{n} X_j^2/\overline{X}^2 \tag{85.2}$$

where $\hat{\delta} = (A - \hat{\xi})/\hat{\sigma}$ (= estimate of lower truncation point as a standardized deviate) and $\overline{X} = n^{-1} \sum_{j=1}^{n} X_j$. Using (85.1), (85.2) can be solved for $\hat{\delta}$, and then $\hat{\sigma}$ is calculated from (85.1).

Cohen and Woodward [47] give tables of the functions

$$Y(\delta) = \left(\frac{Z(\delta)}{1 - \Phi(\delta)} - \delta\right)^{-1} \quad \text{and} \quad \tfrac{1}{2}Y(\delta)[Y(\delta) - \delta]$$

(to 8 significant figures for $\delta = -4.0(0.1)3.0$) to aid in the solution of (85.1) and (85.2). Since (85.2) can be written in the form

$$\tfrac{1}{2}n^{-1} \sum_{j=1}^{n} X_j^2/\overline{X}^2 = \tfrac{1}{2} Y(\hat{\delta})[Y(\hat{\delta}) - \hat{\delta}]$$

inverse interpolation suffices to determine $\hat{\delta}$, and then the table of $Y(\delta)$ is used to calculate $\hat{\sigma}$.

Distributions of functions of truncated normal variables cannot, usually, be expressed in mathematically elegant forms. In Francis [88] there are some tables (calculated by Campbell) of the distribution of the sums of n independent identically distributed truncated normal variables. The tables are not extensive, but they do give a useful idea of the effects of truncation on the distribution of arithmetic means, even for nonnormal distributions. Values of the cumulative distribution, to 4 decimal places, are given for values of the argument at intervals of 0.1, and for $n = 2$ and 4.

As an example of special problems which can sometimes arise, we mention the derivation of the distribution of the sum of two independent random variables, one normal and the other truncated normal. This has been discussed in Weinstein [315].

7.2 *Mixtures*

Compound normal distributions are formed by ascribing a distribution to one, or both, of the parameters ξ, σ in (1). There are two essentially distinct kinds of such distributions. Those obtained by treating ξ and/or σ as continuous random variables are of methodological and theoretical interest, while when (ξ,σ) takes only a finite (usually small) number of possible values, the fitting of corresponding distributions is usually regarded as the 'dissection' of a heterogeneous population into more homogeneous 'parts'.

Using the notation introduced in Chapter 8, we remark that the distribution*

$$\text{Normal}(\xi,\sigma) \bigwedge_{\xi} \text{Normal}(\mu,\sigma')$$

*We use the notation introduced in Chapter 8, so here ξ is a random variable.

is also a normal distribution, with expected value μ and standard deviation $\sqrt{\sigma^2 + \sigma'^2}$. This can be demonstrated by direct integration, or simply by regarding the compound distribution as that of $(\mu + U'\sigma') + U\sigma$ where U, U' are independent unit normal variables.

It can also be shown that

$$\text{Normal }(\xi,\sigma) \underset{\sigma^{-2}}{\wedge} \text{Gamma }(c\chi_\nu^2)$$

is equivalent to a Pearson Type VII distribution (Chapter 12, Section 4.1). In fact

$$(86)\qquad (2c)^{-\frac{1}{2}\nu}[\Gamma(\tfrac{1}{2}\nu)]^{-1}\int_0^\infty [\sqrt{2\pi}\,\sigma]^{-1}(\sigma^{-2})^{\frac{1}{2}\nu-1} \cdot \exp[-(2c\sigma^2)^{-1} - (2\sigma^2)^{-1}(x-\xi)^2]\,d\sigma^{-2} = \frac{c^{\frac{1}{2}}}{B(\frac{1}{2},\frac{1}{2}\nu)}\left[1 + \frac{(x-\xi)^2}{c}\right]^{-\frac{1}{2}(\nu+1)}.$$

(It may be noted that the distribution ascribed to σ^{-2} is of a kind sometimes called *fiducial*, which is obtained by formal inversion of the statement "V is distributed as $\chi_\nu^2\sigma^2$" to become "σ^{-2} is distributed as $V^{-1}\chi_\nu^2$".)

Teichroew [300] has studied the distribution

$$\text{Normal }(0,\sigma^2) \underset{\sigma^2}{\wedge} \text{Gamma }(c\chi_\nu^2).$$

The distribution has a complicated form, although its characteristic function is simply $(1 + ct^2)^{-\frac{1}{2}\nu}$.

The distribution

$$\text{Normal }(\xi,\sigma) \underset{\xi}{\wedge} \text{Rectangular}$$

has been studied by Bhattacharjee *et al.* [11].

Coming now to mixtures of a finite number (k) of normal *components*, a general form for the probability density function is

$$(87)\qquad \sum_{t=1}^{k} \omega_t(\sqrt{2\pi}\,\sigma_t)^{-1}\exp[-\tfrac{1}{2}\{(x-\xi_t)^2/\sigma_t^2\}].$$

The quantities $\omega_1, \omega_2, \ldots, \omega_k$ $(0 < \omega_t;\ \sum_{t=1}^{k}\omega_t = 1)$ are called the *weights* of the component normal distributions. We will consider in detail only the case $k = 2$. With increasing number of components, the general case rapidly becomes extremely complicated, though simplifications (for example, supposing all σ_t's to be equal) can sometimes be used to render the analysis more manageable.

With $k = 2$, and ξ_1 and ξ_2 sufficiently different, it is possible for (87) to represent bimodal distributions (Helguero [139], Prasad [254], Teichroew [300]). A systematic study of conditions under which this is so has been made by Eisenberger [72] (see also Wessels [317]). He summarizes his results as follows:

(1) If

$$(\xi_1 - \xi_2)^2 < \frac{27\sigma_1^2\sigma_2^2}{4(\sigma_1^2 + \sigma_2^2)}$$

the distribution cannot be bimodal. (This is so, in particular, if $\xi_1 = \xi_2$.)

(2) If

$$(\xi_1 - \xi_2)^2 > \frac{8\sigma_1^2\sigma_2^2}{\sigma_1^2 + \sigma_2^2}$$

there are values of ω_1 and ω_2 $(= 1 - \omega_1)$ for which the distribution is bimodal.

(3) For *any* set of values $\xi_1, \xi_2, \sigma_1, \sigma_2$ there are values of ω_1 and ω_2 for which the distribution is unimodal. (This is fairly evident, on considering that if $\omega_1 = 0$ or $\omega_1 = 1$, a normal distribution, which is unimodal, is obtained.)

Tables of moments of (87) with $k = 2$ ((standard deviation) σ^{-1}, α_3 and $(\alpha_4 - 3)$ to 3 decimal places for $\omega_2/\omega_1 = 0.1, 0.9, 1.0$; $\sigma_2/\sigma_1 = 0.0(0.5)3.0$ and $(\xi_2 - \xi_1)/\sigma_1 = 0.0(0.5)3.0$) have been given by Linders [184].

If ω_1 is nearly 1, and so ω_t (for $t > 1$) is small, the mixture distribution (87) is sometimes called a *contaminated* normal distribution (Tukey [309]). It has been used as a model to assess tests for rejection of outlying observations obtained in samples from a supposedly normal distribution.

In the general case (87), the rth moment of X about zero is

$$\mu_r'(X) = \sum_{t=1}^{k} E[X^r \mid \xi_t, \sigma_t] = \sum_{t=1}^{k} E[(\xi_t + U\sigma_t)^r] \tag{88}$$

$$= \sum_{j=0}^{[r/2]} \binom{r}{2j} E[U^{2j}] \sum_{t=1}^{k} \xi_t^{r-2j}\sigma_t^{2j}$$

where U is a unit normal variable (remembering that $E[U^j] = 0$ if j is odd).

Supposing that k is equal to 2, and it is desired to estimate the 5 parameters ω_1 $(= 1 - \omega_2)$, ξ_1, ξ_2, σ_1 and σ_2 by the method of moments. Five moments will be needed. From the equations

$$\begin{aligned}
\mu_1' &= \omega_1\xi_1 + (1 - \omega_1)\xi_2 \\
\mu_2' &= \omega_1(\xi_1^2 + \sigma_1^2) + (1 - \omega_1)(\xi_2^2 + \sigma_2^2) \\
\mu_3' &= \sum_{t=1}^{2} \omega_t(\xi_t^3 + 3\xi_t\sigma_t^2) \\
\mu_4' &= \sum_{t=1}^{2} \omega_t(\xi_t^4 + 6\xi_t^2\sigma_t^2 + 3\sigma_t^4) \\
\mu_5' &= \sum_{t=1}^{2} \omega_t(\xi_t^5 + 10\xi_t^3\sigma_t^2 + 15\xi_t\sigma_t^4)
\end{aligned} \tag{89}$$

we try to find values for ω_1, ξ_1, ξ_2, σ_1 and σ_2.

This problem was considered by Pearson [243] in 1894. Subsequently a number of improvements have been effected (e.g. Charlier and Wicksell [33], Helguero [140]). Recent papers by Molenaar [210] and Cohen [46] give a useful account of this work. The following summary is based on information in these papers.

Putting $\theta_j = \xi_j - \mu_1'$ $(j = 1,2)$ the *central* moments $\mu_2, \mu_3, \mu_4, \mu_5$ are obtained from (89) by replacing ξ_t by θ_t $(t = 1,2)$. From the resulting equations can be derived an equation of ninth degree for $\phi = \theta_1\theta_2$;

$$\sum_{j=0}^{9} a_j\phi^j = 0 \tag{90}$$

where

$$a_0 = -24\mu_3^6;\ a_1 = -96\mu_3^4\kappa_4;\ a_2 = -63\mu_3^2\kappa_4^2 - 72\mu_3^3\kappa_5;$$
$$a_3 = 288\mu_3^4 - 108\mu_3\kappa_4\kappa_5 + 27\kappa_4^3;\ a_4 = 444\mu_3^2\kappa_4 - 18\kappa_5^2;$$
$$a_5 = 90\kappa_4^2 + 72\mu_3\kappa_5;\ a_6 = 36\mu_3^2;\ a_7 = 84\kappa_4;\ a_8 = 0;\ a_9 = 24$$

with

$$\kappa_4 = \mu_4 - 3\mu_2^2;\ \kappa_5 = \mu_5 - 10\mu_2\mu_3$$

In application, values of μ_r are replaced by sample values of these moments.

Since equation (90) might have as many as nine roots, there may be difficulty in choosing the 'right' root. Since μ_1' lies between ξ_1 and ξ_2 it follows (unless $\xi_1 = \xi_2$) that θ_1 and θ_2 are of opposite signs, and so $\phi = \theta_1\theta_2 < 0$. Hence only negative roots of (90) need be considered. For ease of computation, the following method appears to be convenient.

If the value of $\phi' = \theta_1 + \theta_2$ is known then ϕ is a negative root of the cubic equation

$$6\phi^3 - 2\phi'^2\phi^2 + (3\kappa_4 - 4\phi'\mu_3)\phi + \mu_3^2 = 0 \tag{91}$$

and this equation has only one such root. Using ϕ', and the value of ϕ obtained from (91), values of θ_1 and θ_2 can be determined, and from these, in turn ξ_1, ξ_2 are estimated as

$$\xi_j = \overline{X} + \theta_j \qquad (j = 1,2) \tag{92}$$

($\overline{X}$ being the sample mean) and ω_1 as

$$\omega_1 = \theta_2/(\theta_2 - \theta_1). \tag{93}$$

Finally, we have

$$\sigma_j^2 = \tfrac{1}{3}\theta_j(2\phi' - \mu_3/\phi) + \mu_3 - \theta_j^2 \qquad (j = 1,2). \tag{94}$$

Using the parameter values so obtained, a value for μ_5 can be calculated — $\mu_5(\phi')$ say. By inverse interpolation from a series of such values of $\mu_5(\phi')$ a value for ϕ' (and hence a set of values for all five parameters) can be estimated. (This does not exclude the possibility that more than one value of ϕ' may make $\mu_5(\phi')$ equal to the sample value, so it may still be necessary to distinguish among such values. Pearson [243] suggested that the value giving closest agreement between sample and population sixth moments be chosen. It is easy, however, to think of other criteria, e.g. choosing the value giving the least value of χ^2 or some other goodness-of-fit criterion.)

In view of the likely inaccuracy in estimating the sixth central moment, it seems preferable to use the first and third absolute central moments, ν_1 (mean deviation) and ν_3, together with the variance μ_2. From the equations

$$\begin{aligned} \nu_1 &= [\omega_1\sigma_1 + (1 - \omega_1)\sigma_2]\sqrt{2/\pi} \\ \mu_2 &= \omega_1\sigma_1^2 + (1 - \omega_1)\sigma_2^2 \qquad (95) \\ \nu_3 &= [\omega_1\sigma_1^3 + (1 - \omega_1)\sigma_2^3](2\sqrt{2/\pi}) \end{aligned}$$

we obtain σ_1, σ_2 as roots of the equation

$$(\mu_2 - \tfrac{1}{2}\pi\nu_1^2)z^2 - (\tfrac{1}{2}\sqrt{\pi/2}\,\nu_3 - \sqrt{\pi/2}\,\nu_1\mu_2)z + \tfrac{1}{4}\pi\nu_1\nu_3 - \mu_2^2 = 0 \qquad (96)$$

Sometimes simpler procedures can give adequate results. If the difference between the means $|\xi_1 - \xi_2|$ is large enough then the left hand and right hand tails of the distribution come almost entirely from single (and different) components of the mixture. Figure 4 typifies such a situation, with $\xi_1 < \xi_2$ (and $\sigma_1 > \sigma_2$). In such cases a truncated normal distribution may be fitted to each tail separately (as described, for single truncation, in Section 1). This gives estimates of ξ_1, ξ_2, σ_1, and σ_2. Finally ω_1 is determined from the equation

$$\omega_1\xi_1 + (1 - \omega_1)\xi_2 = \overline{X}.$$

A major difficulty in this method is the choice of points of truncation. Some check is possible by moving the points inward, towards the bulk of the distribution, so long as the estimators of ξ_j and σ_j remain 'reasonably' consistent.

A number of graphical methods of estimation have been developed (e.g. Harding [124], Molenaar [210], Taylor [297], and Wiechselberger [314]).

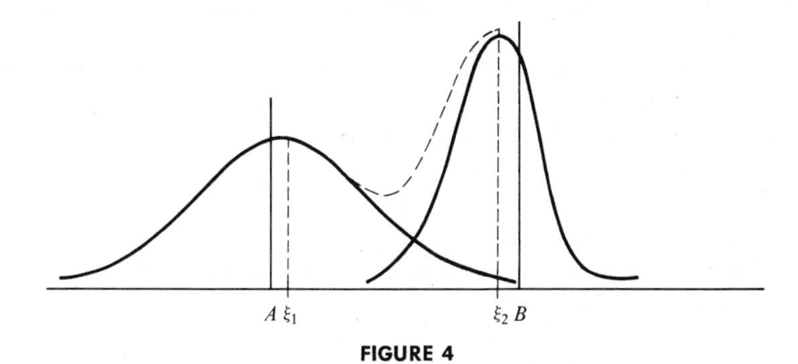

FIGURE 4

The Data in the Region Below A is Used for Estimation of the Component with Expected Value ξ.

The Data above B is Used for Estimation of the Component with Expected Value ξ_2.

If $\xi_1 = \xi_2$ then the distribution is symmetrical for any value of ω_1. A symmetrical distribution is also obtained (even with $\xi_1 \neq \xi_2$) if $\sigma_1 = \sigma_2$ *and* $\omega_1 = \frac{1}{2}$. This can be distinguished from the symmetrical distribution obtained with $\xi_1 = \xi_2$ (and $\sigma_1 \neq \sigma_2$), since in the former case $\kappa_4 < 0$, while in the latter case $\kappa_4 > 0$.

Moment estimators for the case $\xi_1 = \xi_2$ have been discussed in detail by Agard [1]. A maximum likelihood estimation procedure, for the case $\sigma_1 = \sigma_2$, was described by Molenaar [210], who also constructed a computer program for this procedure.

7.3 *Other Related Distributions*

If X_1 and X_2 are independent unit normal variables, and $F = (X_1/X_2)^2$, then F is distributed as $F_{1,1}$ (see Chapter 26). Equivalently, $t = X_1/X_2$ is distributed as t_1 (see Chapter 27), and so has a Cauchy distribution (see Chapter 16).

Among statistics used in testing normality are:

(i) the sample skewness

$$\sqrt{b_1} = \left[n^{-1}\sum_{j=1}^{n}(X_j - \overline{X})^3\right]\left[n^{-1}\sum_{j=1}^{n}(X_j - \overline{X})^2\right]^{-\frac{3}{2}}$$

(ii) the sample kurtosis

$$b_2 = \left[n^{-1}\sum_{j=1}^{n}(X_j - \overline{X})^4\right]\left[n^{-1}\sum_{j=1}^{n}(X_j - \overline{X})^2\right]^{-2}.$$

(Considerable attention has been devoted to obtaining formulas for the higher moments (up to the eighth) of these two statistics. See Fisher [84], Geary [92], [95], Geary and Worlledge [97], Hsu and Lawley [147], and Wilkinson [319]. Tables of approximate percentiles are available in Pearson [237].)

(iii) the ratio (sample mean deviation)/(sample standard deviation) (Geary [93], [94]).

(iv) the ratio (best linear unbiased estimator of σ) $\times$ (sample variance)$^{-1}$ (Shapiro and Wilk [278]).

(v) ratios of symmetrical differences between order statistics

$$(X'_{n-r_1+1} - X'_{r_1+1})/(X'_{n-r_2+1} - X'_{r_2+1})$$

(David and Johnson [58]).

(vi) the ratio (sample range)/(sample standard deviation) (Pearson *et al.* [241], Pearson and Stephens [242]).

In the construction of tests of outlying observations distributions the following statistics have been studied:

(i) $(X'_n - \overline{X})/\sigma$, $(\overline{X} - X'_1)/\sigma$ and the same statistics with σ replaced by the sample standard deviation (David [59], Nair [214], Sarhan [269], Pearson and Chandra Sekar [238]).

(ii) Ratios of order statistics of form

$$\frac{X'_r - X'_1}{X'_{r+1} - X'_1};\quad \frac{X'_n - X'_{n-r-1}}{X'_n - X'_{n-r}}\qquad \text{for } r = 2, 3, 4$$

etc. (Dixon [64]).

If X has a normal distribution as in (2), then $|X|$ is said to have a *folded normal distribution*. The reason for this name is that the distribution can be regarded as being formed by folding the part corresponding to negative X about the vertical axis (see Figure 5), and then adding it to the positive part.

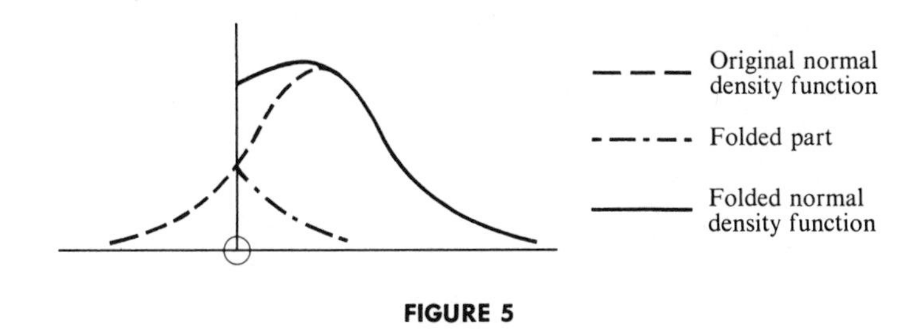

FIGURE 5

Folded Normal Distribution

The folded normal distribution is identical with the distribution of

$\sigma \times$ (noncentral χ with one degree of freedom and noncentrality parameter $(\xi/\sigma)^2$),

and, as such, will be described in Chapter 28. If the 'folding' is about the mean then $\xi = 0$ and a *central* χ is obtained. This is also the *half-normal* distribution referred to earlier in this chapter (Section 7.1).

8. Applications

In view of the common use of normal distributions in model construction it may seem strange that we devote only a short section to this topic. But enumerating the fields of application would be lengthy, and not really informative, therefore we do not attempt the task.

We do emphasize that the normal distribution is almost always used as an approximation — either to a theoretical or an unknown distribution. The normal distribution is well suited to this because its theoretical analysis is fully worked out and often simple in form. Where these conditions are not fulfilled, substitutes for normal distributions should be sought.

Even when normal distributions are not used results corresponding to 'normal theory' are useful as standards of comparison. In this book standardized deviates of percentile points of various distributions are occasionally presented. Comparison of these with the 'normal values' given in Table 1 of this chapter is often interesting and helpful.

It is our opinion that a judgment of the applicability of normal distribution in particular practical cases as well as an assessment of its prominence in the framework of the statistical distribution theory can be performed more successfully and efficiently if other models, given by other distributions, are also thoroughly investigated. It is, indeed, a purpose of these volumes to help a reader in such a judgment.

REFERENCES*

*To compromise between the desire for reasonable length and the desire for completeness and thoroughness, in this chapter a certain number of references included in the list below are not referred to directly in the text. The titles indicate the content of the paper. These references are for the reader who would like to study in greater depth certain topics presented in this chapter.

[1] Agard, J. (1961). Mélange de deux populations normales et étude de quelques fonctions $f(x,y)$ de variables normales x, y, *Revue de Statistique Appliquée*, **9,** No. 4, 53–70.

[2] Aitkin, M. (1966). The correlation between variate–values and ranks in a doubly truncated normal distribution, *Biometrika*, **53,** 281–282.

[3] Archibald, R. C. (1926). A rare pamphlet of Moivre and some of his discoveries, *Isis*, **8,** 671.

[4] Baker, G. A. (1930). Random sampling from non-homogeneous populations, *Metron*, **8,** 67–87.

[5] Barton, D. E. (1961). Unbiased estimation of a set of probabilities. *Biometrika*, **48,** 227–229.

[6] Basu, D. and Laha, R. G. (1954). On some characterizations of the normal distribution, *Sankhyā*, **13,** 359–362. (Addendum, **14,** 180)

[7] Bell, S. (1962). *Approximating the normal distribution with the triangular*, Sandia Corporation Report No. 494.

[8] Benson, F. (1949). A note on the estimation of mean and standard deviation from quantiles, *Journal of the Royal Statistical Society, Series B*, **11,** 91–100.

[9] Bergström, H. (1945). On the central limit theorem in the space R^k, $k > 1$, *Skandinavisk Aktuarietidskrift*, **28,** 106–127.

[10] Berry, A. C. (1941). The accuracy of the Gaussian approximation to the sum of independent variates. *Transactions of the American Mathematical Society*, **49,** 122–136.

[11] Bhattacharjee, G. P., Pandit, S. N. N. and Mohan, R. (1963). Dimensional chains involving rectangular and normal error-distributions, *Technometrics*, **5,** 404–406.

[12] Bhattacharya, N. (1959). An extension of Hald's table for one-sided censored normal distribution, *Sankhyā*, **21,** 377–380.

[13] Billingsley, P. (1963). Limit theorems for randomly selected partial sums, *Annals of Mathematical Statistics*, **33,** 85–92.

[14] Bland, R. P., Gilbert, R. D., Kapadia, C. H. and Owen, D. B. (1966). On the distributions of the range and mean range for samples from a normal distribution, *Biometrika*, **53,** 245–248.

[15] Bliss, C. I. and Stevens, W. L. (1937). The calculation of the time-mortality curve, *Annals of Applied Biology*, **24,** 815–852.

[16] Blum, J. R. (1956). On a characterization of the normal distribution, *Skandinavisk Aktuarietidskrift*, **39,** 59–62.

[17] Bolch, B. W. (1968). More on unbiased estimation of the standard derivation, *American Statistician*, **22,** No. 3, 27.

[18] Bol'shev, L. N. (1959). On transformations of random variables, *Teoriya Veroyatnostei i ee Primeneniya*, **4,** 136–149. (In Russian. English translation, 129–141.)

[19] Bol'shev, L. N. and Smirnov, N. V. (1965). *Tables of Mathematical Statistics*, Moscow: Nauka.

[20] Borenius, G. (1958). On the distribution of extreme values in a sample from a normal distribution, *Skandinavisk Aktuarietidskrift*, **41,** 131–166.

[21] Borenius, G. (1965). On the limit distribution of an extreme value in a sample from a normal distribution, *Skandinavisk Aktuarietidskrift*, **48,** 1–16.

[22] Bose, R. C. and Gupta, S. S. (1959). Moments of order statistics from a normal population, *Biometrika*, **46,** 433–440.

[23] Box, G. E. P. and Muller, M. E. (1958). A note on the generation of random normal deviates, *Annals of Mathematical Statistics*, **29,** 610–611.

[24] Breakwell, J. V. (1953). On estimating both mean and standard deviation of a normal population from the lowest r out of n observations, (Abstract), *Annals of Mathematical Statistics*, **24,** 683.

[25] Burr, I. W. (1967). A useful approximation to the normal distribution function, with application to simulation, *Technometrics*, **9,** 647–651.

[26] Buslenko, N. P., Golenko, D. I., Shreider, Yu. A., Sobol, I. M. and Sragovich, V. G. (1966). *The Monte Carlo Method*, Oxford: Pergamon Press (original Russian edition published 1962).

[27] Cacoullos, T. (1967). Some characterizations of normality, *Sankhyā, Series A*, **29,** 399–404.

[28] Cacoullos, T. (1967). Characterizations of normality by constant regression of linear statistics on another linear statistic, *Annals of Mathematical Statistics*, **38,** 1894–1898.

[29] Cadwell, J. H. (1951). The bivariate normal integral, *Biometrika*, **38,** 475–479.

[30] Cadwell, J. H. (1953). Approximating to the distributions of measures of dispersion by a power of χ^2, *Biometrika*, **40,** 336–346.

[31] Cadwell, J. H. (1953). The distribution of quasi-ranges in samples from a normal population, *Annals of Mathematical Statistics*, **24,** 603–613.

[32] Cadwell, J. H. (1954). The statistical treatment of mean deviation, *Biometrika*, **41,** 12–18.

[33] Charlier, C. V. L. and Wicksell, S. D. (1924). On the dissection of frequency functions, *Arkiv för Matematik, Astronomi och Fysik*, **18,** No. 6, 1–64.

[34] Chernoff, H., Gastwirth, J. L. and Johns, M. V. (1967). Asymptotic distribution of linear combinations of order statistics with application to estimation, *Annals of Mathematical Statistics*, **38,** 52–71.

[35] Chew, V. (1968). Some useful alternatives to the normal distribution, *American Statistician*, **22,** No. 3, 22–24.

[36] Chu, J. T. (1955). On bounds for the normal integral, *Biometrika*, **42,** 263–265.

[37] Chu, J. T. and Ya'Coub, K. (1966). Quadratic order estimates and moments of normal order statistics, *Annals of the Institute of Statistical Mathematics, Tokyo*, **18,** 337–341.

[38] Clark, C. E. (1966). *Random Numbers in Uniform and Normal Distribution*, San Francisco: Chandler.

[39] Cohen, A. C. (1949). On estimating the mean and standard deviation of truncated normal distributions, *Journal of the American Statistical Association*, **44,** 518–525.

[40] Cohen, A. C. (1950). Estimating the mean and variance of normal populations from singly truncated and doubly truncated samples, *Annals of Mathematical Statistics*, **21,** 557–569.

[41] Cohen, A. C. (1950). On estimating the mean and variance of singly truncated normal frequency distributions from the first three sample moments, *Annals of the Institute of Statistical Mathematics, Tokyo*, **3,** 37–44.

[42] Cohen, A. C. (1955). Censored samples from truncated normal distributions, *Biometrika*, **42,** 516–519.

[43] Cohen, A. C. (1957). On the solution of estimating equations for truncated and censored samples from normal populations, *Biometrika*, **44,** 225–236.

[44] Cohen, A. C. (1961). Tables for maximum likelihood estimates: Singly truncated and singly censored samples, *Technometrics*, **3,** 535–540.

[45] Cohen, A. C. (1963). Progressively censored samples in life testing, *Technometrics*, **5,** 327–339.

[46] Cohen, A. C. (1967). Estimation in mixtures of two normal distributions, *Technometrics*, **9,** 15–28.

[47] Cohen, A. C. and Woodward, J. (1953). Tables of Pearson-Lee-Fisher functions of singly truncated normal distributions, *Biometrics*, **9,** 489–497.

[48] Cox, D. R. (1949). Use of range in sequential analysis, *Journal of the Royal Statistical Society, Series B*, **11,** 101–114.

[49] Cox, D. R. (1960). Notes on the analysis of mixed frequency distributions, *British Journal of Mathematical and Statistical Psychology*, **19,** 39–47.

[50] Craig, C. C. (1928). An application of Thiele's semi-invariants to the sampling problem, *Metron*, **7,** (No. 4), 3–74.

[51] Cramér, H. (1928). On the composition of elementary errors, *Skandinavisk Aktuarietidskrift*, **11,** 13–74, 141–180.

[52] Cramér, H. (1936). Über eine Eigenschaft der normalen Verteilungsfunktion, *Mathematische Zeitschrift*, **41,** 405–414.

[53] Cureton, E. E. (1968). Unbiased estimation of the standard deviation, *American Statistician*, **22,** No. 1, 22. (Priority correction, *Ibid* **22,** No. 3, 27).

[54] Daly, J. F. (1946). On the use of the sample range in an analogue of Student's *t*-test, *Annals of Mathematical Statistics*, **17,** 71–74.

[55] Daniel, C. (1959). Use of half-normal plots in interpreting factorial two-level experiments, *Technometrics*, **1,** 311–341.

[56] Darmois, G. (1951). Sur une propriété caractéristique de la loi de probabilité de Laplace, *Comptes Rendus de l'Académie des Sciences, Paris*, **232,** 1999–2000.

[57] Darmois, G. (1953). Analyse générale des liasons stochastiques. Étude particuliaire de l'analyse factorielle linéaire, *Revue de l'Institut Internationale de Statistique*, **21,** 2–8.

[58] David, F. N. and Johnson, N. L. (1954). Tests for skewness and kurtosis with ordered variables, *Proceedings of the International Congress of Mathematics*, **2,** 286.

[59] David, H. A. (1956). Revised upper percentage points of the extreme studentized deviate from the sample mean, *Biometrika*, **43,** 449–451.

[60] David, H. A. (1957). Estimation of means of normal populations from observed minima, *Biometrika*, **44,** 282–286.

[61] David, H. A. (1963). The sample mean among the extreme normal order statistics, *Annals of Mathematical Statistics*, **34,** 33–55.

[62] David, H. A. (1968). Gini's mean difference rediscovered, *Biometrika*, **55,** 573–575.

[63] Daw, R. H. (1966). Why the normal distribution, *Journal of the Institute of Actuaries Students Society*, **18,** 2–15.

[64] Dixon, W. J. (1950). Analysis of extreme values, *Annals of Mathematical Statistics*, **21,** 488–506.

[65] Dixon, W. J. (1960). Simplified estimation from censored normal samples, *Annals of Mathematical Statistics*, **31,** 385–391.

[66] Dixon, W. J. (1962). Rejection of observations (pp. 299–342, in *Contributions to Order Statistics* (Ed. A. E. Sarhan and B. G. Greenberg), New York: John Wiley & Sons, Inc.).

[67] Doeblin, W. (1937). *Sur les propriétés asymptotiques de mouvements regis par certains types de chaînes simples*, Thesis, University of Paris.

[68] Doornbos, R. (1956). Significance of the smallest of a set of estimated normal variances, *Statistica Neerlandica*, **10,** 117–126.

[69] D'Ortenzio, R. J. (1965). Approximating the normal distribution function, *Systems Design*, **9,** 4–7.

[70] Eaton, M. L. (1966). Characterization of distributions by the identical distribution of linear forms, *Journal of Applied Probability*, **3,** 481–494.

[71] Edwards, A. W. F. (1963). A linkage for drawing the normal distribution, *Applied Statistics*, **12,** 44–45.

[72] Eisenberger, I. (1964). Genesis of bimodal distributions, *Technometrics*, **6,** 357–363.

[73] Eisenberger, I. and Posner, E. C. (1965). Systematic statistics used for data compression in space telemetry, *Journal of the American Statistical Association*, **60,** 97–133.

[74] Elfving, G. (1947). The asymptotical distribution of range in samples from a normal population, *Biometrika*, **34,** 111–119.

[75] Emersleben, O. (1951). Numerische Werte des Fehlerintegrals für $\sqrt{n\pi}$, *Zeitshrift für Angewandte Mathematik und Mechanik*, **31,** 393–394.

[76] Esseen, C.–G. (1942). On the Liapounoff limit of error in the theory of probability, *Arkiv för Matematik, Astronomi och Fysik*, **28A,** 1–19.

[77] Esseen, C.–G. (1945). Fourier analysis of distribution functions. A mathematical study of the Laplace-Gauss law, *Acta Mathematica*, **77,** 1–125.

[78] Esseen, C.–G. (1958). On mean central limit theorems, *Kunglige Tekniska Högskolans Handlingar*, No. 121, 1–30.

[79] Evans, I. G. (1964). Bayesian estimation of the variance of a normal distribution, *Journal of the Royal Statistical Society, Series B*, **26,** 63–68.

[80] Feller, W. (1935). Über den zentralen Grenzwertsatz der Wahrscheinlichkeitsrechnung, *Mathematische Zeitschrift*, **40,** 521–559.

[81] Feller, W. (1945). The fundamental limit theorems in probability, *Bulletin of the American Mathematical Society, Series 2*, **51,** 800–832.

[82] Fields, R. I., Kramer, C. Y. and Clunies-Ross, C. W. (1962). Joint estimation of the parameters of two normal populations, *Journal of the American Statistical Association*, **57,** 446–454.

[83] Fisher, R. A. (1925). Theory of statistical estimation, *Proceedings of the Cambridge Philosophical Society*, **22,** 700–706.

[84] Fisher, R. A. (1930). The moments of the distributions for normal samples of measures of departure from normality, *Proceedings of the Royal Society of London*, **130A,** 16–28.

[85] Fisher, R. A. (1931). The truncated normal distribution, *British Association for the Advancement of Science, Mathematical Tables*, **5,** xxxiii–xxxiv.

[86] Fisher, R. A. and Yates, F. (1963). *Statistical Tables for Biological, Agricultural and Medical Research*, London and Edinburgh: Oliver & Boyd.

[87] Fox, C. (1965). A family of distributions with the same ratio property as normal distribution, *Canadian Mathematical Bulletin*, **8,** 631–636.

[88] Francis, V. J. (1946). On the distribution of the sum of n sample values drawn from a truncated normal population, *Journal of the Royal Statistical Society, Series B*, **8,** 223–232.

[89] Gauss, C. F. (1809). *Theoria Motus Corporum Coelestium*, Hamburg: Perthes & Besser. (English translation by C. H. Davis, published 1857, Boston: Little, Brown, Co.)

[90] Gauss, C. F. (1816). Bestimmung der Genauigkeit der Beobachtungen, *Zeitschrift Astronomi*, **1,** 185–197.

[91] Gautschi, W. (1964). Error function and Fresnel integrals (pp. 295–309 in *Handbook of Mathematical Functions* (Ed. M. Abramowitz and I. A. Stegun), U.S. Department of Commerce, *Applied Mathematics Series*, **55**).

[92] Geary, R. C. (1933). A general expression for the moments of certain symmetrical functions of normal samples, *Biometrika*, **25,** 184–186.

[93] Geary, R. C. (1935). The ratio of the mean deviation to the standard deviation as a test of normality, *Biometrika*, **27,** 310–332.

[94] Geary, R. C. (1936). Moments of the ratio of the mean deviation to the standard deviation for normal samples, *Biometrika*, **28,** 295–305.

[95] Geary, R. C. (1947). The frequency distribution of $\sqrt{b_1}$ for samples of all sizes drawn at random from a normal population, *Biometrika*, **34,** 68–97.

[96] Geary, R. C. (1947). Testing for normality, *Biometrika*, **34,** 209–242.

[97] Geary, R. C. and Worlledge, J. P. G. (1947). On the computation of universal moments of tests of statistical normality derived from samples drawn at random from a normal universe. Application to the calculation of the seventh moment of b_2, *Biometrika*, **34,** 98–110. (Correction **37,** 189)

[98] Geisser, S. (1956). A note on the normal distribution, *Annals of Mathematical Statistics*, **27,** 858–859.

[99] Gjeddebaek, N. F. (1949). Contribution to the study of grouped observations. Application of the method of maximum likelihood in case of normally distributed observations, *Skandinavisk Aktuarietidskrift*, **42,** 135–150.

[100] Gnedenko, B. V. (1948). On a theorem of S. N. Bernstein, *Izvestiya Akademii Nauk SSSR, Seria Matematichiskaya*, **12,** 97–100. (In Russian)

[101] Gnedenko, B. V. and Kolmogorov, A. N. (1954). *Limit distributions for sums of independent random variables*, Reading, Mass.: Addison-Wesley.

[102] Godwin, H. J. (1945). On the distribution of the estimate of mean deviation obtained from samples from a normal population, *Biometrika*, **33,** 254–256.

[103] Godwin, H. J. (1948). A further note on the mean deviation, *Biometrika*, **35,** 304–309.

[104] Godwin, H. J. (1949). On the estimation of dispersion by linear systematic statistics, *Biometrika*, **36,** 92–100.

[105] Godwin, H. J. (1949). Some low moments of order statistics, *Annals of Mathematical Statistics*, **20,** 279–285.

[106] Godwin, H. J. and Hartley, H. O. (1945). Tables of the probability integral and the percentage points of the mean deviation in samples from a normal population, *Biometrika*, **33,** 254–265.

[107] Gosslee, D. G. and Bowman, K. O. (1967). *Evaluation of maximum likelihood estimates of parameters in mixtures of normal distributions*, Oak Ridge National Laboratory, Report ORNL-TM-2110.

[108] Govindarajulu, Z. (1963). On moments of order statistics and quasi-ranges from normal populations, *Annals of Mathematical Statistics*, **34,** 633–651.

[109] Govindarajulu, Z. (1966). Characterization of normal and generalized truncated normal distributions using order statistics, *Annals of Mathematical Statistics*, **37,** 1011–1015.

[110] Greenwood, J. A. and Hartley, H. O. (1962). *Guide to Tables in Mathematical Statistics*, Princeton, N. J.: Princeton University Press.

[111] Grubbs, F. E. (1950). Sample criteria for testing outlying observations, *Annals of Mathematical Statistics*, **21,** 27–58.

[112] Grubbs, F. E. and Weaver, C. L. (1947). The best unbiased estimate of population standard deviation based on group ranges, *Journal of the American Statistical Association*, **42,** 224–241.

[113] Grundy, P. M. (1952). The fitting of grouped truncated and grouped censored normal distributions, *Biometrika*, **39,** 252–259.

[114] Gupta, A. K. (1952). Estimation of the mean and standard deviation of a normal population from a censored sample, *Biometrika*, **39,** 260–273.

[115] Gupta, S. S. (1961). Percentage points and modes of order statistics from the normal distribution, *Annals of Mathematical Statistics*, **32,** 888–893.

[116] Gupta, S. S. (1962). Life-test sampling plans for normal and lognormal distributions, *Technometrics*, **4,** 151–175.

[117] Guttmann, I. (1960). Tests for the scale parameter of the truncated normal, *Canadian Mathematical Bulletin*, **3,** 225–236.

[118] Hald, A. (1949). Maximum likelihood estimation of the parameters of a normal distribution which is truncated at a known point, *Skandinavisk Aktuarietidskrift*, **32,** 119–134.

[119] Hald, A. (1952). *Statistical Tables and Formulas*, New York: John Wiley & Sons, Inc.

[120] Halperin, M. (1952). Maximum likelihood estimation in truncated samples, *Annals of Mathematical Statistics*, **23,** 226–238.

[121] Halperin, M. (1952). Estimation in the truncated normal distribution, *Journal of the American Statistical Association*, **47,** 457–465.

[122] Halperin, M. (1961). Confidence intervals from censored samples, *Annals of Mathematical Statistics*, **32,** 828–837.

[123] Halperin, M. (1966). Confidence intervals from censored samples, **II**, *Technometrics*, **8,** 291–301.

[124] Harding, J. P. (1949). The use of probability paper for the graphical analysis of polynomial frequency distributions, *Journal of Marine Biology Association, United Kingdom*, **28,** 141–153.

[125] Harris, H. and Smith, C. A. B. (1949). The sib-sib age of onset correlation among individuals suffering from a hereditary syndrome produced by more than one gene, *Annals of Eugenics, London*, **14,** 309–318.

[126] Hart, R. G. (1966). A close approximation related to the error function, *Mathematics of Computation*, **20,** 600–602.

[127] Harter, H. L. (1960). Tables of range and studentized range, *Annals of Mathematical Statistics*, **31,** 1122–1147.

[128] Harter, H. L. (1961). Expected values of normal order statistics, *Biometrika*, **48,** 151–157.

[129] Harter, H. L. (1963). Percentage points of the ratio of two ranges and power of the associated test, *Biometrika*, **50,** 187–194.

[130] Harter, H. L. (1964). Criteria for best substitute interval estimators, with an application to the normal distribution, *Journal of the American Statistical Association*, **59,** 1133–1140.

[131] Harter, H. L. and Moore, A. H. (1966). Iterative maximum likelihood estimation of the parameters of normal populations from singly and doubly censored samples, *Biometrika*, **53,** 205–213.

[132] Hartley, H. O. (1942). Numerical evaluation of the probability integral (*of range*), *Biometrika*, **32,** 309–310. (See also E. S. Pearson [**234**].)

[133] Hartley, H. O. (1945). Note on the calculation of the distribution of the estimate of mean deviation in normal samples, *Biometrika*, **33,** 257–258.

[134] Hartley, H. O. and Pearson, E. S. (1951). Moment constants for the distribution of range in small samples, *Biometrika*, **38,** 463–464.

[135] Hasselblad, V. (1966). Estimation of parameters for a mixture of normal distributions, *Technometrics*, **8,** 431–444, (Discussion by A. C. Cohen, pp. 445–446).

[136] Hastings, C. (1955). *Approximations for Digital Computers*, Princeton: Princeton University Press.

[137] Hastings, C., Mosteller, F., Tukey, J. W. and Winsor, C. P. (1947). Low moments for small samples: A comparative study of order statistics, *Annals of Mathematical Statistics*, **18,** 413–426.

[138] Heath, D. F. (1967). Normal or lognormal: Appropriate distributions, (Letter in) *Nature, London*, **213,** 1159–1160.

[139] Helguero, F. de (1904). Sui massima delle curve dimorfiche, *Biometrika*, **3,** 84–98.

[140] Helguero, F. de (1905). Per la risoluzione delle curve dimorfiche, *Biometrika*, **4,** 230–231.

[141] Helmert, F. R. (1875). Über die Berechnung des wahrscheinlichen Fehlers aus einer endlichen Anzahl wahrer Beobachtungsfehler, *Zeitschrift für angewandte Mathematik und Physik*, **20,** 300–303.

[142] Helmert, F. R. (1876). Über die Wahrscheinlichkeit der Potenzsummen der Beobachtungsfehler und über einige damit im Zusammenhänge stehende Fragen, *Zeitschrift für Angewandte Mathematik und Physik*, **21,** 192–218.

[143] Herrey, Erna M. J. (1965). Confidence intervals based on the mean absolute deviation of a normal sample, *Journal of the American Statistical Association*, **60,** 257–269.

[144] Hill, B. M. (1963). Information for estimating the proportions in mixtures of exponential and normal distributions, *Journal of the American Statistical Association*, **58,** 918–932.

[145] Hojo, T. (1931). Distribution of the median, quartiles and interquartile distance in samples from a normal population, *Biometrika*, **23,** 315–360.

[146] Hoyt, J. P. (1968). A simple approximation to the standard normal probability density function, *American Statistician*, **22,** No. 2, 25–26.

[147] Hsu, C. T. and Lawley, D. N. (1940). The derivation of the fifth and sixth moments of the distribution of b_2 in samples from a normal population, *Biometrika*, **31,** 238–248.

[148] Hyrenius, H. (1949). Sampling distributions from a compound normal parent population, *Skandinavisk Aktuarietidskrift*, **32,** 180–187.

[149] Ibragimov, I. A. (1965). On the rate of convergence to normality, *Doklady Akademii Nauk SSSR*, **161,** 1267–1269. (In Russian)

[150] Irwin, J. O. (1925). The further theory of Francis Galton's individual-difference problem, *Biometrika*, **17,** 100–128.

[151] Jarrett, R. F. (1968). A minor exercise in history, *American Statistician*, **22,** No. 3, 25–26.

[152] Jílek, M. and Líkař, O. (1960). Tolerance limits of the normal distribution with known variance and unknown mean, *Australian Journal of Statistics*, **2,** 78–83.

[153] Jones, A. E. (1946). A useful method for the routine estimation of dispersion in large samples, *Biometrika*, **33,** 274–282.

[154] Jones, G. M., Kapadia, C. H., Owen, D. B. and Bland, R. P. (1969). *On the distribution of the quasi-range and mid-range for samples from a normal population*, Technical Report No. 20, THEMIS Contract, Department of Statistics, Southern Methodist University, Dallas, Texas.

[155] Jones, H. L. (1948). Exact lower moments of order statistics in small samples from a normal distribution, *Annals of Mathematical Statistics*, **19,** 270–273.

[156] Kac, M. (1939). On a characterization of the normal distribution, *American Journal of Mathematics*, **61,** 726–728.

[157] Kagan, A. M., Linnik, Yu. V. and Rao, C. R. (1965). On a characterization of the normal law based on a property of the sample average, *Sankhyā, Series A*, **27,** 405–406.

[158] Kagan, A. M. and Shalayevskiĭ, O. V. (1967). Characterization of the normal law by a property of the non-central χ^2-distribution (In Russian), *Lietuvos Matematikos Rinkinys*, **7,** 57–58.

[159] Kamat, A. R. (1953). The third moment of Gini's mean difference, *Biometrika*, **40,** 451–452.

[160] Kamat, A. R. (1954). Distribution theory of two estimates for standard deviation based on second variate differences, *Biometrika*, **41,** 1–11.

[161] Kamat, A. R. (1954). Moments of the mean deviation, *Biometrika*, **41,** 541–542.

[162] Kamat, A. R. and Sathe, Y. S. (1957). Approximations to the distributions of some measures of dispersion based on successive differences, *Biometrika*, **44,** 349–359.

[163] Kaplansky, I. (1943). A characterization of the normal distribution, *Annals of Mathematical Statistics*, **14,** 197–198.

[164] Kelley, T. L. (1948). *The Kelley Statistical Tables*, Cambridge: Harvard University Press.

[165] Kendall, M. G. and Babington Smith, B. (1942). Random Sampling Numbers, *Tracts for Computers*, **xxiv.** London: Cambridge University Press.

[166] Keyfitz, N. (1938). Graduation by a truncated normal, *Annals of Mathematical Statistics*, **9,** 66–67.

[167] King, E. P. (1953). Estimating the standard deviation of a normal population, *Industrial Quality Control*, **10,** No. 2, 30–33.

[168] Kondo, T. and Elderton, Ethel M. (1931). Table of functions of the normal curve to ten decimal places, *Biometrika*, **22,** 368–376.

[169] Koopmans, L. H., Owen, D. B. and Rosenblatt, J. I. (1964). Confidence intervals for the coefficient of variation for the normal and log-normal distributions, *Biometrika*, **51,** 25–32.

[170] Kotlarski, I. (1966). On characterizing the normal distribution by Student's law, *Biometrika*, **53,** 603–606.

[171] Krutchkoff, R. G. (1966). The correct use of the sample mean absolute deviation in confidence intervals for a normal variate, *Technometrics*, **8,** 663–673.

[172] Kulldorff, G. (1958). Maximum likelihood estimation of the mean of a normal random variable when the sample is grouped, *Skandinavisk Aktuarietidskrift*, **41,** 1–17.

[173] Kulldorff, G. (1958). Maximum likelihood estimation of the standard deviation of a normal random variable when the sample is grouped, *Skandinavisk Aktuarietidskrift*, **41,** 18–36.

[174] Kulldorff, G. (1963,1964). On the optimum spacing of sample quantiles from a normal distribution, Part I, *Skandnavisk Aktuarietidskrift*, **46,** 143–161. Part II, *Skandinavisk Aktuarietidskrift*, **47,** 71–87.

[175] Laha, R. G. (1957). On a characterization of the normal distribution from properties of suitable linear statistics, *Annals of Mathematical Statistics*, **28,** 126–139.

[176] Laplace, P. S. (1774). Determiner le milieu que l'on doit prendre entre trois obsevations données d'un même phénomené, *Mémoires de Mathématique et Physique presentées à l'Académie Royale des Sciences par divers Savans*, **6,** 621–625.

[177] Laplace, P. S. (1778). Mémoire sur les probabilités, *Histoire de l'Academie Royale de Sciences, Année 1778* (published 1781), 227–332.

[178] Lee, Alice (1914). Table of the Gaussian 'tail' functions when the 'tail' is longer than the body, *Biometrika*, **10,** 208–214.

[179] Leslie, R. T. and Brown, B. M. (1966). Use of range in testing heterogeneity of variance, *Biometrika*, **53,** 221–227.

[180] Levy, H. and Roth, L. (1936). *Elements of Probability*, Oxford: Oxford University Press.

[181] Lévy, P. (1935). Propriétés asymptotiques des sommes de variables aléatoires independantes ou enchaînées, *Journal de Mathématiques Pures et Appliquées*, **14,** 347–402.

[182] Lifsey, J. D. (1965). *Auxiliary estimating functions for doubly truncated normal samples*, NASA Technical Memo. TM X-53221.

[183] Lindeberg, J. W. (1922). Eine neue Herleitung des Exponentialgesetzes in der Wahrscheinlichkeitsrechnung, *Mathematische Zeitschrift*, **15,** 211–225.

[184] Linders, F. J. (1930). On the addition of two normal frequency curves, *Nordic Statistical Journal*, **2,** 63–73.

[185] Linnik, Yu. V. (1952). Linear statistics and the normal distribution law, *Doklady Akademii Nauk SSSR*, **83,** 353–355. (In Russian. English translation published by American Mathematical Society, 1961.)

[186] Lloyd, E. H. (1952). Least squares estimation of location and scale parameters using order statistics, *Biometrika*, **39,** 88–95.

[187] Loève, M. (1950). Fundamental limit theorems of probability theory, *Annals of Mathematical Statistics*, **21,** 321–338.

[188] Loève, M. (1963). *Probability Theory*, 3rd edition, New York: D. Van Nostrand.

[189] Lord, E. (1947). The use of range in place of standard deviation in the *t*-test, *Biometrika*, **34,** 41–67. (Correction **39,** 442)

[190] Lukacs, E. (1942). A characterization of the normal distribution, *Annals of Mathematical Statistics*, **13,** 91–93.

[191] Lukacs, E. (1955). Applications of Faà di Bruno's formula in mathematical statistics, *American Mathematical Monthly*, **62,** 340–348.

[192] Lukacs, E. (1956). Characterization of populations by properties of suitable statistics, *Proceedings of the 3rd Berkeley Symposium on Mathematical Statistics and Probability*, **2,** 195–214.

[193] Lukacs, E. (1960). *Characteristic Functions*, London: Griffin.

[194] Lukacs, E. (1962). On tube statistics and characterization problems, *Zeitschrift für Wahrscheinlichkeitstheorie und Verwandte Gebiete*, **1,** 116–125.

[195] Lukacs, E. and King, E. P. (1954). A property of the normal distribution, *Annals of Mathematics Statistics*, **25,** 389–394.

[196] Lyapunov, A. (1900). Sur une proposition de la théorie des probabilités, *Izvestiya Akademii Nauk SSSR, Series V*, **13,** 359–386. (Also Lyapunov, A. (1954). *Collected Works*, Vol. I, 125–151, Moscow: Akademia Nauk SSSR, (In Russian).)

[197] Lyapunov, A. (1901). Nouvelle forme du théorème sur la limite de probabilité, *Mémoire Académie St. Petersbourg, Series VII*, **12,** 1–24. (Also Lyapunov, A.,

Collected Works, Vol. I, 157–176. Moscow: Akademia Nauk SSSR, (In Russian).)

[198] Mahalanobis, P. C., Bose, S. S., Roy, P. R. and Banerjee, S. K. (1934). Tables of random samples from a normal distribution, *Sankhyā*, **1**, 289–328.

[199] Marcinkiewicz, J. (1939). Sur une propriété de la loi de Gauss, *Mathematische Zeitschrift*, **44**, 612–618.

[200] Markowitz, E. (1968). Minimum mean-square-error estimation of the standard deviation of the normal distribution, *American Statistician*, **22**, No. 3, 26.

[201] Marsaglia, G. and Bray, T. A. (1964). A convenient method for generating normal variables, *SIAM Review*, **6**, 260–264.

[202] McGillivray, W. R. and Kaller, C. L. (1966). A characterization of deviation from normality under certain moment assumptions, *Canadian Mathematical Bulletin*, **9**, 509–514.

[203] McKay, A. T. (1933). The distribution of $\sqrt{\beta_1}$ in samples of four from a normal universe, *Biometrika*, **25**, 204–210.

[204] McKay, A. T. (1933). The distribution of β_2 in samples of four from a normal universe, *Biometrika*, **25**, 411–415.

[205] McKay, A. T. (1935). The distribution of the difference between the extreme observation and the sample mean in samples of n from a normal universe, *Biometrika*, **27**, 466–471.

[206] Mead, R. (1966). A quick method of estimating the standard deviation, *Biometrika*, **53**, 559–564.

[207] Mills, J. F. (1926). Table of the ratio: area to bounding ordinate for any portion of normal curve, *Biometrika*, **18**, 395–400.

[208] Moivre, A. de (1733). Approximatio ad Summam Ferminorum Binomii $(a + b)^n$ in Seriem expansi, *Supplementum II to Miscellanae Analytica*, 1–7.

[209] Moivre, A. de (1738). *The Doctrine of Chances* (2nd edition). [1st edition 1718, 3rd edition 1756; new impression of 2nd edition, with some additional material, 1967. London: Frank Cass and Co.]

[210] Molenaar, W. (1965). Survey of separation methods for two normal distributions, *Statistica Neerlandica*, **19**, 249–263. (In Dutch)

[211] Monk, D. T. and Owen, D. B. (1957). *Tables of the Normal Probability Integral*, Sandia Corporation, (Publications of the U.S. Department of Commerce, Office of Technical Services).

[212] Moore, P. G. (1956). The estimation of the mean of a censored normal distribution by ordered variables, *Biometrika*, **43**, 482–485.

[213] Muller, M. E. (1959). A comparison of methods for generating normal deviates on digital computers, *Journal of the Association of Computing Machinery*, **6**, 376–383.

[214] Nair, K. R. (1947). The distribution of the extreme deviate from the sample mean and its studentized form, *Biometrika*, **34**, 118–144.

[215] Nair, K. R. (1949). The efficiency of Gini's mean difference, *Bulletin of the Calcutta Statistical Association*, **2**, 129–130.

[216] Nair, K. R. (1950). Efficiencies of certain linear systematic statistics for estimating dispersion from normal samples, *Biometrika*, **37**, 182–183.

[217] Nair, K. R. (1952). Tables of percentage points of the studentized extreme deviate from the sample mean, *Biometrika*, **39,** 189–191.

[218] Nair, U. S. (1936). The standard error of Gini's mean difference, *Biometrika*, **28,** 428–438.

[219] National Bureau of Standards (1953). *Tables of Normal Probability Functions*, U.S. Department of Commerce, *Applied Mathematics Series*, **23,** (Original edition, 1942).

[220] National Bureau of Standards, (1952). *A Guide to Tables of the Normal Probability Integral*, U.S. Department of Commerce, *Applied Mathematics Series*, **21.**

[221] Newman, D. (1939). The distribution of range in samples from a normal population, expressed in terms of an independent estimate of standard deviation, *Biometrika*, **31,** 20–30.

[222] Oderfeld, J. and Pleszczyńska, E. (1961). Liniowy estymator odchylenia średniego w populacji normalnej, *Zastosowania Matematyki*, **6,** 111–117.

[223] Ogawa, J. (1951). Contributions to the theory of systematic statistics, I, *Osaka Mathematics Journal*, **3,** 175–213.

[224] Ogawa, J. (1962). Determination of optimum spacings in the case of normal distributions (pp. 272–283 in *Contributions to Order Statistics* (Ed. A. E. Sarhan and B. G. Greenberg), New York: John Wiley & Sons, Inc.).

[225] Owen, D. B. (1962). *Handbook of Statistical Tables*, Reading, Massachusetts: Addison-Wesley.

[226] Pachares, J. (1959). Tables of the upper 10% points of the studentized range, *Biometrika*, **46,** 461–466.

[227] Paskevich, V. S. (1953). On a property of control charts used in quality control, *Prikladnaya Matematika i Mekhanika*, **17,** 49–57. (In Russian)

[228] Patil, G. P. (1964). On certain compound Poisson and compound binomial distributions, *Sankhyā, Series A*, **26,** 293–294.

[229] Patil, S. A. (1961). Monotonicity property of indices of information functions of samples censored below and samples censored above from normal distribution, *Bulletin of the Calcutta Statistical Association*, **10,** 153–162.

[230] Pearson, E. S. (1930). A further development of tests for normality, *Biometrika*, **22,** 239–249.

[231] Pearson, E. S. (1931). Note on tests for normality, *Biometrika*, **22,** 423–424.

[232] Pearson, E. S. (1932). The percentage limits for the distribution of range in samples from a normal population, *Biometrika*, **24,** 404–417.

[233] Pearson, E. S. (1936). Note on probability levels for $\sqrt{b_1}$, *Biometrika*, **28,** 306–307.

[234] Pearson, E. S. (1942). The probability integral of the range in samples of n observations from a normal population: I. Foreward and tables, *Biometrika*, **32,** 301–308. (See also H. O. Hartley (1942).)

[235] Pearson, E. S. (1945). The probability integral of the mean deviation, (Editorial Note), *Biometrika*, **33,** 252–253.

[236] Pearson, E. S. (1952). Comparison of two approximations to the distribution of the range in small samples from normal populations, *Biometrika*, **39,** 130–136.

[237] Pearson, E. S. (1965). Tables of percentage points of $\sqrt{b_1}$ and b_2 in normal samples; a rounding off, *Biometrika*, **52,** 282–285.

[238] Pearson, E. S. and Chandra Sekar, C. (1936). The efficiency of statistical tools and a criterion for the rejection of outlying observations, *Biometrika*, **28,** 308–320.

[239] Pearson, E. S. and Hartley, H. O. (1943). Tables of the probability integral of the 'Studentized' range, *Biometrika*, **33,** 89–99.

[240] Pearson, E. S. and Hartley, H. O. (1948). *Biometrika Tables for Statisticians*, **1,** London: Cambridge University Press. (2nd edition)

[241] Pearson, E. S., Hartley, H. O. and David, H. A. (1954). The distribution of the ratio, in a single normal sample, of range to standard deviation, *Biometrika*, **41,** 482–493.

[242] Pearson, E. S. and Stephens, M. A. (1964). The ratio of range to standard deviation in the same normal sample, *Biometrika*, **51,** 484–487.

[243] Pearson, K. (1894). Contributions to the mathematical study of evolution, *Philosophical Transactions of the Royal Society of London, Series A*, **185,** 71–110.

[244] Pearson, K. (1920). On the probable errors of frequency constants, Part III, *Biometrika*, **13,** 112–132.

[245] Pearson, K. (1924). Historical note on the origin of the normal curve of errors, *Biometrika*, **16,** 402–404.

[246] Pearson, K. (Ed.) (1931). *Tables for Statisticians and Biometricians*, Part II, London: Cambridge University Press.

[247] Pearson, K. (1931). Historical note on the distribution of the standard deviations of samples of any size drawn from an indefinitely large normal parent population, *Biometrika*, **23,** 416–418.

[248] Pearson, K. and Lee, Alice (1908). On the generalized probable error in multiple normal correlation, *Biometrika*, **6,** 59–68.

[249] Pepper, J. (1932). The sampling distribution of the third moment coefficient — an experiment, *Biometrika*, **24,** 55–64.

[250] Pillai, K. C. S. (1952). On the distribution of 'studentized' range, *Biometrika*, **39,** 194–195.

[251] Pleszczyńska, E. (1963). Tabela wag liniowego estymatora odchylenia średniego w populacji normalnej, *Zastosowania Matematyki*, **7,** 117–124.

[252] Plucinska, A. (1965). O pewnych zagadieniach zwiazanych z podzialem populacji normalnej na czesci, *Zastoswania Matematyki*, **8,** 117–125.

[253] Pólya, G. (1945). Remarks on computing the probability integral in one and two dimensions, *Proceedings of the 1st Berkeley Symposium on Mathematical Statistics and Probability*, 63–78.

[254] Prasad, A. (1955). Bi-modal distributions derived from the normal distribution, *Sankhyā*, **14,** 369–374.

[255] Prescott, P. (1968). A simple method of estimating dispersion from normal samples, *Applied Statistics*, **17,** 70–74.

[256] Proschan, F. (1962). Confidence and tolerance intervals for the normal distribution, *Journal of the American Statistical Association*, **48,** 550–564.

[257] Raab, D. H. and Green, E. H. (1961). A cosine approximation to the normal distribution, *Psychometrika*, **26,** 447–450.

[258] RAND Corporation (1955). *A Million Random Digits with 100,000 Normal Deviates*, Glencoe, Illinois: Free Press.

[259] Rao, C. R., Mitra, S. K. and Matthai, A. (1966). *Formulae and Tables for Statistical Work*, Calcutta: Statistical Publishing Society.

[260] Rao, C. R. (1967). On some characterizations of the normal law, *Sankhyā, Series A*, **29,** 1–14.

[261] Rao, J. N. K. (1958). A characterization of the normal distribution, *Annals of Mathematical Statistics*, **29,** 914–919.

[262] Rao, M. M. (1963). Characterizing normal law and a nonlinear integral equation, *Journal of Mathematics and Mechanics*, **12,** 869–880.

[263] Riffenburgh, R. H. (1967). *Transformation for statistical distribution approximately normal but of finite sample range*, Report NUWC-TP-19, Naval Undersea Warfare Center, San Diego, California. (Abbreviated version *Technometrics*, **11,** (1969), 47–59.)

[264] Ruben, H. (1956). On the moments of the range and product moments of extreme order statistics in normal samples, *Biometrika*, **43,** 458–460.

[265] Ruben, H. (1960). On the geometrical significance of the moments of order statistics, and of deviations of order statistics from the mean in samples from Gaussian populations, *Journal of Mathematics and Mechanics*, **9,** 631–638.

[266] Ruben, H. (1962). The moments of the order statistics and of the range in samples from normal populations, (pp. 165–190 in *Contributions to Order Statistics* (Ed. A. E. Sarhan and B. G. Greenberg), New York: John Wiley & Sons, Inc.)

[267] Sadikova, S. M. (1966). On two-dimensional analogs of an inequality of C.–G. Esseen and their application to the central limit theorem, *Teoriya Veroyatnostei i ee Primeneniya*, **11,** 370–380. (In Russian)

[268] Sandelius, M. (1957). On the estimation of the standard deviation of a normal distribution from a pair of percentiles, *Skandinavisk Aktuarietidskrift*, **40,** 85–88.

[269] Sarhan, A. E. (1954). Estimation of the mean and standard deviation by order statistics, I, *Annals of Mathematical Statistics*, **25,** 317–328.

[270] Sarhan, A. E. and Greenberg, G. B. (1962). The best linear estimates for the parameters of the normal distribution (pp. 206–269 in *Contributions to Order Statistics* (Ed. A. E. Sarhan and B. G. Greenberg), New York: John Wiley & Sons, Inc.).

[271] Sarkadi, K. (1960). On testing for normality, *Matematikai Kutato Intezetenet Közlemenyei, Hungarian Academy of Science*, **5,** 269–275.

[272] Saw, J. G. and Chow, B. (1966). The curve through the expected values of ordered variates and the sum of squares of normal scores, *Biometrika*, **53,** 252–255.

[273] Sazanov, V. V. (1967). On the rate of convergence in the multidimensional central limit theorem, *Teoriya Veroyatnostei i ee Primeneniya*, **12,** 82–95. (In Russian)

[274] Schay, G. (1965). Approximation of sums of independent random variables by normal random variables, *Zeitschrift für Wahrscheinlichkeitstheorie und verwandte Gebiete*, **4,** 209–216.

[275] Schucany, W. R. and Gray, H. L. (1968). A new approximation related to the error function, *Mathematics of Computation*, **22,** 201–202.

[276] Sengupta, J. M. and Bhattacharya, N. (1958). Tables of random normal deviates, *Sankhyā*, **20,** 250–286.

[277] Shah, S. M. and Jaiswal, M. C. (1966). Estimation of parameters of doubly truncated normal distribution from first four sample moments, *Annals of the Institute of Statistical Mathematics, Tokyo*, **18,** 107–111.

[278] Shapiro, S. S. and Wilk, M. B. (1965). An analysis of variance test for normality (complete samples), *Biometrika*, **52,** 591–611.

[279] Shepp, L. (1964). Normal functions of normal random variables, *SIAM Review*, **6,** 459–460.

[280] Sheppard, W. F. (1903). New tables of the probability integral, *Biometrika*, **2,** 174–190.

[281] Sheppard, W. F. (1907). Table of deviates of the normal curve, *Biometrika*, **5,** 404–406.

[282] Sheppard, W. F. (1939). The Probability Integral, *British Association for the Advancement of Science, Mathematical Tables*, **7.**

[283] Shimizu, R. (1961). A characterization of the normal distribution, *Annals of the Institute of Statistical Mathematics, Tokyo*, **13,** 53–56.

[284] Shimizu, R. (1962). Characterization of the normal distribution, II *Annals of the Institute of Statistical Mathematics, Tokyo*, **14,** 173–178.

[285] Skitovich, V. P. (1953). On a property of the normal distribution, *Doklady Akademii Nauk SSSR*, **89,** 217–219. (In Russian)

[286] Smirnov, N. V. (Ed.) (1960). *Tables of the Normal Integral, Normal Density and its Normalized Derivatives*, Moscow: Akademia Nauk SSSR. (In Russian)

[287] Smirnov, N. V. (Ed.) (1965). *Tables of the Normal Probability Integral, the Normal Density and its Normalized Derivatives*, New York: Macmillan. (Translation of [286])

[288] Steffensen, J. F. (1937). On the semi-normal distribution, *Skandinavisk Aktuarietidskrift*, **20,** 60–74.

[289] Strecock, A. J. (1968). On the calculation of the inverse of the error function, *Mathematics of Computation*, **22,** 144–158.

[290] Subramanya, M. T. (1965). Tables of two-sided tolerance intervals for normal distribution $N(\mu, \sigma)$ with μ unknown and σ known, *Journal of the Indian Statistical Association*, **3,** 195–201.

[291] Swamy, P. S. (1963). On the amount of information supplied by truncated samples of grouped observations in the estimation of the parameters of normal populations, *Biometrika*, **50,** 207–213.

[292] *Tables of the Error Function and its First Twenty Derivatives* (1952). Cambridge: Harvard University Computation Laboratory.

[293] Tables of Normal Distribution (1961). *Advanced Series of Mathematics and Engineering Tables*, **3,** Tokyo: Corona.

[294] *Tables of Probability Functions* (1959). Vol. 2, Computing Center, Moscow: Akademia Nauk SSSR.

[295] Tallis, G. M. and Young, S. S. M. (1962). Maximum likelihood estimation of parameters of the normal, log-normal, truncated normal and bivariate normal distributions from grouped data, *Australian Journal of Statistics*, **4,** 49–54.

[296] Tamhankar, M. V. (1967). A characterization of normality, *Annals of Mathematical Statistics*, **38,** 1924–1927.

[297] Taylor, B. J. R. (1965). The analysis of polymodal frequency distributions, *Journal of Animal Ecology*, **34,** 445–452.

[298] Teicher, H. (1961). Maximum likelihood characterization of distributions, *Annals of Mathematical Statistics*, **32,** 1214–1222.

[299] Teichroew, D. (1956). Tables of expected values of order statistics and products of order statistics for samples of size twenty and less from the normal distribution, *Annals of Mathematical Statistics*, **27,** 410–426.

[300] Teichroew, D. (1957). The mixture of normal distributions with different variances, *Annals of Mathematical Statistics*, **28,** 510–512.

[301] Teichroew, D. (1962). Tables of lower moments of order statistics for samples from the normal distribution, (pp. 190–205 in *Contributions to Order Statistics* (Ed. A. E. Sarhan and B. G. Greenberg), New York: John Wiley & Sons, Inc.)

[302] Thompson, W. R. (1935). On a criterion for the rejection of observations and the distribution of the ratio of deviation to sample standard deviation, *Annals of Mathematical Statistics*, **6,** 214–219.

[303] Tiku, M. L. (1967). Estimating the mean and standard deviation from a censored normal sample, *Biometrika*, **54,** 155–165.

[304] Tippett, L. H. C. (1925). On the extreme individuals and the range of samples taken from a normal population, *Biometrika*, **17,** 364–387.

[305] Tippett, L. H. C. (1927). *Random Sampling Numbers* (Tracts for Computers XV), London: Cambridge University Press.

[306] Tranquilli, G. B. (1966). Sul teorema di Basu-Darmois, *Giornale dell'Istituto Italiano degli Attuari*, **29,** 135–152.

[307] Trotter, H. F. (1959). An elementary proof of the central limit theorem, *Archiv der Mathematik*, **10,** 226–234.

[308] Truax, D. R. (1953). An optimum slippage test for the variances of k normal distributions, *Annals of Mathematical Statistics*, **24,** 669–674.

[309] Tukey, J. W. (1949). Comparing individual means in the analysis of variance, *Biometrics*, **5,** 99–114.

[310] Tukey, J. W. (1962). The future of data analysis, *Annals of Mathematical Statistics*, **33,** 1–67.

[311] Wald, A. and Wolfowitz, J. (1946). Tolerance limits for a normal distribution, *Annals of Mathematical Statistics*, **17,** 208–215.

[312] Walker, Helen M. (1924). *Studies in the History of Statistical Methods*, Baltimore: William and Wilkins.

[313] Wallace, D. (1959). A corrected computation of Berry's bound for the central limit theorem error, *Statistics Research Center, University of Chicago.*

[314] Weichselberger, K. (1961). Über ein graphisches Verfahren zur Trennung von Mischverteilungen und zur Identifikation kapierter Normalverteilungen bei grossem Stichprobenumfang, *Metrika*, **4,** 178–229.

[315] Weinstein, M. A. (1964). The sum of values from a normal and a truncated normal distribution (Answer to Query), *Technometrics*, **6,** 104–105. (See also answers by M. Lipow, N. Mantel and J. W. Wilkinson, *Technometrics*, **6,** 469–471.)

[316] Weissberg, A. and Beatty, G. H. (1960). Tables of tolerance-limit factors for normal distributions, *Technometrics*, **2,** 483–500.

[317] Wessels, J. (1964). Multimodality in a family of probability densities, with application to a linear mixture of two normal densities, *Statistica Neerlandica*, **18,** 267–282.

[318] Wetherill, G. B. (1965). An approximation to the inverse normal function suitable for the generation of random normal deviates on electronic computers, *Applied Statistics*, **14,** 201–205.

[319] Wilkinson, G. N. (1961). The eighth moment and cumulant of the distribution for normal samples of the skewness coefficient, $\gamma = k_3 k_2^{-\frac{3}{2}}$, *Australian Journal of Statistics*, **3,** 108–109.

[320] Wishart, J. (1930). The derivation of certain high order sampling product moments from a normal population, *Biometrika*, **22,** 224–238.

[321] Wold, H. (1948). *Random Normal Deviates* (Tracts for Computers, XXV) London: Cambridge University Press. (Also *Statistica Uppsala*, **3.**)

[322] Wolfowitz, J. (1946). Confidence limits for the fraction of a normal population which lies between two given limits, *Annals of Mathematical Statistics*, **17,** 483–488.

[323] Zackrisson, U. (1959). The distribution of "Student's" t in samples from individual nonnormal populations, *Publications of the Statistical Institute, University of Gothenburg, Sweden*, No. 6.

[324] Zacks, S. (1966). Unbiased estimation of the common mean of two normal distributions based on small samples of equal size, *Journal of the American Statistical Association*, **61,** 467–476.

[325] Zahl, S. (1966). Bounds for the central limit theorem error, *SIAM Journal of Applied Mathematics*, **14,** 1225–1245.

[326] Zeigler, R. K. (1965). A uniqueness theorem concerning moment distributions, *Journal of the American Statistical Association*, **60,** 1203–1206.

[327] Zelen, M. and Severo, N. C. (1964). Probability functions (pp. 925–995 in *Handbook of Mathematical Functions* (Ed. M. Abramowitz and I. A. Stegun), U.S. Department of Commerce, *Applied Mathematics Series*, **55**).

[328] Zinger, A. A. (1958). New results on independent statistics, *Trudy Soveshchaniya Teorii Veroyatnostei i Matematicheskoi Statistike, Erevan, AN ArmSSR*, 103–105. (In Russian)

[329] Zinger, A. A. and Linnik, Yn. V. (1964). On a characterization of the normal distribution, *Teoriya Veroyatnostei i ee Primeneniya*, **9,** 692–695. (In Russian)

[330] Zolotarev, V. M. (1967). A sharpening of the inequality of Berry-Esseen, *Zeitschrift für Wahrscheinlichkeitstheorie und verwandte Gebiete*, **8,** 332–342.

14

Lognormal Distributions

1. Introduction and Discussion

The idea of a transformation such that the transformed variable is normally distributed has been encountered in Chapter 12. In Section 4.3 of that chapter some specific transformations were introduced. Of these, the most commonly used, and the only one of sufficient importance to merit a separate chapter, is the simple logarithmic transformation.

If there is a number, θ, such that $Z = \log(X - \theta)$ is normally distributed, the distribution of X is said to be *lognormal*. For this to be the case it is clearly necessary that X can take any value exceeding θ, but has zero probability of taking any value less than θ. The name 'lognormal' can also be applied to the distribution of X if $\log(\theta - X)$ is normally distributed, X having zero probability of exceeding θ. However, since replacement of X by $-X$, (and θ by $-\theta$) reduces this situation to the first, we will consider only the first case.

The distribution of X can be defined by the equation

$$U = \gamma + \delta \log(X - \theta) \tag{1}$$

where U is a unit normal variable and γ, δ and θ are parameters. From (1) it follows that the probability density function of X is

$$p_X(x) = \delta[(x - \theta)\sqrt{2\pi}]^{-1} \exp[-\tfrac{1}{2}\{\gamma + \delta \log(x - \theta)\}^2] \qquad (x > \theta). \tag{2}$$

An alternative, more fashionable, notation replaces γ and δ by the expected value ζ and standard deviation σ of $Z = \log(X - \theta)$. The two sets of parameters are related by the equations

$$\zeta = -\gamma/\delta, \qquad \sigma = \delta^{-1}$$

so that (1) becomes

(1)′ $$U = \{\log (X - \theta) - \zeta\}/\sigma$$

and (2) becomes

(2)′ $$p_X(x) = [(x - \theta)\sqrt{2\pi}\,\sigma]^{-1} \exp[-\tfrac{1}{2}\{\log (x - \theta) - \zeta\}^2/\sigma^2] \quad (x > \theta).$$

The lognormal distribution is sometimes called the *antilognormal* distribution. This name has some logical basis, in that it is not the distribution of the logarithm of a normal variable (this is not even always real) but of an exponential — that is, antilogarithmic — function of such a variable. However, 'lognormal' is most commonly used, and we shall follow this practice. The minor variants, *logarithmic-* or *logarithmico-*normal, have been used, as have the names of pioneers in its development, notably *Galton* and *McAlister*, *Kapteyn* and *Gibrat* (see Section 2). When applied to economic data, particularly production functions, it is often called the *Cobb-Douglas* distribution (e.g., Dhrymes [16]).

It can be seen that a change in the value of the parameter θ affects only the location of the distribution. It does not affect the variance or the shape (or any property depending only on differences between values of the variable and its expected value). It is convenient to assign θ a particular value for ease of algebra, realizing that many of the results so obtained can be transferred to the more general distribution. In many applications, θ is 'known' to be zero (so that $\Pr[X \leq 0] = 0$ or X is a 'positive random variable'). This important case has been given the name *two-parameter lognormal distribution* (parameters γ, δ or ζ,σ). For this distribution (1) becomes

(3) $$U = \gamma + \delta \log X$$

and (1)′ becomes

(4) $$U = (\log X - \zeta)/\sigma.$$

The general case (with θ not necessarily zero) can be called the *three-parameter lognormal distribution* (parameters γ,δ,θ or ζ,σ,θ).

2. Genesis and Historical Remarks

In 1879, Galton [26] pointed out that if $X_1, X_2, \ldots, X_n$ are independent positive random variables and

$$T_n = \prod_{j=1}^{n} X_j$$

then

$$\log T_n = \sum_{j=1}^{n} \log X_j$$

and so, if the independent random variables, $\log X_j$, are such that a central limit type of result applies, then the standardized distribution of $\log T_n$ would tend to a unit normal distribution as n tends to infinity. The limiting distribution of T_n would then be (two-parameter) lognormal. In an accompanying paper, McAlister [54] obtained expressions for the mean, median, mode, variance, and certain percentiles of the distribution.

Subsequent to this, little material was published relating to the lognormal distribution until 1903 when Kapteyn [45] again considered its genesis, on the lines described above. (Fechner [20], in 1897, mentioned the use of the distribution in the description of psychophysical phenomena, but gave little emphasis to this topic.) In 1916, Kapteyn and van Uven [46] gave a graphical method (based on the use of sample quantiles) for estimating parameters, and in the following years there was a considerable increase in published information on the lognormal and related distributions. In 1917, Wicksell [79] obtained formulas for the higher moments, while van Uven ([75], [76]) considered transformations to normality from a more general point of view; and, in 1919, Nydell [59] obtained approximate formulas for the standard deviations of estimators obtained by the method of moments. Estimation from percentile points was described by Davies [14], [15] in 1925 and 1929, and tables to facilitate estimation from sample moments were published by Yuan [87] in 1933. Unbiased estimators based on sample moments were constructed by Finney [23] in 1941.

From 1930 onwards, fields of application of the lognormal distribution have increased steadily. Gibrat [27], [28] found the distribution usefully represented the distribution of size for varied kinds of 'natural' economic units. Soon after this, Gaddum [24] and Bliss [6] found that distribution of critical dose (dose just causing reaction) for a number of forms of drug application could be represented with adequate accuracy by a (two-parameter) lognormal distribution. On the basis of these observations, a highly developed method of statistical analysis of 'quantal' ('all-or-none') response data — probit analysis — has been elaborated. Contemporaneously, lognormal distributions were found to be applicable to distributions of particle size in naturally occurring aggregates (Hatch [34], Hatch and Choute [35], Krumbein [51]; also Bol'shev *et al.* [7], Herdan [37], Kalinske [44], Kolmogorov [48], Kottler [50], Wise [83]). It is possible to give some general theoretical basis for this application, on lines similar to Galton's argument. Consider a quantity A subject to a large number of independent subdivisions each of which results in division in two parts, the proportions being uniformly distributed between 0 and 1. After n subdivisions, the size of a randomly chosen item will be distributed as $AX_1 X_2 \cdots X_n$, where $X_1, \ldots, X_n$ are independent random variables, rectangularly distributed over the interval 0 to 1. For n large, the distribution of this quantity will be approximately lognormal. (See Halmos [32] and Herdan [37] for more complete accounts of this argument.)

A little later (1937–1940) further applications, in agricultural, entomological and even literary, research were described (Cochran [11], Williams [80], [81] also Grundy [30], Herdan [36], [38], Pearce [61]).

This increase in practical usefulness of the distribution was followed by (and presumably associated with) renewed interest in problems of estimation of the parameters. (Technical aspects of these problems will be discussed in Section 4 of this chapter.) As late as 1957, there were so many outstanding doubtful points in this regard that Aitchison and Brown [3], directed 'a substantial part' of their book to discussion of problems of estimation, on account of 'unresolved difficulties'. (We have found this book by Aitchison and Brown very useful in organizing material for the present chapter. It can be recommended for supplementary reading.)

Wu [86] has shown that lognormal distributions can arise as limiting distributions of order statistics, if sample size and order increase in certain relationships.

3. Moments and Other Properties

Most of the following discussion will be in terms of the two-parameter distribution, and using (4) rather than (3).

The rth moment of X about zero is

$$\begin{aligned}\mu_r' = E[X^r] &= E[\exp r(\zeta + U\sigma)] \\ &= \exp(r\zeta + \tfrac{1}{2}r^2\sigma^2)\end{aligned} \tag{5}$$

μ_r' increases very rapidly with r. Heyde [40] has shown that the moment sequence $\{\mu_r'\}$ does not belong *only* to lognormal distributions — i.e., the distribution cannot be defined by its moments.

The expected value is

$$\mu_1' = \exp(\zeta + \tfrac{1}{2}\sigma^2) \tag{6.1}$$

and the lower order central moments are

$$\mu_2 = e^{2\zeta}e^{\sigma^2}(e^{\sigma^2} - 1) = \omega(\omega - 1)e^{2\zeta}, \tag{6.2}$$

so that

$$\sigma(X) = e^{\zeta}\sqrt{\omega(\omega - 1)}$$

$$\mu_3 = \omega^{\frac{3}{2}}(\omega - 1)^2(\omega + 2)e^{3\zeta} \tag{6.3}$$

$$\mu_4 = \omega^2(\omega - 1)^2(\omega^4 + 2\omega^3 + 3\omega^2 - 3)e^{4\zeta} \tag{6.4}$$

where $\omega = \exp(\sigma^2)$.

Wartmann [77] has given the general formula

$$\mu_r = \frac{\mu_2^{\frac{1}{2}n}}{(\omega - 1)^{\frac{1}{2}n}} \sum_{j=0}^{r} (-1)^j \binom{r}{j} \omega^{\frac{1}{2}(r-j)(r-j-1)}.$$

The shape factors are

$$\alpha_3 = \sqrt{\beta_1} = (\omega - 1)^{\frac{1}{2}}(\omega + 2) \tag{7.1}$$

$$\alpha_4 = \beta_2 = \omega^4 + 2\omega^3 + 3\omega^2 - 3. \tag{7.2}$$

(Note that $\alpha_3 > 0$; and neither α_3 nor α_4 depends on ζ.)

Equations (7.1) and (7.2) may be regarded as parametric equations of a curve in the (β_1,β_2) plane. This curve is called the *lognormal line* and is shown in

TABLE 1

Standardized 100α% Points of Lognormal Distributions

Note: $\sigma = \delta^{-1}$.

σ \ α		99.95	99.9	99.75	99.5	99	97.5	95	90	75	50 (median)
	(Upper tail) (Lower tail)	0.05	0.1	0.25	0.5	1	2.5	5	10	25	
0.02		3.39	3.18	2.88	2.63	2.37	1.99	1.66	1.29	0.669	−0.010
		−3.19	−3.01	−2.74	−2.52	−2.28	−1.93	−1.63	−1.27	−0.680	
0.04		3.49	3.27	2.95	2.69	2.42	2.02	1.68	1.29	0.663	−0.020
		−3.10	−2.92	−2.67	−2.46	−2.24	−1.90	−1.61	−1.27	−0.680	
0.06		3.60	3.36	3.02	2.75	2.46	2.04	1.69	1.30	0.657	−0.030
		−3.01	−2.84	−2.61	−2.41	−2.20	−1.87	−1.59	−1.26	−0.689	
0.08		3.71	3.45	3.09	2.81	2.50	2.07	1.71	1.30	0.650	−0.040
		−2.29	−2.76	−2.54	−2.36	−2.15	−1.85	−1.57	−1.25	−0.693	
0.10		3.82	3.54	3.17	2.87	2.55	2.10	1.72	1.31	0.643	−0.050
		−2.83	−2.69	−2.48	−2.30	−2.11	−1.82	−1.56	−1.24	−0.697	
0.12		3.93	3.64	3.24	2.93	2.60	2.13	1.74	1.31	0.635	−0.060
		−2.75	−2.61	−2.42	−2.25	−2.07	−1.79	−1.54	−1.23	−0.701	
0.14		4.05	3.74	3.32	2.99	2.64	2.15	1.75	1.31	0.628	−0.069
		−2.67	−2.54	−2.36	−2.20	−2.03	−1.76	−1.52	−1.23	−0.704	
0.16		4.17	3.84	3.40	3.05	2.69	2.18	1.77	1.32	0.620	−0.079
		−2.59	−2.47	−2.30	−2.15	−1.98	−1.73	−1.50	−1.22	−0.706	
0.18		4.29	3.95	3.48	3.11	2.73	2.21	1.78	1.32	0.611	−0.089
		−2.51	−2.40	−2.24	−2.10	−1.94	−1.70	−1.48	−1.21	−0.708	
0.20		4.42	4.05	3.56	3.17	2.78	2.23	1.79	1.32	0.603	−0.098
		−2.44	−2.33	−2.18	−2.05	−1.90	−1.67	−1.46	−1.14	−0.714	
0.30		5.10	4.61	3.97	3.49	3.00	2.35	1.84	1.32	0.555	−0.143
		−2.10	−2.03	−1.92	−1.82	−1.71	−1.53	−1.36	−1.14	−0.714	
0.40		5.86	5.23	4.41	3.81	3.22	2.45	1.88	1.30	0.502	−0.185
		−1.81	−1.76	−1.68	−1.61	−1.53	−1.39	−1.25	−1.07	−0.709	
0.50		6.71	5.89	4.86	4.13	3.42	2.54	1.89	1.27	0.444	−0.220
		−1.56	−1.52	−1.47	−1.42	−1.36	−1.25	−1.15	−1.00	−0.695	
1.00		11.66	9.41	6.90	5.32	3.98	2.52	1.63	0.904	0.145	−0.300
		−0.746	−0.742	−0.735	−0.728	−0.718	−0.698	−0.674	−0.634	−0.527	

Figure 1 of Chapter 12.

The coefficient of variation is $(\omega - 1)^{\frac{1}{2}}$ and also does not depend on ζ.

The distribution of X is unimodal, and the *mode* is at

$$\text{mode}\,(X) = \exp\,(\zeta - \sigma^2). \tag{8}$$

From (4) it is clear that the value X_α such that $\Pr[X \leq X_\alpha] = \alpha$ is related to the corresponding percentile, U_α, of the unit normal distribution by the relation:

$$X_\alpha = \exp\,(\zeta + U_\alpha \sigma). \tag{9.1}$$

In particular the median X (corresponding to $\alpha = \frac{1}{2}$) is

$$\text{median}\,(X) = e^{\zeta}. \tag{9.2}$$

Comparison of (6.1), (8) and (9.2) shows that

$$E[X] > \text{median}\,(X) > \text{mode}\,(X)$$

and

$$\frac{\text{mode}\,(X)}{E[X]} = e^{-3\sigma^2/2} = \left[\frac{\text{median}\,(X)}{E[X]}\right]^3. \tag{9.3}$$

The *standardized 100α% deviate* is

$$X'_\alpha = \frac{X_\alpha - E[X]}{\sigma[X]} = \frac{e^{U_\alpha \sigma} - e^{\frac{1}{2}\sigma^2}}{[e^{\sigma^2}(e^{\sigma^2} - 1)]^{\frac{1}{2}}} \tag{10}$$

$$= \frac{\exp\,(U_\alpha \sigma - \frac{1}{2}\sigma^2) - 1}{(e^{\sigma^2} - 1)^{\frac{1}{2}}}.$$

Some values of X'_α are shown in Table 1.

As σ tends to zero (or δ to infinity) the standardized lognormal distribution tends to a unit normal distribution, and $X'_\alpha \to U_\alpha$, as can be seen from Table 1. For σ small

$$X'_\alpha \doteqdot U_\alpha + \tfrac{1}{2}(U_\alpha^2 - 1)\sigma + \tfrac{1}{12}(2U_\alpha^3 - 9U_\alpha)\sigma^2. \tag{11}$$

It is to be expected that

(i) X'_α will change relatively slowly with σ (for σ small) if $U_\alpha^2 \doteqdot 1$, i.e., $\alpha \doteqdot 0.84$ or 0.16, and

(ii) the *standardized inter-100α% distance*

$$X'_\alpha - X'_{1-\alpha} \doteqdot 2U_\alpha + \tfrac{1}{6}(2U_\alpha^3 - 9U_\alpha)\sigma^2$$

will change relatively slowly if $U_\alpha^2 \doteqdot \frac{9}{2}$; i.e., $\alpha \doteqdot 0.983$ or 0.017. These features, also, are indicated from a study of Table 1. It is also clear from this table that as σ increases the lognormal distribution rapidly becomes markedly nonnormal. Table 2 of α_3 and α_4 also indicates how rapidly the skewness and kurtosis increase with σ.

TABLE 2

Values of α_3 and α_4 for Distribution (2)′

σ	α_3	α_4
0.1	0.30	3.16
0.2	0.61	3.68
0.3	0.95	4.64
0.4	1.32	6.26
0.5	1.75	8.90
0.6	2.26	13.27
0.7	2.89	20.79
0.8	3.69	34.37
0.9	4.75	60.41
1.0	6.18	113.94

It is for this reason that only relatively small values of σ are used in Table 1. It can be seen that for larger σ there is a high probability density below the expected value, leading to small numerical standardized deviations. Conversely there is a long positive tail with large standardized deviations for upper percentiles. Figure 1 shows some typical probability density functions (standardized in each case, so that expected value is zero and standard deviation is 1).

Wise [84] has shown that the probability density function of the two-parameter distribution has two points of inflection at

$$x = \exp\left[\zeta - \frac{3\sigma^2}{2} \pm \sigma\sqrt{1 + \tfrac{1}{4}\sigma^2}\right].$$

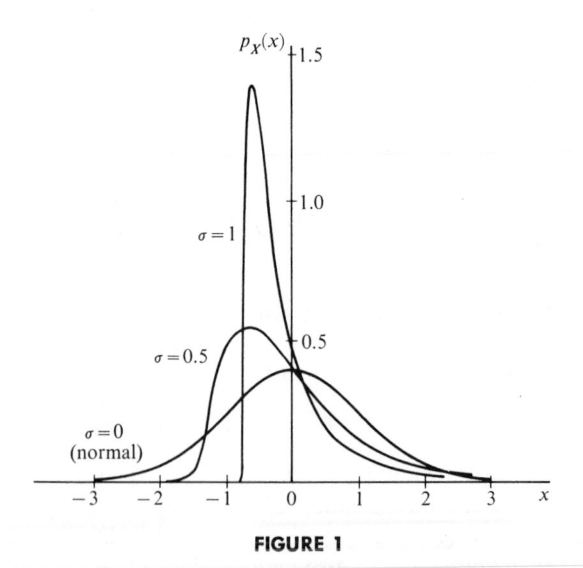

FIGURE 1

Standardized Lognormal Distribution

Just as the sum of two independent normal variables is also normally distributed, so the product of two independent (two-parameter) lognormally distributed variables is also lognormally distributed. If

$$U_j = \gamma_j + \delta_j \log X_j \qquad (j = 1,2)$$

are independent unit normal variables, then

$$\log (X_1 X_2) = \delta_1^{-1} U_1 + \delta_2^{-1} U_2 - (\gamma_1 \delta_1^{-1} + \gamma_2 \delta_2^{-1})$$

and so

$$(\delta_1^{-2} + \delta_2^{-2})^{-\frac{1}{2}}[\log (X_1 X_2) + \gamma_1 \delta_1^{-1} + \gamma_2 \delta_2^{-1}]$$

is a unit normal variable.

Introducing the notation $\Lambda(\zeta,\sigma^2)$ to denote a distribution defined by (4), then if X_j has the distribution $\Lambda(\zeta_j,\sigma_j^2)$ (for any j) and X_1, X_2 are mutually independent then $X_1 X_2$ has the distribution $\Lambda(\zeta_1 + \zeta_2,\sigma_1^2 + \sigma_2^2)$.

We also note that the distribution of $a_j X_j$, where a_j is any positive constant, is $\Lambda(\zeta_j + \log a_j,\sigma_j^2)$. Evidently, if $X_1, X_2, \ldots, X_k$ are a mutually independent set, then $\prod a_j X_j$ is distributed as $\Lambda\left(\sum_{j=1}^{k} \{\zeta_j + \log a_j\}, \sum_{j=1}^{k} \sigma_j^2\right)$.

The preceding discussion has been limited to variables with *two-parameter* lognormal distributions. The extra parameter, θ, in (2) moves ('translates') the whole distribution an amount θ in the direction of increasing X. Many properties — particularly variance and shape — remain unchanged. However the reproductive property just discussed does not hold if $\theta \neq 0$ for either X_1 or X_2.

It is also clear that if X has a two-parameter lognormal distribution then so do X^b and bX (provided $b \neq 0$).

The information generating function corresponding to (2) is

$$(\delta/\sqrt{2\pi})^{u-1} \exp [\tfrac{1}{2}(u - 1)^2(u\delta^2)^{-1} + (u - 1)\gamma/\delta].$$

The entropy is

$$\tfrac{1}{2} - (\gamma/\delta) - \log (\delta/\sqrt{2\pi}).$$

4. Estimation

(i) θ *known*: If the value of the parameter θ is known (it is often possible to take θ equal to zero) then estimation of the parameters ζ and σ (or γ and δ) presents few difficulties beyond those already discussed in connection with the normal distribution. In fact, by using values of $Z_i = \log (X_i - \theta)$, the problem is reduced to that of estimation of parameters of a normal distribution. Many specialized problems, such as those relating to truncated or censored distributions, or the use of order statistics, may be reduced to corresponding problems (already discussed in Chapter 13) for the normal distribution. Maximum likelihood estimation is exactly equivalent to maximum likelihood estimation for normal distributions, so that the maximum likelihood estimators $\hat{\zeta}$, $\hat{\sigma}$ of ζ and σ, respectively, are

(12.1) $$\hat{\zeta} = \bar{Z}$$

(12.2) $$\hat{\sigma} = \left[n^{-1} \sum_{j=1}^{n} (Z_j - \bar{Z})^2\right]^{\frac{1}{2}}$$

where $\bar{Z} = n^{-1} \sum_{j=1}^{n} Z_j$ (assuming, of course, that $X_1, X_2, \ldots, X_n$ are independent random variables each having the same lognormal distribution with θ known).

There are a few situations in which techniques appropriate to normal distributions cannot be applied directly. One is the construction of best *linear* unbiased estimators, using order statistics of the original variables X_j. Another is that in which the original values $\{X_i\}$ are not available, but the data are in groups of equal width. While the results can immediately be written down in the form of grouped observations from a normal distribution, the groups appropriate to the transformed variables will not be of equal width. In such circumstances methods of the kind described in Chapter 12, Section 2, might be applied.

Sometimes it may be desired to estimate not ζ or σ, but rather the expected value $\exp(\zeta + \frac{1}{2}\sigma^2)$ and the variance $e^{\sigma^2}(e^{\sigma^2} - 1)e^{2\zeta}$ of the variable X, since $\bar{Z}$ and $S^2 = (n-1)^{-1} \sum_{j=1}^{n} (Z_j - \bar{Z})^2$ are jointly complete and sufficient statistics for ζ and σ. If a function $f(\bar{Z}, S^2)$ has expected value $h(\zeta, \sigma)$ it is the minimum variance unbiased estimator of $h(\zeta, \sigma)$. Finney [23] obtained such estimators of the expected value and variance in the form of infinite series

(13.1) $$M = \exp(\bar{Z}) g(\tfrac{1}{2}S^2)$$

(13.2) $$V = \exp(2\bar{Z})\{g(2S^2) - g((n-2)S^2/(n-1))\}$$

respectively, where

(14) $$g(t) = 1 + \frac{n-1}{n} t + \sum_{j=2}^{\infty} \frac{(n-1)^{2j-1}}{n^j(n+1)(n+3)\cdots(n+2j-1)} \frac{t^j}{j!}.$$

Unfortunately the series in (14) converges slowly (except for very small values of t). Finney recommends using the approximation

(15) $$g(t) \doteqdot e^t\left\{1 - \frac{t(t+1)}{n} + \frac{t^2(3t^2 + 22t + 21)}{6n^2}\right\}$$

which (he states) should be safe for $n > 50$ in (13.1) and $n > 100$ in (13.2) provided the coefficient of variation $((e^{\sigma^2} - 1)^{\frac{1}{2}})$ is less than 1 (corresponding to $\sigma < 0.83$).

Finney showed that (with $\omega = \exp(\sigma^2)$)

(16.1) $$n \operatorname{var}(M) \doteqdot \sigma^2 e^{2\zeta} \omega (1 + \tfrac{1}{2}\sigma^2)$$

(16.2) $$n \operatorname{var}(V) \doteqdot 2\sigma^2 e^{\zeta} \omega^2 \{2(\omega - 1)^2 + \sigma^2(2\omega - 1)^2\}.$$

By comparison, for the unbiased estimators of the expected value and variance

$\overline{X}$ and $S_X^2 = (n-1)^{-1} \sum_{j=1}^{n} (X_j - \overline{X})^2$, respectively,

$$n \operatorname{var}(\overline{X}) = e^{2\zeta}\omega(\omega - 1) \tag{17.1}$$

$$n \operatorname{var}(S_X^2) \doteqdot e^{4\zeta}\omega^2(\omega - 1)^2(\omega^4 + 2\omega^3 + 3\omega^2 - 4). \tag{17.2}$$

Figure 2 (taken from Finney [23]) shows approximate values of the 'efficiency' ratios $100 \operatorname{var}(M)/\operatorname{var}(\overline{X})$ and $100 \operatorname{var}(V)/\operatorname{var}(S_X^2)$ as a function of σ^2. It will be noted that while $\overline{X}$ is reasonably efficient compared with M, considerable reduction in variance is achieved by using V in place of S_X^2.

Similar investigations have been published more recently (Oldham [60]). Peters [62] has constructed best *quadratic* estimators of $\log \mu_r' = r\zeta + \frac{1}{2}r^2\sigma^2$, using order statistics based on $\log X_j$'s.

Confidence limits for ζ and/or σ can, of course, be constructed from the transformed values $Z_1, Z_2, \ldots, Z_n$ using techniques described in Chapter 13. In particular, (when $\theta = 0$), since the coefficient of variation of X is $(e^{\sigma^2} - 1)^{\frac{1}{2}}$, confidence limits (with confidence coefficient $100\alpha\%$) for this quantity are $(\exp[(n-1)S^2/\chi^2_{n-1,1-\alpha/2}] - 1)^{\frac{1}{2}}$ and $(\exp[(n-1)S^2/\chi^2_{n-1,\alpha/2}] - 1)^{\frac{1}{2}}$ where $\chi^2_{\nu,\epsilon}$ denotes the $100\epsilon\%$ point of the χ^2 distribution with ν degrees of freedom, to be described in Chapter 17. See Koopmans *et al.* [49].

Similar arguments could be used to construct confidence limits for any other monotonic function of σ, such as $\sqrt{\beta_1}$ and β_2.

(ii) *θ unknown*: Estimation problems present considerable difficulty when θ is not known. As might be expected, estimation of θ is particularly inaccurate. This parameter is a 'threshold value', below which the cumulative distribution function is zero, and above which it is positive. Such values are often — one might say 'usually' — difficult to estimate.

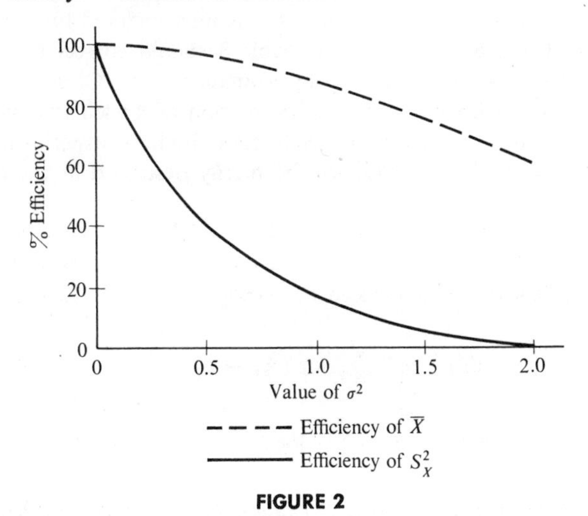

FIGURE 2

The Efficiency of $\overline{X}$ and of S_X^2 in Large Samples

However, estimation of parameters is often not as important as estimation of probabilities — in particular, of cumulative distribution functions. Table 3 shows the median and upper and lower 10% and 25% points of lognormal distributions with widely differing values of θ, but with ζ and σ so chosen that each distribution is standardized (that is, has zero expected value and unit standard deviation). (See also Table 1.)

TABLE 3

Percentile Points of Standardized Lognormal Distributions

σ	θ	Lower 10%	Lower 25%	Median	Upper 25%	Upper 10%
0.02	−49.995	−1.275	−0.680	−0.010	0.669	1.288
0.04	−24.990	−1.268	−0.685	−0.020	0.663	1.293
0.06	−16.652	−1.260	−0.689	−0.030	0.657	1.299
0.08	−12.480	−1.252	−0.693	−0.040	0.650	1.303
0.10	−9.975	−1.244	−0.697	−0.050	0.643	1.307
0.12	−8.303	−1.235	−0.700	−0.060	0.635	1.311
0.14	−7.108	−1.225	−0.703	−0.069	0.628	1.314
0.16	−6.210	−1.216	−0.706	−0.079	0.620	1.316
0.18	−5.511	−1.206	−0.708	−0.089	0.611	1.318
0.20	−4.950	−1.195	−0.710	−0.098	0.603	1.320

Note: The mode is at $(e^{\sigma^2} - 1)^{-\frac{1}{2}}(-1 + e^{-3\sigma^2/2}) = \theta(1 - e^{-3\sigma^2/2})$.

There can be considerable variation in θ with little effect on the percentiles, and little effect on values of the cumulative distribution functions for fixed values of X. Insensitivity to variation in θ is most marked for large negative values of θ. These observations and Table 3 should correct any feelings of depression caused by the inaccuracy of estimation of θ. Of course, there can be situations in which the accuracy of estimation of θ itself is a primary consideration. In these cases special techniques, both of experimentation and analysis, may be needed. These will be briefly described at the end of this section.

Maximum likelihood estimation of θ, ζ and σ might be expected to be attained by the following tedious, but straightforward, procedure. Since for given θ, the likelihood function is maximized by taking

$$\begin{aligned} \zeta &= \hat{\zeta}(\theta) = n^{-1} \sum_{j=1}^{n} \log (X_i - \theta) \\ \sigma &= \hat{\sigma}(\theta) = \left[n^{-1} \sum_{j=1}^{n} \{\log (X_i - \theta) - \hat{\zeta}(\theta)\}^2 \right]^{\frac{1}{2}} \end{aligned} \tag{18}$$

one might take a sequence of values of θ, and calculate the maximized likelihood corresponding to each, and then try to estimate, numerically, the value, $\hat{\theta}$, of θ maximizing the maximized likelihood. However, Hill [41] has shown that as θ tends to min $(X_1,X_2,\ldots,X_n)$* the maximized likelihood tends to infinity. Formally this would seem to indicate that we should accept the 'estimates': $\hat{\theta} = \min (X_1,X_2,\ldots,X_n)$, $\hat{\zeta} = -\infty$, $\hat{\sigma} = \infty$. Hill resolved this difficulty by introducing a prior joint distribution for θ, ζ and σ, and using Bayes' theorem. This leads to the conclusion that solutions of the *formal* maximum likelihood equations should be used, with $\hat{\theta}$ satisfying

$$\sum_{j=1}^{n} (X_j - \hat{\theta})^{-1} + [\hat{\sigma}(\hat{\theta})]^{-1} \left\{\sum_{j=1}^{n} (X_j - \hat{\theta})^{-1} Z_j'\right\} = 0 \tag{19}$$

where $Z_j' = [\log (X_i - \hat{\theta}) - \hat{\zeta}(\hat{\theta})]/\hat{\sigma}(\hat{\theta})$ and $\hat{\zeta}(\hat{\theta})$, $\hat{\sigma}(\hat{\theta})$ satisfy (18), with θ replaced by $\hat{\theta}$.

Formal calculation of limiting variances gives:

$$n \operatorname{var}(\hat{\theta}) \doteqdot \sigma^2 e^{2\zeta} \omega^{-1}[\omega(1 + \sigma^2) - 2\sigma^2 - 1]^{-1} \tag{20.1}$$

$$n \operatorname{var}(\hat{\zeta}) \doteqdot \sigma^2[\omega(1 + \sigma^2) - 2\sigma^2][\omega(1 + \sigma^2) - 2\sigma^2 - 1]^{-1} \tag{20.2}$$

$$n \operatorname{var}(\hat{\sigma}) \doteqdot \sigma^2[\omega(1 + \sigma^2) - 1][\omega(1 + \sigma^2) - 2\sigma^2 - 1]^{-1}. \tag{20.3}$$

Formulas equivalent to these have been given by Cohen [12] and Hill [41]. The latter, however, points out that in his opinion the formulas have not been rigorously established, and he seems to be right.

We may note that Harter and Moore [33] carried out a sampling experiment, with $\zeta = 4$, $\sigma = 2$, $\theta = 10$. Their results can be represented approximately, for sample size n, in the range 50 to 200 by

$$n \operatorname{var}(\hat{\zeta}) \doteqdot 4.2;$$
$$n \operatorname{var}(\hat{\sigma}) \doteqdot 3.2;$$
$$n^2 \operatorname{var}(\hat{\theta}) \doteqdot 1800$$

(note the factor n^2 in the last formula). If θ is known, their results give $n \operatorname{var}(\hat{\zeta}) \doteqdot 4.1$ — not much less than if θ is not known — but $n \operatorname{var}(\hat{\sigma})$ is now only about 2.1.

Tiku [72] has suggested an approximate linearization of the maximum likelihood equations based on the approximate formula

$$Z(x)[1 - \Phi(x)]^{-1} \doteqdot \alpha + \beta x$$

for Mills' ratio. Appropriate values of α and β depend on the range of values of x. (See discussion of a similar technique in Chapter 13, Section 6.4.)

Uncertainties and difficulties in the use of (18) and (19) lead us to consider other methods.

If the first three sample moments are equated to the corresponding population values we obtain the formulas:

*It is clear that we must take $\theta \leq \min (X_1,X_2,\ldots,X_n)$.

(21.1) $$\bar{X} = \tilde{\theta} + \exp(\tilde{\xi} + \tfrac{1}{2}\tilde{\sigma}^2)$$

(21.2) $$m_2 = n^{-1}\sum_{j=1}^{n}(X_j - \bar{X})^2 = e^{2\tilde{\xi}+\tilde{\sigma}^2}(e^{\tilde{\sigma}^2} - 1)$$

(21.3) $$m_3 = n^{-1}\sum_{j=1}^{n}(X_j - \bar{X})^3 = e^{3\tilde{\xi}+3\tilde{\sigma}^2/2}(e^{\tilde{\sigma}^2} - 1)^2(e^{\tilde{\sigma}^2} + 2)$$

whence

(22) $$\begin{aligned} b_1 = m_3^2 m_2^{-3} &= (e^{\tilde{\sigma}^2} - 1)(e^{\tilde{\sigma}^2} + 2)^2 \\ &= (\tilde{\omega} - 1)(\tilde{\omega} + 2)^2. \end{aligned}$$

From (22), $\tilde{\omega}$ and so $\tilde{\sigma}$ can be determined. Then, from (21.2),

$$\tilde{\xi} = \tfrac{1}{2}\log[m_2\tilde{\omega}^{-1}(\tilde{\omega} - 1)^{-1}]$$

and finally $\tilde{\theta}$ can be determined from (21.1).

Yuan [87] provided a table of values of the right-hand side of (22) to aid in the solution of that equation. Aitchison and Brown [3] give a table for $\tilde{\sigma}^2$ directly to four decimal places for $\sqrt{b_1} = 0.0(0.2)10.0(1)24$. Without using such tables, an iterative method using the following equation

(22)′ $$\tilde{\omega} = 1 + (\tilde{\omega} + 2)^{-2}b_1$$

(equivalent to (22)) is quite easy to apply, or the explicit solution

$$\tilde{\omega} = [1 + \tfrac{1}{2}b_1 + \sqrt{(1 + \tfrac{1}{2}b_1)^2 - 1}]^{\frac{1}{3}} + [1 + \tfrac{1}{2}b_1 - \sqrt{(1 + \tfrac{1}{2}b_1)^2 - 1}]^{\frac{1}{3}} - 1$$

can be used.

This method is easy to apply but liable to be inaccurate. Even when θ is known, estimation of population variance by sample variance is relatively inaccurate (see Figure 2). Approximate formulas for the variances of the estimators were obtained by Nydell [59]. They are

$$n\,\mathrm{var}(\tilde{\xi}) \doteqdot \tfrac{1}{6}\{4u^{-1} + 54 + 306u + 895u^2 + 1598u^3 + 1753.5u^4 + 1242.7u^5\}$$
$$n\,\mathrm{var}(\tilde{\sigma}) \doteqdot \tfrac{1}{2}u\sigma^{-2}\{8 + 106u + 332u^2 + 479u^3 + 404u^4 + 216u^5\}$$

where $u = \omega - 1 = e^{\sigma^2} - 1 =$ (square of coefficient of variation).

A modification of the method of moments is to estimate θ by this method and then use Finney's estimators M, V (see (13.1) and (13.2)) of expected value and variance, applied to variables $\{X_i - \tilde{\theta}\}$.

Alternatively, if one is prepared to accept some loss in efficiency, it might be possible to use a relatively simple method of calculating estimates. Certain special forms of the method of percentile points give very simple formulas for the estimators. Using the relationship (see (1) and (9.1))

(23) $$U_{\alpha_j} = \gamma + \delta\log(X_{\alpha_j} - \theta) \qquad (j = 1,2,3)$$

the following formula can be obtained:

$$(24) \qquad \frac{U_{\alpha_1} - U_{\alpha_2}}{U_{\alpha_2} - U_{\alpha_3}} = \left\{\log\left[\frac{X_{\alpha_1} - \theta}{X_{\alpha_2} - \theta}\right]\right\} \Big/ \left\{\log\left[\frac{X_{\alpha_2} - \theta}{X_{\alpha_3} - \theta}\right]\right\}.$$

If estimated values, $\hat{X}_{\alpha_j}$, of the X_{α_j}'s are inserted in (24), solution of the resulting equation for θ must be effected by trial and error. However if we choose $\alpha_2 = \frac{1}{2}$ (corresponding to the median) and $\alpha_1 = 1 - \alpha_3$ then (noting that $U_{\frac{1}{2}} = 0$, and $U_{\alpha_3} = -U_{\alpha_1}$) the equations are (asterisk denoting 'estimated value'):

$$(25.1) \qquad \hat{X}_{\alpha_1} = \theta^* + e^{-\gamma^*/\delta^*} \exp [U_{\alpha_1}/\delta^*]$$

$$(25.2) \qquad \hat{X}_{\frac{1}{2}} = \theta^* + e^{-\gamma^*/\delta^*}$$

$$(25.3) \qquad \hat{X}_{\alpha_3} = \theta^* + e^{-\gamma^*/\delta^*} \exp [-U_{\alpha_1}/\delta^*]$$

whence

$$(\hat{X}_{\alpha_1} - \hat{X}_{\frac{1}{2}})/(\hat{X}_{\frac{1}{2}} - \hat{X}_{\alpha_3}) = \exp [U_{\alpha_1}/\delta^*]$$

and

$$(26) \qquad \delta^* = U_{\alpha_1}\left[\log \frac{\hat{X}_{\alpha_1} - \hat{X}_{\alpha_2}}{\hat{X}_{\alpha_2} - \hat{X}_{\alpha_3}}\right]^{-1}$$

so that no solution by trial and error is needed. Aitchison and Brown [3] suggest taking $\alpha_1 = 0.95$, but it is likely that any value between 0.90 and 0.95 will give about the same accuracy. Lambert [52] suggested using values of the parameters obtained by this method as a starting point for an iterative formal solution of the maximum likelihood equations. He used the value $\alpha_1 = \frac{61}{64} \doteqdot 0.953$.

It is possible for (26) to give a negative value of δ^*, but this happens only if

$$\hat{X}_{\alpha_1} - \hat{X}_{\frac{1}{2}} < \hat{X}_{\frac{1}{2}} - \hat{X}_{1-\alpha_1}$$

which is very unlikely to be the case if the distribution has substantial positive skewness.

If the sample median, $\hat{X}_{\frac{1}{2}}$, be replaced by the sample arithmetic mean, $\overline{X}$, and equation (25.2) is replaced by

$$(25.4) \qquad \overline{X} = \theta^* + e^{-\gamma^*/\delta^*} \exp (\tfrac{1}{2}\delta^{*-2})$$

then from (25.1), (25.3) and (25.4)

$$(27) \qquad \frac{\overline{X} - \hat{X}_{1-\alpha_1}}{\hat{X}_{\alpha_1} - \overline{X}} = \frac{\exp (\frac{1}{2}\delta^{*-2}) - \exp (-U_{\alpha_1}\delta^{*-2})}{\exp (U_{\alpha_1}\delta^{*-2}) - \exp (\frac{1}{2}\delta^{*-2})}$$

From this equation δ^* can be found numerically. Aitchison and Brown [3] give graphs of the function on the right-hand side of (27), for $\alpha_1 = 0.95$, 0.90 and 0.80, which helps to give an initial value for the solution.

This method of estimation is called *Kemsley's method* [47]. It, also, can give a negative value for δ^*, though this is unlikely to occur.

Aitchison and Brown [3] made an experimental comparison of results of using the method of moments (22), the method of quantiles (26) and Kemsley's method (27) to estimate θ. They came to the conclusion that (26) with $\alpha_1 = 0.95$ is slightly better than (27), and that both are considerably better than the method of moments.

Other methods of estimating θ have been suggested. We mention here only *Cohen's method* [12], which is based on the idea that in a large sample at least one observed value of X should be not much greater than the threshold value θ. The least sample value $\hat{X}_{\min}$ is the equated to the formula for the $100(n+1)^{-1}\%$ point of the distribution, giving the relationship

$$\hat{X}_{\min} = \theta^* + e^{-\gamma^*/\delta^*} \exp [U_{(n+1)^{-1}}/\delta^*]$$

or

$$\hat{X}_{\min} = \theta^* + e^{\zeta^*} \exp [U_{(n+1)^{-1}}\, \sigma^*].$$

This is then combined with the first two maximum likelihood equations (18).

There are natural modifications of this method, to allow for cases when there are a number of equal (or indistinguishable — as in grouped data) minimum values. Using a variant of this method, an initial value of θ (for use in some iterative process) may be chosen as $\hat{X}_{\min}$ *minus* some arbitrary (usually rather small) value.

Wise [83] has put forward the following interesting suggestion for a graphical method of estimating the parameters. It is particularly useful when the 'rate of increase' of probability (i.e., probability density) is observed, rather than actual frequencies, though it can also be used in the latter case, with sufficiently extensive data. He starts by observing that the tangents at the two points of inflection of the probability density function remain close to the curve for a considerable length, and so should be estimable graphically with fair accuracy. The modulus of the ratio of the slopes (lower/upper) is

$$\exp [2 \sinh^{-1} \tfrac{1}{2}\sigma + \sigma\sqrt{1 + \tfrac{1}{4}\sigma^2}].$$

Wise [84] provides tables of the logarithm of this quantity to four decimal places for $\sigma = 0(0.01)0.80$, to aid in estimating σ from the observed slopes of the inflection tangents. The initial point θ is estimated from the values X_1, X_2 of X at the points where these tangents cut the horizontal axis (see Figure 3). If $x_1 < x_2$, then θ is estimated from the formula

$$\tfrac{1}{2}(x_1 + x_2) - \tfrac{1}{2}(x_2 - x_1)L^{-1}$$

where

$$L = [1 + \phi \tanh (\sigma\sqrt{1 + \tfrac{1}{4}\sigma^2})][\tanh (\sigma\sqrt{1 + \tfrac{1}{4}\sigma^2}) + \phi]^{-1}$$

with

$$\phi = (1 + \tfrac{1}{2}\sigma^2)[\sigma\sqrt{1 + \tfrac{1}{4}\sigma^2}]^{-1}.$$

Wise [84] also provides values of L to four decimal places for $\sigma = 0(0.01)0.80$.

Finally ζ is estimated from the formula

$$\zeta = \tfrac{3}{2}\sigma^2 + \tfrac{1}{2} \log [(x_1 - \theta)(x_2 - \theta)].$$

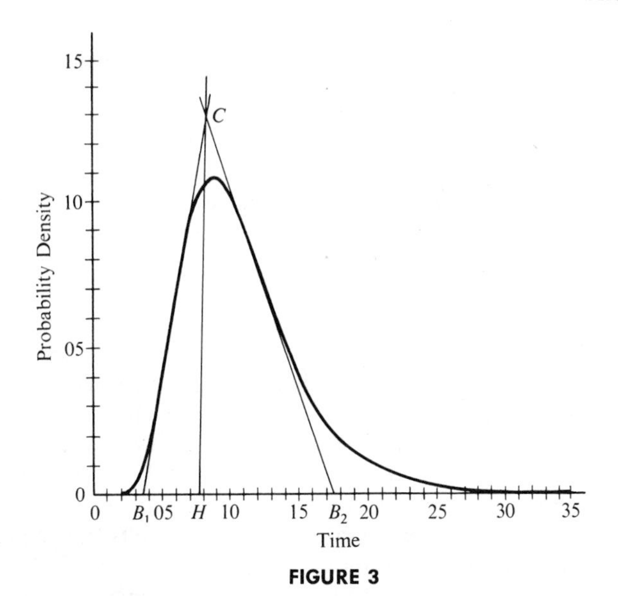

FIGURE 3

Geometrical Method of Estimating Lognormal Parameters

Sequential estimation and testing have been discussed by Zacks [88] and Tomlinson [74].

5. Tables and Graphs

Aitchison and Brown [3] give values of the coefficient of variation, α_3, $(\alpha_4 - 3)$, the ratios (from (9.3)) of mean to median $(e^{\frac{1}{2}\sigma^2})$ and mean to mode $(e^{3\sigma^2/2})$, and the probability that X does not exceed $E[X]$ for

$$\sigma = 0.00(0.005)1.00(0.1)3.0.$$

(When $\sigma = 0$, values appropriate to the unit normal distribution are shown.) All values (except for coefficient of variation and $(\alpha_4 - 3)$ when $\sigma = 0.05$) are given to four significant figures.

Tables of percentile points of (two-parameter) lognormal distributions have been published by Moshman [57]: Upper and lower 10%, 5%, $2\frac{1}{2}$%, 1% and 0.5% values to 4 decimal places, for $\alpha_3 = 0(.05)3.00$ and Broadbent [8]: ratios of upper and lower 5% and 1% values to the expected value, to 4 decimal places, for coefficient of variation $((e^{\sigma^2} - 1)^{\frac{1}{2}})$ equal to 0(0.001)0.150. (Note that this corresponds to $0 \leq \sigma < 0.15$ approximately.)

Tables of *random lognormal deviates* have been published by Hyrenius and Gustafsson [42]. These were derived from tables of random normal deviates (see Chapter 13) and are given to 2 decimal places for distributions with skewness $\alpha_3^2 = 0.2$, 0.5, 1.0 and 2.0 (corresponding to $\sigma = 0.006$, 0.16, 0.31 and 0.55 (approximately) respectively).

Graph paper with the horizontal (abscissa) scale logarithmic and the vertical (ordinate) scale normal (i.e. marked with proportion P at a distance from the origin equal to y where

$$P = (\sqrt{2\pi})^{-1} \int_{-\infty}^{y} e^{-\frac{1}{2}u^2}\, du)$$

is called *lognormal probability* paper. If X has a two-parameter lognormal distribution, plotting $P = \Pr[X \leq x]$ as ordinate against x as abscissa will give a straight line plot. The slope of the line is $\delta(= \sigma^{-1})$; it meets the horizontal axis at $-\gamma/\delta(= \zeta)$. From sample plots these parameters may be estimated by observing the slope of a fitted line and its intersection with the horizontal axis. If a third parameter (θ) is needed, the plotted points tend to curve up, away from a straight line, as x decreases. It is possible, with some patience, to obtain a graphical estimate of θ by trial and error (as suggested by Gibrat [27]), changing the value of θ until a plot using $(x - \theta)$ in place of x can be most closely fitted by a straight line. For a subjective method of this kind, it is not possible to obtain even an approximate formula for the standard deviation of the estimator of θ. However the method seems to give quite useful results and can certainly be used to obtain initial values for use in iterative processes.

Graph paper of the kind described above is useful in probit analysis of quantal response data.

6. Applications

Section 2 of this chapter indicates several fields in which lognormal distributions have been found applicable, and the references to this chapter contain further examples. Application of the distribution is not only based on empirical observation, but can in some cases be supported by theoretical argument — for example, in the distribution of particle sizes in natural aggregates (see Section 2), and in the closely related distribution of dust concentration in industrial atmospheres (Kolmogorov [48], Tomlinson [74], Oldham [60]). Geological applications have been described by Ahrens [2], Chayes [10], Miller and Goldberg [55] and Prohorov [63]. Thébault [70] gives a number of examples of such applications and includes some further references. Further applications, mentioned by Oldham [60], include duration of sickness absence and physicians' consultation time. Wise [85] has described application to dye-dilution curves representing concentration of indicator as a function of time. Hermanson and Johnson [39] found that it gives a good representation of flood flows although extreme value distributions (see Chapter 21) are more generally associated with this field.

The lognormal distribution has also been found to be a serious competitor to the Weibull distribution (Chapter 20) in representing life-time distributions (for manufactured products). Among our references, Adams [1], Epstein [18] [19], Feinlieb [21], Goldthwaite [29], Gupta [31] and Nowick and Berry [58] refer to this topic. Other applications in quality control are described by Ferrell [22], Morrison [56] and Rohn [65]. It is quite likely that the lognormal distribution will be one of the most widely applied distributions in practical statistical work in the near future.

The two-parameter distribution is in at least one important respect a more realistic representation of distributions of characters like weight, height, density than is the normal distribution. These quantities cannot take negative values; but a normal distribution ascribes positive probability to such events, while the two-parameter lognormal distribution does not. Furthermore, by taking σ small enough, it is possible to construct a lognormal distribution closely resembling any normal distribution. Hence, even if a normal distribution is felt to be really appropriate, it might be replaced by a suitable lognormal distribution. Such a replacement is convenient when obtaining confidence limits for the coefficient of variation. Koopmans *et al.* [49] pointed out that if the normal distribution is replaced by a lognormal distribution, then confidence limits for the coefficient of variation are easily constructed (as described in Section 4 of this chapter). Wise [85] has pointed out marked similarities in shape between appropriately chosen lognormal distributions and random walk (Chapter 15) and gamma (Chapter 17) distributions.

The lognormal distribution is also applied, in effect, when certain approximations to the distribution of Fisher's $z = \frac{1}{2} \log F$ are used (see Chapter 26). It is well known that the distribution of z is much closer to normality than is that of F (see, for example, Aroian [4]). Logarithmic transformations, also, are often used in attempts to 'equalize variances' (Curtiss [13], Pearce [61]). If the standard deviation of a character is expected to vary (e.g. from locality to locality) in such a way that the coefficient of variation remains roughly constant (i.e. with standard deviation roughly proportional to expected value) then application of the method of statistical differentials (as described in Chapter 1) indicates that by use of logarithms of observed values, the dependence of standard deviation on expected value should be substantially reduced. Very often, the transformed variables also have a distribution more nearly normal than that of the original variables.

7. Truncated Lognormal and Other Related Distributions

As pointed out in Section 4, estimation for the two-parameter lognormal distribution, and the three-parameter distribution with known value for θ, presents no special difficulties beyond those already encountered with normal distributions in Chapter 13. The same is true of censored or truncated forms of these distributions.

We may note that if $\log X$ has a normal distribution truncated from below at $\log X_0$, with expected value ζ and standard deviation σ, then the rth moment of X about zero is

$$(r\text{th moment if not truncated})[(1 - \Phi(U_0 - r\sigma)]/(1 - \Phi(U_0))$$

with $U_0 = (\log X_0 - \zeta)/\sigma$ (Quensel [64]).

For the three-parameter distribution, with θ not known, when censoring or truncation has been applied, estimation presents considerable difficulty. Only recently has published material contained detailed information on these problems.

Tiku [72] has used his approximate linearization formula (described in Section 4) to simplify the maximum likelihood equations for the truncated lognormal; Tallis and Young [69] have considered estimation from grouped data.

Thompson [71] gives details of fitting a 'truncated lognormal distribution,' which is a mixed distribution defined by

$$\Pr[\log x = a] = \Phi((a - \xi)/\sigma)$$
$$\Pr[a < \log x \leq z] = \Phi((z - \xi)/\sigma) - \Phi((a - \xi)/\sigma).$$

Tables are provided to assist estimation by moments.

In [33], Harter and Moore gave the results of some sampling experiments, in which values of parameters were estimated for lognormal distributions censored by omission of proportions q_1 at the lower, and q_2 at the upper limits of the range of variation, with $q_1 = 0$ or 0.01, $q_2 = 0$ or 0.5. (The case $q_1 = q_2 = 0$ corresponds to a complete, untruncated lognormal distribution. These results have already been referred to in Section 4).

Taking $\sigma = 2$ and $\theta = 10$, Table 5 shows the values which were obtained for the variances of estimators obtained by solving the maximum likelihood equations.

TABLE 5

Variances and Covariances of 'Maximum Likelihood' Estimators for Censored Three-parameter Lognormal Distribution

q_1	q_2	var($\hat{\zeta}$)	var($\hat{\sigma}$)	var($\hat{\theta}$)	cov($\hat{\sigma}$,$\hat{\theta}$)	cov($\hat{\zeta}$,$\hat{\theta}$)	cov($\hat{\zeta}$,$\hat{\sigma}$)
0.00	0.5	0.0600	0.0961	0.2109	0.0628	0.0146	0.0296
0.01	0.0	0.0428	0.0351	0.2861	0.0064	0.0126	−0.0047
0.01	0.5	0.0628	0.1244	0.3011	0.0199	0.0239	0.0375
0.00*	0.0*	0.0416	0.0312	0.1733	0.0232	−0.0015	−0.0032

Note: $\zeta = 4$; $\sigma = 2$; $\theta = 10$ in Sampling Experiment with Samples of Size 100 (Value of θ not Used in Analysis).

*(No censoring)

Although the results are applicable to a *censored* lognormal distribution they should give a useful indication of the accuracy to be expected with *truncated* lognormal distributions.

The substantial increase in var($\hat{\theta}$) with even a small amount of censoring ($q_1 = 0.01$) at the lower limits is notable. It would seem reasonable to suppose variances and covariances approximately inversely proportional to sample size, if estimates are needed for sample sizes other than 100.

The arithmetic means of the estimators were also calculated. They indicated a positive bias in $\hat{\theta}$ of about 0.8–0.9 when $q_1 = 0.01$ (whether $q_2 = 0$ or not), and a positive bias of about 0.3 when $q_2 = 0.5$ with $q_1 = 0$. There was also a positive bias of about 0.5 in $\hat{\sigma}$ when $q_1 = 0.01$ and $q_2 = 0.5$.

It is of interest to compare the figures in Table 5 with corresponding values in Table 6, where the value (10) of θ was supposed known, and used in the analysis.

TABLE 6

As Table 5, Except That Value of θ is Supposed Known

q_1	q_2	$\text{var}(\hat{\zeta} \mid \theta)$	$\text{var}(\hat{\sigma} \mid \theta)$	$\text{cov}(\hat{\zeta},\hat{\sigma} \mid \theta)$
0.00	0.5	0.0576	0.0460	0.0198
0.01	0.0	0.0406	0.0224	0.0000
0.01	0.5	0.0574	0.0473	0.0197
0.00*	0.0*	0.0416	0.0218	0.0002

*(No censoring)

The variance of $\hat{\sigma}$ is considerably reduced (by comparison with Table 5), but that of $\hat{\zeta}$ is not greatly changed. The effect of q_1 is much smaller.

The S_U and S_B systems of distributions (Johnson [43]) discussed in Chapter 12, are related to the lognormal distribution, especially in that the lognormal distribution is a limiting form of either system, and also the 'lognormal line' is the border between the regions in the (β_1,β_2) plane corresponding to the two systems.

The distribution corresponding to the S_B transformation, defined by

$$U = \gamma + \delta \log \left(\frac{X - \zeta}{\zeta + \lambda - X} \right) \tag{28}$$

which has four parameters γ, δ, ζ, γ was considered by van Uven [75] as early as 1917. It has been termed the *four-parameter lognormal* distribution by Aitchison and Brown [3]. This name is not so well recognized as the two-parameter or three-parameter lognormal nomenclature. In view of the existence of other four-parameter transformations (such as the S_U described in Chapter 12) the usefulness of this name is doubtful.

The discretized form of the (truncated) lognormal distribution has been found to offer a competitive alternative to logarithmic series distributions in some practical situations (see Chapter 6).

Quensel [64] has described a *logarithmic Gram-Charlier* distribution in which $\log X$ has a Gram-Charlier distribution.

Multivariate distributions with lognormal marginal distributions will be discussed in Chapter 34. These distributions are usually formed by supposing the transformed (and normally distributed) variables to have a joint multinormal distribution.

REFERENCES

[1] Adams, J. D. (1962). Failure time distribution estimation, *Semiconductor Reliability*, **2,** 41–52.

[2] Ahrens, L. H. (1954–7). The lognormal distribution of the elements, *Geochimica et Cosmochimica Acta*, (1) **5,** 49–73, (2) **6,** 121–131, (3) **11,** 205–212.

[3] Aitchison, J. and Brown, J. A. C. (1957). *The Lognormal Distribution*, London: Cambridge University Press.

[4] Aroian, L. A. (1939). A study of Fisher's *z* distribution, *Annals of Mathematical Statistics*, **12,** 429–448.

[5] Bain, A. D. (1964). *The Growth of Television Ownership in the United Kingdom since the War*, London: Cambridge University Press.

[6] Bliss, C. I. (1934). The method of probits, *Science*, **79,** 38–39.

[7] Bol'shev, L. N., Prohorov, Yu. V. and Rodinov, D. A. (1963). On the logarithmic-normal law in geology, *Teoriya Veroyatnostei i ee Primeneniya*, **8,** 114 (Abstract). (In Russian. English translation, 107.)

[8] Broadbent, S. R. (1956). Lognormal approximation to products and quotients, *Biometrika*, **43,** 404–417.

[9] Camp, B. H. (1938). Notes on the distribution of the geometric mean, *Annals of Mathematical Statistics*, **9,** 221–226.

[10] Chayes, F. (1954). The lognormal distribution of the elements: a discussion, *Geochimica et Cosmochimica Acta*, **6,** 119–120.

[11] Cochran, W. G. (1938). Some difficulties in the statistical analysis of replicated experiments, *Empire Journal of Experimental Agriculture*, **6,** 157–163.

[12] Cohen, A. C. (1951). Estimating parameters of logarithmic-normal distributions by maximum likelihood, *Journal of the American Statistical Association*, **46,** 206–212.

[13] Curtiss, J. H. (1943). On transformations used in the analysis of variance, *Annals of Mathematical Statistics*, **14,** 107–122.

[14] Davies, G. R. (1925). The logarithmic curve of distribution, *Journal of the American Statistical Association*, **20,** 467–480.

[15] Davies, G. R. (1929). The analysis of frequency distributions, *Journal of the American Statistical Association*, **24,** 349–366.

[16] Dhrymes, P. J. (1962). On devising unbiased estimators for the parameters of the Cobb-Douglas production function, *Econometrica*, **30,** 297–304.

[17] Éltetö, Ö. (1965). Large-sample lognormality tests based on new inequality measures, *Bulletin of the International Statistical Institute*, **41,** (**1**) 382–385.

[18] Epstein, B. (1947). The mathematical description of certain breakage mechanisms leading to the logarithmico-normal distribution, *Journal of the Franklin Institute*, **244,** 471–477.

[19] Epstein, B. (1948). Statistical aspects of fracture problems, *Journal of Applied Physics*, **19,** 140–147.

[20] Fechner, G. T. (1897). *Kollektivmasslehre*, Leipzig: Engelmann.

[21] Feinlieb, M. (1960). A method of analyzing log-normally distributed survival data with incomplete follow-up, *Journal of the American Statistical Association*, **55,** 534–545.

[22] Ferrell, E. B. (1958). Control charts for log-normal universes, *Industrial Quality Control*, **15,** No. 2, 4–6.

[23] Finney, D. J. (1941). On the distribution of a variate whose logarithm is normally distributed, *Journal of the Royal Statistical Society, Series B*, **7,** 155–161.

[24] Gaddum, J. H. (1933). Reports on biological standards. III: Methods of biological assay depending on a quantal response, *Special Report Series, Medical Research Council*, London, No. **183.**

[25] Gaddum, J. H. (1945). Lognormal distributions, *Nature*, London, **156,** 463–466.

[26] Galton, F. (1879). The geometric mean in vital and social statistics, *Proceedings of the Royal Society of London*, **29,** 365–367.

[27] Gibrat, R. (1930). Une loi des répartitions économiques: l'effet proportionnel, *Bulletin de Statistique Géneral, France* **19,** 469ff.

[28] Gibrat, R. (1931). *Les Inégalités Economiques*, Paris: Libraire du Recueil Sirey.

[29] Goldthwaite, L. R. (1961). Failure rate study for the lognormal lifetime model, *Proceedings of the 7th National Symposium on Reliability and Quality Control in Electronics*, 208–213.

[30] Grundy, P. M. (1951). The expected frequencies in a sample of an animal population in which the abundances of species are log-normally distributed, Part I, *Biometrika*, **38,** 427–434.

[31] Gupta, S. S. (1962). Life test sampling plans for normal and lognormal distributions, *Technometrics*, **4,** 151–175.

[32] Halmos, P. R. (1944). Random alms, *Annals of Mathematical Statistics*, **15,** 182–189.

[33] Harter, H. L. and Moore, A. H. (1966). Local-maximum-likelihood estimation of the parameters of three-parameter lognormal populations from complete and censored samples, *Journal of the American Statistical Association*, **61,** 842–851.

[34] Hatch, T. (1933). Determination of average particle size from the screen-analysis of non-uniform particulate substances, *Journal of the Franklin Institute*, **215,** 27–37.

[35] Hatch, T. and Choute, S. P. (1929). Statistical description of the size properties of non-uniform particulate substances, *Journal of the Franklin Institute*, **207,** 369–388.

[36] Herdan, G. (1958). The relation between the dictionary distribution and the occurrence distribution of word length and its importance for the study of quantitative linguistics, *Biometrika*, **45,** 222–228.

[37] Herdan, G. (1960). *Small Particle Statistics*, 2nd Edition, London: Butterworth's.

[38] Herdan, G. (1966). *The Advanced Theory of Language as Choice and Chance*, New York: Springer.

[39] Hermanson, R. E. and Johnson, H. P. (1967). Generalized flood-frequency relationships, *Iowa State Journal of Science*, **41,** 247–268.

[40] Heyde, C. C. (1963). On a property of the lognormal distribution, *Journal of the Royal Statistical Society, Series B*, **25,** 392–393.

[41] Hill, B. M. (1963). The three-parameter lognormal distribution and Bayesian analysis of a point-source epidemic, *Journal of the American Statistical Association*, **58,** 72–84.

[42] Hyrenius, H. and Gustafsson, R. (1962). *Tables of Normal and Log-normal Random Deviates: I, II*, Stockholm: Almqvist and Wiksell.

[43] Johnson, N. L. (1949). Systems of frequency curves generated by methods of translation, *Biometrika*, **36,** 149–176.

[44] Kalinske, A. A. (1946). On the logarithmic probability law, *Transactions of the American Geophysical Union*, **27,** 709–711.

[45] Kapteyn, J. C. (1903). *Skew Frequency Curves in Biology and Statistics*, Astronomical Laboratory, Groningen: Noordhoff.

[46] Kapteyn, J. C. and van Uven, M. J. (1916). *Skew Frequency Curves in Biology and Statistics*, Groningen: Hotsema Brothers, Inc.

[47] Kemsley, W. F. F. (1952). Body weight at different ages and heights, *Annals of Eugenics, London*, **16,** 316–334.

[48] Kolmogorov, A. N. (1941). Über das logarithmisch normale Verteilungsgesetz der Dimensionen der Teilchen bei Zerstückelung, *Doklady Akademii Nauk* SSSR, **31,** 99–101.

[49] Koopmans, L. H., Owen, D. B. and Rosenblatt, J. I. (1964). Confidence intervals for the coefficient of variation for the normal and lognormal distributions, *Biometrika*, **51,** 25–32.

[50] Kottler, F. (1950). The distribution of particle sizes, I, *Journal of the Franklin Institute*, **250,** 339–356.

[51] Krumbein, W. C. (1936). Application of logarithmic moments to size frequency distributions of sediments, *Journal of Sedimentary Petrology*, **6,** 35–47.

[52] Lambert, J. A. (1964). Estimation of parameters in the three-parameter lognormal distribution, *Australian Journal of Statistics*, **6,** 29–32.

[53] Laurent, A. G. (1963). The lognormal distribution and the translation method: description and estimation problems, *Journal of the American Statistical Association*, **58,** 231–235 (Correction, **58,** 1163).

[54] McAlister, D. (1879). The law of the geometric mean, *Proceedings of the Royal Society of London*, **29,** 367–375.

[55] Miller, R. L. and Goldberg, E. D. (1955). The normal distribution in geochemistry, *Geochimica et Cosmochimica Acta*, **8,** 53–62.

[56] Morrison, J. (1958). The lognormal distribution in quality control, *Applied Statistics*, **7,** 160–172.

[57] Moshman, J. (1953). Critical values of the log-normal distribution, *Journal of the American Statistical Association*, **48,** 600–609.

[58] Nowick, A. S. and Berry, B. S. (1961). Lognormal distribution function for describing anelastic and other relaxation processes, *IBM Journal of Research and Development*, **5,** 297–311, *Ibid*, 312–320.

[59] Nydell, S. (1919). The mean errors of the characteristics in logarithmic-normal distribution, *Skandinavisk Aktuarietidskrift*, **2,** 134–144.

[60] Oldham, P. D. (1965). On estimating the arithmetic means of lognormally-distributed populations, *Biometrics*, **21,** 235–239.

[61] Pearce, S. C. (1945). Lognormal distributions, (Letter in) *Nature*, *London*, **156,** 747.

[62] Peters, S. (1963). *Multi-quantile estimates of the moments of a lognormal distribution*, *Working Memo No.* 142, Arthur D. Little, Inc.

[63] Prohorov, Yu. V. (1963). On the lognormal distribution in geo-chemistry, *Teoriya Veroyatnostei i ee Primeneniya*, **10,** 184–187. (In Russian)

[64] Quensel, C.-E. (1945). Studies of the logarithmic normal curve, *Skandinavisk Aktuarietidskrift*, **28,** 141–153.

[65] Rohn, W. B. (1959). Reliability prediction for complex systems, *Proceedings of the 5th National Symposium on Reliability and Quality Control in Electronics*, 381–388.

[66] Sartwell, P. E. (1950). The distribution of incubation periods of infectious diseases, *American Journal of Hygiene*, **51,** 310–318.

[67] Severo, N. C. and Olds, E. G. (1956). A comparison of tests on the mean of a logarithmico-normal distribution with known variance, *Annals of Mathematical Statistics*, **27,** 670–686.

[68] Sinnott, E. W. (1937). The relation of gene to character in quantitative inheritance, *Proceedings of the National Academy of Science*, *Washington*, **23,** 224–227.

[69] Tallis, G. M. and Young, S. S. Y. (1962). Maximum likelihood estimation of parameters of the normal, the log-normal, truncated normal and bivariate normal distributions from grouped data, *Australian Journal of Statistics*, **4,** 49–54.

[70] Thébault, J. Y. (1961). Distribution lognormale de certains caractères de quelques phénomènes géologiques et ses applications, *Revue de Statistique Appliquée*, **9,** No. 2, 37–87.

[71] Thompson, H. R. (1951). Truncated lognormal distributions. I. Solution by moments, *Biometrika*, **38,** 414–422.

[72] Tiku, M. L. (1968). Estimating the parameters of log-normal distribution from censored samples, *Journal of the American Statistical Association*, **63,** 134–140.

[73] Tokoko, K. (1966). On the mode and median of Gibrat distribution, *Bulletin of the Faculty of Arts and Sciences, Ibaraki University*, (*Natural Science*), **17,** 11–15.

[74] Tomlinson, R. C. (1957). A simple sequential procedure to test whether average conditions achieve a certain standard, *Applied Statistics*, **6,** 198–207.

[75] Uven, M. J. van (1917). Logarithmic frequency distributions, *Proceedings of the Royal Academy of Sciences*, *Amsterdam*, **19,** 533–546.

[76] Uven, M. J. van (1917). Skew frequency curves, *Proceedings of the Royal Academy of Sciences*, *Amsterdam*, **19,** 670–684.

[77] Wartmann, R. (1956). Anwendung der logarithmischen Normalverteilung, *Mitteilungsblatt für Mathematische Statistik*, **8,** 83–91.

[78] Weiss, L. L. (1957). A nomogram for log-normal frequency analysis, *Transactions of the American Geophysical Union*, **38,** 33–37.

[79] Wicksell, S. D. (1917). On the genetic theory of frequency, *Arkiv för Mathematik, Astronomi och Fysik*, **12,** (No. 20), 1–56.

[80] Williams, C. B. (1937). The use of logarithms in the interpretation of certain entomological problems, *Annals of Applied Biology*, **24,** 404–414.

[81] Williams, C. B. (1940). A note on the statistical analysis of sentence length, *Biometrika*, **31,** 356–361.

[82] Wilson, E. B. and Worcester, J. (1945). The normal logarithmic transformation, *Review of Economics and Statistics*, **27,** 17–22.

[83] Wise, M. E. (1952). Dense random packing of unequal spheres, *Philips Research Reports*, **7,** 321–343.

[84] Wise, M. E. (1966). The geometry of log-normal and related distributions and an application to tracer-dilution curves, *Statistica Neerlandica*, **20,** 119–142.

[85] Wise, M. E. (1966). Tracer dilution curves in cardiology and random walk and lognormal distributions, *Acta Physiologica Pharmacologica Neerlandica*, **14,** 175–204.

[86] Wu, Chuan-yi (1966). The types of limit distributions for some terms of variational series, *Scientia Sinica*, **15,** 749–762.

[87] Yuan, P. T. (1933). On the logarithmic frequency distribution and the semi-logarithmic correlation surface, *Annals of Mathematical Statistics*, **4,** 30–74.

[88] Zacks, S. (1966). Sequential estimation of the mean of a log-normal distribution having a prescribed proportional closeness, *Annals of Mathematical Statistics*, **37,** 1688–1696.

15

Inverse Gaussian (Wald) Distributions

1. Introduction

The name "inverse Gaussian" was applied to a certain class of distributions by Tweedie [11], who noted the inverse relationship between the cumulant generating functions of these distributions and those of Gaussian distributions.

The same class of distributions was derived by Wald [16] as an asymptotic form of distribution of average sample number in sequential analysis. The name *Wald distribution* is also used for members of this class. Wasan [18] calls these distributions *first passage time distributions of Brownian motion with positive drift* and uses the abbreviation *T.B.M.P. distribution.*

2. Genesis

Suppose a particle moving along a line tends to move with a uniform velocity v. Suppose also, that the particle is subject to linear Brownian motion which causes it to take a variable amount of time to cover a fixed distance (d). It can be shown that the time (X) required to cover the distance is a random variable with probability density function

$$(1) \qquad p_X(x) = \frac{1}{\sqrt{2\pi\beta x^3}}\, d e^{-(d-vx)^2/2\beta x} \qquad (0 < x)$$

where β is a diffusion constant (Schrödinger [8]).

Alternatively, when the time (x) is fixed, the distance (D) over which the particle travels is a random variable with the normal distribution:

$$(2) \qquad p_D(d) = \frac{1}{\sqrt{2\pi\beta x}}\, e^{-(d-vx)^2/(2\beta x)}.$$

While examining the cumulant generating functions of (1) and (2), Tweedie [12], noticed the inverse relationship between these functions, and suggested the name "Inverse Gaussian" for the distribution (1).

Distributions of this kind were derived by Wald [16] as a limiting form of distribution of sample size in certain sequential probability ratio tests. In a more general form, the distribution can be obtained as a solution to the following problem.

Given that $Z_1, Z_2, \ldots$ are independent random variables each having the same distribution, with finite expected value $E[Z] > 0$ and nonzero variance $V(Z)$, what is the limiting distribution of the random variable N, defined by

$$\left[\bigcap_{j=1}^{N-1}\left(\sum_{i=1}^{j} Z_i < K\right)\right] \cap \left(\sum_{i=1}^{N} Z_i \geq K\right) \qquad \text{with } K > 0$$

(i.e. $\sum_{j=1}^{N} Z_i$ is the first of the sums $Z_1, Z_1 + Z_2, Z_1 + Z_2 + Z_3, \ldots$, to be not less than K, with $K > 0$)?

It can be shown that $E[N] = K/E[Z]$ and

$$\lim_{E(N)\to\infty} \Pr[N \leq xE(N)]$$

is given by the integral up to x of

$$\text{(3)} \qquad p_X(x) = \sqrt{\frac{\phi}{2\pi}}\, e^{\phi} x^{-3/2} \exp\left[-\tfrac{1}{2}\phi(x + x^{-1})\right] \qquad (x > 0)$$

with

$$\phi = K\,E[Z]/V(Z) > 0.$$

This is the *standard form* of probability density function of the *Wald distribution.*

3. Definition

On substituting $v = \dfrac{d}{\mu}$ and $\beta = \dfrac{d^2}{\lambda}$ into (1) we obtain

$$\text{(4.1)} \quad p_X(x \mid \mu,\lambda) = \sqrt{\lambda/(2\pi x^3)}\, e^{-\lambda(x-\mu)^2/2\mu^2 x} \qquad (x > 0;\ \lambda > 0;\ \mu > 0).$$

It may be shown (see below) that μ is a measure of location (and is, in fact, the population mean) while λ is a reciprocal measure of dispersion. We will treat formula (4.1) as a standard form of the *inverse Gaussian* distribution. Our distinction between standard Wald (one parameter) and standard inverse Gaussian (two parameters) is arbitrary, but seems convenient.

Alternatively the distribution can be written in any of the following equivalent forms:

$$\text{(4.2)} \qquad p_X(x \mid \mu,\phi) = [\mu\phi/2\pi x^3]^{\frac{1}{2}} \exp\left\{-\frac{\phi x}{2\mu} + \phi - \frac{\mu\phi}{2x}\right\}$$

$$\text{(4.3)} \qquad p_X(x \mid \phi,\lambda) = [\lambda/2\pi x^3]^{\frac{1}{2}} \exp\left\{-\frac{\phi^2 x}{2\lambda} + \phi - \frac{\lambda}{2x}\right\}$$

(4.4) $$p_X(x \mid \alpha,\lambda) = [\lambda/2\pi x^3]^{\frac{1}{2}} \exp\{-\alpha\lambda x + \lambda(2\alpha)^{\frac{1}{2}} - \lambda/2x\}$$

where $\phi = \dfrac{\lambda}{\mu}$, and $\alpha = \sqrt{2\mu}$, and where μ, λ, ϕ, α are each positive (Tweedie [13]).

Some graphs of $p_X(x \mid \mu,\lambda)$ are shown in Figure 1. The standard form of Wald distribution is obtained by putting $\mu = 1$ in (4.2).

4. Moments

From (4.1) the cumulant generating function

(5.1) $$\Psi_X(t;\mu,\lambda) = \frac{\lambda}{\mu}\left\{1 - \left(1 - \frac{2\mu^2 t}{\lambda}\right)^{\frac{1}{2}}\right\}$$

is obtained. The cumulant generating functions corresponding to (4.2), (4.3) and (4.4) are

(5.2) $$\Psi_X(t;\mu,\phi) = \phi\left\{1 - \left(1 - \frac{2\mu t}{\phi}\right)^{\frac{1}{2}}\right\}$$

(5.3) $$\Psi_X(t;\phi,\lambda) = \phi\left\{1 - \left(1 - \frac{2\lambda t}{\phi^2}\right)^{\frac{1}{2}}\right\}$$

(5.4) $$\Psi_X(t;\alpha,\lambda) = \lambda\left\{(2\alpha)^{\frac{1}{2}} - 2^{\frac{1}{2}}\left(\alpha - \frac{t}{\lambda}\right)^{\frac{1}{2}}\right\}$$

respectively.

The first four cumulants corresponding to (5.1) are

(6.1) $$\kappa_1 = \mu$$

(6.2) $$\kappa_2 = \mu^3/\lambda$$

(6.3) $$\kappa_3 = 3\mu^5/\lambda^2$$

(6.4) $$\kappa_4 = 15\mu^7/\lambda^3$$

and generally, when $r \geq 2$.

(7) $$\kappa_r = 1 \cdot 3 \cdot 5 \cdot \cdots \cdots (2r - 3)\mu^{2r-1}\lambda^{1-r}.$$

These results were obtained by Tweedie [13].

The positive integral moments about zero may be derived from the cumulants, or by direct integration. The integration may be performed by making use of modified Bessel functions of the second kind (Tweedie [13]).

Note that, for the Wald distribution, κ_1 is 1, whatever the value of ϕ. For this distribution

$$\kappa_r = 1 \cdot 3 \cdot 5 \cdot \cdots \cdot (2r - 3)\phi^{-(r-1)}.$$

From (6.1)–(6.4), the central moments are (Wald distribution values, with $\mu = 1$, in parentheses)

$$\mathrm{var}(X) = \mu_2 = \mu^3/\lambda \qquad (= \phi^{-1})$$

$$\mu_3(X) = 3\mu^5/\lambda^2 \qquad (= 3\phi^{-2})$$

$$\mu_4(X) = 15(\mu^7/\lambda^3) + 3(\mu^6/\lambda^2) \qquad (= 15\phi^{-3} + 3\phi^{-2})$$

and the first two moment-ratios are

(8.1) $$\sqrt{\beta_1} = \alpha_3 = 3\sqrt{\mu/\lambda} \qquad (= 3\phi^{-\frac{1}{2}})$$

(8.2) $$\beta_2 = \alpha_4 = 3 + 15\mu/\lambda \qquad (= 3 + 15\phi^{-1}).$$

The first three negative moments about zero of the inverse Gaussian distribution are

(9.1) $$\mu'_{-1} = \mu^{-1} + \lambda^{-1}$$

(9.2) $$\mu'_{-2} = \mu^{-2} + 3\mu^{-1}\lambda^{-1} + 3\lambda^{-2}$$

(9.3) $$\mu'_{-3} = \mu^{-3} + 6\mu^{-2}\lambda^{-1} + 15\mu^{-1}\lambda^{-2} + 15\lambda^{-3}.$$

From (4.2), the density function of $Y = X^{-1}$ is

$$p_Y(y \mid \mu,\phi) = \left[\frac{\mu\phi}{2\pi y^3}\right]^{-\frac{1}{2}} y \exp\left\{-\frac{\mu\phi y}{2} + \phi - \frac{\phi}{2\mu y}\right\}.$$

It follows that

$$E[X^{-r} \mid \mu,\phi] = E[Y^r \mid \mu,\phi]$$
$$= \mu^{-1}E[X^{r+1} \mid \mu^{-1},\phi].$$

In particular for the standard Wald distribution ($\mu = 1$) we have the remarkable relation between negative and positive moments

(10) $$\mu'_{-r} = \mu'_{r+1}.$$

For the standard Wald distribution, the mean deviation is

(11) $$4e^{2\phi}\Phi(-2\sqrt{\phi}).$$

For ϕ large (whatever be the value of μ)

(12) $$\frac{\text{mean deviation}}{\text{standard deviation}} = 4\sqrt{\phi}\, e^{2\phi}\Phi(-2\sqrt{\phi})$$

$$\doteqdot 4\sqrt{\phi}\, e^{2\phi}[(\sqrt{2\pi})^{-1}e^{-2\phi}(2\sqrt{\phi})^{-1}\{1 - (2\sqrt{\phi})^{-2}\}]$$

$$\doteqdot \sqrt{\frac{2}{\pi}}\,(1 - \tfrac{1}{4}\phi^{-1}).$$

5. Properties

As ϕ tends to infinity (with μ fixed) the *standardized* distribution tends to a unit normal distribution (see Figure 1).

More precisely, there is the following relation between corresponding quantiles of the Wald distribution, $X_p(\phi)$, and the normal distribution, U_p, obtained by Sigangirov [10]:

(13)
$$1 + \frac{U_{p-\epsilon}}{\sqrt{\phi}} + \frac{U_{p-\epsilon}^2}{2\phi} + \frac{U_{p-\epsilon}^3}{8\phi\sqrt{\phi}} - \frac{U_{p-\epsilon}^5}{128\phi^2\sqrt{\phi}}$$
$$< X_p(\phi)$$
$$< 1 + \frac{U_p}{\sqrt{\phi}} + \frac{U_p^2}{2\phi} + \frac{U_p^3}{8\phi\sqrt{\phi}},$$

where

$$\epsilon = \Phi(-2\sqrt{\phi})e^{2\phi} < \frac{1}{2\sqrt{2\pi\phi}}.$$

For small ϕ ($\phi < U_p^2/16$)

(14)
$$\frac{\phi}{U_{\frac{1}{2}p}^2 e^{-2\phi}} < X_p(\phi) < \frac{\phi}{U_{\frac{1}{2}p}^2}.$$

As μ tends to infinity (λ remaining fixed) the distribution of $Y = X^{-1}$ tends to the gamma distribution:

$$p_Y(y) = \lambda(2\pi)^{-\frac{1}{2}}y^{-\frac{1}{2}} \exp(-\tfrac{1}{2}\lambda y) \qquad (y > 0)$$

(Wasan and Roy [19]).

Krapivin [5] notes the following equations, satisfied by the Wald probability density function $p_X(x)$ (as given in (3)):

(15)
$$W_\phi(x) = F_X(x) = \int_0^x p_X(t)\,dt = \int_{1/x}^\infty t p_X(t)\,dt$$
$$= 1 - \int_0^{1/x} t p_X(t)\,dt.$$

This enables one to calculate the expected value of a Wald distribution from tables of $W_\phi(x)$.

The cumulative distribution function is, in fact

(16)
$$\Pr[X \le x] = \Phi((x-1)\sqrt{\phi/x}) + e^{2\phi}\Phi(-(x+1)\sqrt{\phi/x}).$$

For x large

(17)
$$\Pr[X \le x] \doteqdot 1 - e^{-\frac{1}{2}\phi x}e^{\phi}(\log x).$$

It may be shown for the more general case (Tweedie [13]) when $X_1, X_2, \ldots, X_n$ are independent random variables, and X_i is distributed according to (4.1) with $\mu = \mu_i$, $\lambda = \lambda_i$ ($i = 1, \ldots, n$), that the distribution of $\sum_{i=1}^{n} (\mu_i^{-2}\lambda_i X_i)$ is also of form (4.1), with $\phi = \mu = \sum_{i=1}^{n} \phi_i$, $\lambda = \left(\sum_{i=1}^{n} \phi_i\right)^2$.

The inverse Gaussian distribution may be characterized by the fact that if $X_1, \ldots, X_n$ are independently distributed random variables each distributed according to (4.1) then $\overline{X} = \left(\sum_{i=1}^{n} X_i\right)/n$ and $n^{-1} \sum_{i=1}^{n} (X_i^{-1} - \overline{X}^{-1})$ are statistically independent (Tweedie [13]). The converse is also true, that is, if the expected values of X, X^2, $1/X$, and $\left(1/\sum_{i=1}^{n} X_i\right)$ exist and are different from zero, and if $\overline{X}$ and $n^{-1} \sum_{i=1}^{n} (X^{-1} - \overline{X}^{-1})$ are independently distributed, then the distribution of each variate X_i is inverse Gaussian (Khatri [4]).

The inverse Gaussian distributions have the property of reproducibility. Thus the distribution of the arithmetic mean $\overline{X}$ of n independent random variables with distribution (4) has the same form as (4) with the same values for μ and α, but with ϕ and λ replaced by $n\phi$ and $n\lambda$ respectively. This follows from the form of the cumulant generating function (5) (Tweedie [13]).

The location of the mode is

$$X_{\text{mode}} = \mu\left\{\left(1 + \frac{9}{4\phi^2}\right)^{1/2} - \frac{3}{2\phi}\right\}. \tag{18}$$

The distribution is unimodal and its shape depends only on the value of $\phi = \lambda/\mu$ (see Figure 1, taken from [13]).

Shuster [9] has pointed out that if X has distribution (3.1) then $\lambda(X-\mu)^2(\mu^2 X)^{-1}$ has a χ^2 distribution with one degree of freedom (Chapter 17). Some further details are given by Wasan [17], [18].

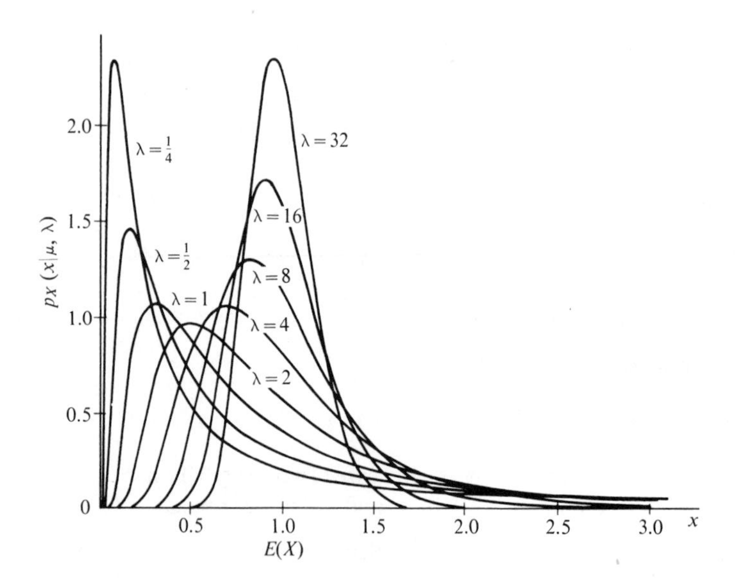

FIGURE 1

Inverse Gaussian Density Functions
$[E(x) = \mu = 1]$

6. Estimation of Parameters

Maximum likelihood estimators (MLE) have been obtained (Tweedie [13]) for the following general case. Let $X_1, \ldots, X_n$ be a series of observations on n distinct inverse Gaussian distributions of the form (4.1) whose unknown parameters are μ (for all i) and λ_i $(i = 1, \ldots, n)$. Also, let $\lambda_i = w_i\lambda_0$, where λ_0 is unknown but w_i is a known positive value. The MLE's of μ and λ_0 satisfy the equations

$$\hat{\mu} = \overline{X} = \left[\sum_{i=1}^{n} w_i X_i\right]\left[\sum_{i=1}^{n} w_i\right]^{-1} \tag{19.1}$$

$$1/\hat{\lambda}_0 = \frac{1}{n}\sum_{i=1}^{n} w_i (X_i^{-1} - \overline{X}^{-1}). \tag{19.2}$$

When the w_i's are all equal to unity, (19.1) and (19.2) become

$$\hat{\mu} = \overline{X} = \left(\sum_{i=1}^{n} X_i\right)\Big/ n \tag{20.1}$$

$$1/\hat{\lambda}_0 = \left(n^{-1}\sum_{i=1}^{n} (X_i^{-1} - \overline{X}^{-1})\right). \tag{20.2}$$

It is apparent that in this case $\overline{X}$ is a sufficient statistic for μ.

The distribution of $1/\hat{\lambda}_0$ can be shown to be of chi-square type. In fact

(21) $\quad \hat{\lambda}_0^{-1}$ is distributed as $(\lambda_0 n)^{-1} \times (\chi^2$ with $(n-1)$ degrees of freedom).

Combining (19.2) and (21) the following relation is obtained

(22) $\quad \sum_{i=1}^{n} w_i(X_i^{-1} - \overline{X}^{-1})$ is distributed as

$\lambda_0^{-1} \times (\chi^2$ with $(n-1)$ degrees of freedom).

Confidence intervals for λ_0 may be constructed using (22) (Tweedie [13]). We also note that

$$(n-1)^{-1}\sum_{i=1}^{n} w_i(X_i^{-1} - \overline{X}^{-1}) \tag{23}$$

is an unbiased estimator of $1/\lambda_0$. It is, in fact, the uniform minimum variance unbiased estimator (Roy and Wasan [7]). Note that the distribution of this estimator is the same form as that of the usual unbiased estimator of variance of a normal distribution (Tweedie [13]).

An *approximately* unbiased estimator of $1/\lambda_0$ is $S^2/\overline{X}^3$ where

$$S^2 = \frac{\sum_{i=1}^{n} (X_i - \overline{X})^2}{n-1}. \tag{24}$$

It is a consistent estimator and has asymptotic efficiency of $\phi/(\phi + 3)$. Some further notes on estimators of the cumulants will be found in Section 9.

If the value of μ is known, we may, by a suitable choice of scale, take it as being equal to 1. We are then in the situation of having to estimate the single parameter, ϕ, of a standard Wald distribution.

If $X_1, X_2, \ldots, X_n$ are independent random variables each having distribution (3), then the maximum likelihood estimator $\hat{\phi}$ of ϕ satisfies the equation

$$\frac{n}{2}\hat{\phi}^{-1} + n - \frac{1}{2}\sum_{i=1}^{n}(X_i + X_i^{-1}) = 0$$

whence

$$\hat{\phi} = \left[\frac{1}{n}\sum_{i=1}^{n}(X_i + X_i^{-1}) - 2\right]^{-1}.$$

For n large

$$n\ \mathrm{var}(\hat{\phi}) \doteqdot 2\phi^2.$$

We note that if ϕ be estimated by equating observed and expected values of $m_2 = n^{-1}\sum_{i=1}^{n}(X_i - \bar{X})^2$, the resultant estimator is

$$\tilde{\phi} = m_2^{-1}. \tag{25}$$

For n large,

$$n\ \mathrm{var}(\tilde{\phi}) \doteqdot 2\phi^2 + 15\phi. \tag{26}$$

The asymptotic efficiency of $\tilde{\phi}$, relative to $\hat{\phi}$, as an estimator of ϕ is

$$(1 + 7.5\phi^{-1})^{-1}.$$

From (10) it follows that

$$\begin{aligned} E(\hat{\phi}^{-1}) &= E[X + X^{-1} - 2] \\ &= 1 + (1 + \phi^{-1}) - 2 \\ &= \phi^{-1} \end{aligned}$$

i.e. $\hat{\phi}^{-1}$ is an unbiased estimator of ϕ^{-1}; m_2 is also an unbiased estimator of ϕ^{-1}. Further

$$\mathrm{var}(\hat{\phi}^{-1}) = 2\phi^{-2}n^{-1} \tag{27.1}$$

$$\mathrm{var}(m_2) = (2\phi^{-2} + 15\phi^{-3})n^{-1}. \tag{27.2}$$

It is of interest to compare these results with those when μ is unknown (as in (4.2)).

The maximum likelihood estimators are then

$$\hat{\mu} = n^{-1}\sum_{i=1}^{n} X_i = \bar{X} \tag{28.1}$$

(28.2) $$\hat{\phi} = \left[n^{-2}\left(\sum_{i=1}^{n} X_i\right)\left(\sum_{i=1}^{n} X_i^{-1}\right) - 1\right]^{-1}.$$

Evidently

(29.1) $$\operatorname{var}(\hat{\mu}) = \mu^2\phi^{-1}n^{-1}.$$

Also

(29.2) $$n \operatorname{var}(\hat{\phi}) \doteqdot 2\phi^2 + \phi$$

and

(29.3) $$\operatorname{corr}(\hat{\mu},\hat{\phi}) \doteqdot -(1 + 2\phi)^{-\frac{1}{2}}.$$

The increase in variance of $\hat{\phi}$, arising from ignorance of the value of μ, should be noted.

Moment estimators are

(30.1) $$\tilde{\mu} = \bar{X}$$

and

(30.2) $$\tilde{\phi} = \frac{n\bar{X}^2}{\sum_{i=1}^{n} (X_i - \bar{X})^2} = (\text{sample coefficient of variation})^{-2}.$$

(Note that $\phi^{-\frac{1}{2}}$ is the coefficient of variation of distribution (4).) The variance of $\tilde{\mu}$ is, of course, the same as that of $\hat{\mu}$ and

(31) $$n \operatorname{var}(\tilde{\phi}) \doteqdot 10\phi^2 + 19\phi \quad (\text{cf. Equation (26)}).$$

7. Truncated Distributions — Estimation of Parameters

It is frequently desirable to estimate the parameters of a truncated inverse Gaussian distribution. Estimators of the parameters and their asymptotic variances have been obtained (Patel [6]) for both the singly and doubly truncated cases.

7.1 *Doubly Truncated Distribution*

The density function of the doubly truncated distribution may be written as

(32) $$p_X(x \mid \mu,\lambda;x_1,x_2) = Kx^{-\frac{3}{2}} \exp\left\{-\frac{\lambda x}{2\mu^2} - \frac{\lambda}{2x}\right\}$$

where

$$K = \left[\int_{x_1}^{x_2} x^{-\frac{3}{2}} \exp\left\{-\frac{\lambda x}{2\mu^2} - \frac{\lambda}{2x}\right\} dx\right]^{-1} \qquad (x_1 \leq x \leq x_2).$$

The values of x_1 and x_2 denote the lower and upper points of truncation, respectively, of the distribution of X $(0 < x_1 < x_2)$; the parameters μ and λ are positive.

A recurrence relation for the moments may be obtained by differentiating (32) partially with respect to x and multiplying both sides by x^r to give, after integration by parts:

$$a\mu'_r - \lambda\mu'_{r-2} - 2(x_1^r p_1 - x_2^r p_2) = (2r-3)\mu'_{r-1} \tag{33}$$

where

$$\begin{aligned} a &= \lambda/\mu^2 \\ p_i &= p_X(x_i \mid \mu,\lambda;x_1,x_2) \qquad (i = 1,2) \\ \mu'_r &= \int_{x_1}^{x_2} x^r p_X(x \mid \mu,\lambda;x_1,x_2)\,dx \qquad (r = 0,\pm 1,\pm 2,\ldots) \end{aligned}$$

p_1 and p_2 are, of course, values of the density at the truncation points x_1 and x_2.

If the population moments (μ'_r) are replaced by the sample moments (m'_r) in the recurrence relation (33) the following set of equations result:

$$\mathbf{M}\hat{\mathbf{h}} = \mathbf{c} \tag{34}$$

where

$$\hat{\mathbf{h}} = \begin{pmatrix} \hat{a} \\ \hat{\lambda} \\ \hat{p}_1 \\ \hat{p}_2 \end{pmatrix}, \mathbf{M} = \begin{pmatrix} 1 & -m'_{-2} & -2 & 2 \\ m'_1 & -m'_{-1} & -2x_1 & 2x_2 \\ m'_2 & -1 & -2x_1^2 & 2x_2^2 \\ m'_3 & -m'_1 & -2x_1^3 & 2x_2^3 \end{pmatrix}, \text{ and } \mathbf{c} = \begin{pmatrix} -3m'_{-1} \\ -1 \\ m'_1 \\ 3m'_2 \end{pmatrix}.$$

($\hat{\mathbf{h}}$ is a vector of estimators of functions of the parameters.) Providing $\mathbf{M}$ is non-singular, (34) may be solved, yielding

$$\hat{\mathbf{h}} = \mathbf{M}^{-1} \cdot \mathbf{c} \tag{35}$$

from which $\hat{\mu}$ may be derived as $\sqrt{\hat{\lambda}/\hat{a}}$.

The asymptotic variances and covariances of the estimators ($\hat{\mathbf{h}}$) of the parameters may be obtained in matrix form by using a differential method (Patel [6]). The elements of this symmetric matrix are as follows:

$$n\,\text{var}(\hat{a}) \doteqdot (14 + \lambda a)\mu'_{-2} + a\mu'_{-1} - P_{-1} - \lambda P_{-2} \tag{36.1}$$

$$n\,\text{cov}(\hat{a},\hat{\lambda}) \doteqdot 2\lambda\mu'_{-2} - \lambda P_{-1} + 3a \tag{36.2}$$

$$n\,\text{cov}(\hat{a},\hat{p}_1) \doteqdot \lambda\mu'_{-1} + 3a\mu'_1 - a\lambda - 3 \tag{36.3}$$

$$n\,\text{cov}(\hat{a},p_2) \doteqdot \lambda^2\mu'_{-1} - (9 + a\lambda)\mu'_1 + 3a\mu'_2 \tag{36.4}$$

$$n\,\text{var}(\hat{\lambda}) \doteqdot \lambda^2\mu'_{-2} + a^2\mu'_2 - 2a\lambda \tag{36.5}$$

$$n\,\text{cov}(\hat{\lambda},\hat{p}_1) \doteqdot a^2\mu'_3 - a\mu'_2 - a\lambda\mu'_1 + \lambda \tag{36.6}$$

$$n\,\text{cov}(\hat{\lambda},\hat{p}_2) \doteqdot 2a\mu'_3 - a\lambda\mu'_2 + 3\lambda\mu'_1 + aP_4 + \lambda^2 \tag{36.7}$$

(36.8) $$n \operatorname{var}(\hat{p}_1) \doteqdot 3a\mu_3' + (1 + a\lambda)\mu_2' + aP_4$$

(36.9) $$n \operatorname{cov}(\hat{p}_1,\hat{p}_2) \doteqdot 18\mu_3' + \mu_2' + 3P_4 + aP_5$$

(36.10) $$n \operatorname{var}(\hat{p}_2) \doteqdot \frac{30}{a}(5\mu_3' + \lambda\mu_2' + P_4) + 4\lambda\mu_3' - \lambda P_4 + 3P_5 + aP_6$$

where $P_r = 2(x_1^r p_1 - x_2^r p_2)$.

Two other useful general relations are:

(37) $$\operatorname{var}(\hat{\mu}) = \frac{1}{4\mu^2 a^4}[a^2 \operatorname{var}(\hat{\lambda}) + \lambda^2 \operatorname{var}(\hat{a}) - 2a\lambda \operatorname{cov}(\hat{a},\hat{\lambda})]$$

(38) $$\operatorname{cov}(\hat{\lambda},\hat{\mu}) = [a \operatorname{var}(\hat{\lambda}) - \lambda \operatorname{cov}(\hat{a},\hat{\lambda})]/2\mu a^2.$$

7.2 *Truncation of the Lower Tail Only*

If the truncation point is at x_0 where $x_0 > 0$, then the estimators $\hat{\mathbf{h}}_1$ of the parameters are

(39) $$\hat{\mathbf{h}}_1 = \mathbf{M}_1^{-1}\mathbf{c}_1$$

where $\mathbf{M}_1$ is a matrix of order 3 (obtained by deleting the last row and column of $\mathbf{M}$ and putting $x_1 = x_0$). The vectors $\hat{\mathbf{h}}_1$ and $\mathbf{c}_1$ are obtained from $\hat{\mathbf{h}}$ and $\mathbf{c}$ respectively by deleting the last element and putting $p_1 = p_0$.

The asymptotic variance of $\hat{\lambda}$ can be obtained from (36) and that of $\hat{\mu}$ and the covariance between $\hat{\lambda}$ and $\hat{\mu}$ from (37) and (38) respectively.

7.3 *Truncation of the Upper Tail Only*

This case may be dealt with in a similar manner. The general relations (34) of the first case are again applicable with $x_1 = 0$, $x_0 = x_2$, $p_0 = p_2$, and omitting the last row and column of $\mathbf{M}$.

The asymptotic variances and covariances of the estimators are obtained in the same way as in the second case.

8. Conditional Expectations of the Estimators of the Cumulants

We now return to study the untruncated distribution. This section is of less general interest than Section 7, but contains some useful results. It may be shown that, given the arithmetic mean ($\overline{X}$), the conditional expectation of any unbiased estimator $\bar{\kappa}_r$ of the rth cumulant is

(40) $$E(\bar{\kappa}_r \mid \overline{X}) = 2\overline{X}(\tfrac{1}{2}\lambda n^2)^{r-1} \exp(\tfrac{1}{2}G)\int_1^\infty (u-1)^{2r-3} \exp(-\tfrac{1}{2}Gu^2)\,du/(r-2)!$$

where $G = \lambda n/\overline{X}$. Since the distribution of $\overline{X}$ has the same form as (4.3) with ϕ replaced by ϕn then the probability density function of G is:

(41) $$p_G(g) = \exp\left(-\frac{\theta^2}{2g} + \theta - \frac{g}{2}\right) \Big/ \sqrt{2\pi g} \qquad (g > 0)$$

where $\theta = \phi n$. The moments of G may be obtained using the formulas which were developed for the inverse Gaussian variate X (see (7) and (10)).

The first few moments of G are

(42)
$$\begin{aligned}
E(G) &= \theta + 1 \\
E(G^2) &= \theta^2 + 3\theta + 3 \\
E(G^3) &= \theta^3 + 6\theta^2 + 15\theta + 15 \\
E(G^4) &= \theta^4 + 10\theta^3 + 45\theta^2 + 105\theta + 105 \\
E(G^5) &= \theta^5 + 15\theta^4 + 105\theta^3 + 420\theta^2 + 945\theta + 945 \\
E(G^6) &= \theta^6 + 21\theta^5 + 210\theta^4 + 1260\theta^3 + 4725\theta^2 + 10395\theta + 10395
\end{aligned}$$

It can be shown (Tweedie [14]) that conditional expectations of the estimators of the cumulants, calculated from (40), are

(43.1) $$E[\bar{\kappa}_2 \mid \bar{X}; \lambda, n] = \lambda^2 n^3 G^{-2} J$$

(43.2) $$E[\bar{\kappa}_3 \mid \bar{X}; \lambda, n] = \tfrac{1}{2} n^2 \bar{X}^3 \{(G + 3)J - 1\}$$

(43.3) $$E[\bar{\kappa}_4 \mid \bar{X}; \lambda, n] = \tfrac{1}{8} n^3 \bar{X}^4 \{(G^2 + 10G + 15)J - G - 7\}$$

where

(44) $$J = 1 - Ge^{\frac{1}{2}G} \int_1^\infty e^{-\frac{1}{2}Gu^2}\, du.$$

J has a maximum value of 1 and decreases monotonically to 0 as G increases to ∞.

An asymptotic series expansion of (44) yields

(45) $$J = G^{-1} - 3G^{-2} + \cdots + (-1)^{r+1} 3 \cdot 5 \cdots (2r - 1) G^{-r} + (-1) 3 \cdot 5 \cdots (2r - 1)(2r + 1) G^{-r} e^{-\frac{1}{2}G} \int_1^\infty u^{-2r-2} e^{-\frac{1}{2}Gu^2}\, du$$

(Tweedie [14]).

A relatively large number of significant figures is required in J in order to yield accurate results when evaluating (43) numerically.

Tables of J and $e^{\frac{1}{2}G} \int_1^\infty e^{-Gu^2/2}\, du (= \text{“}I\text{”})$ are available (National Bureau of Standards [15]). Tweedie [14] gives tables of I and J for $G = 1(1)10$. Tweedie also gives “a useful expansion for moderately large values of G” as the continued fraction

(46) $$J = \cfrac{1}{G + \cfrac{3}{1 + \cfrac{2}{G + \cfrac{5}{1 + \cfrac{4}{G + \cdots}}}}}$$

where the sequence of positive integers in the partial numerators is 1; 3, 2; 5, 4; 7, 6; 9, 8; etc.

For large values of G, asymptotic expansions of Hermite polynomials may be used, giving

$$E(\bar{\kappa}_r \mid \bar{X};\lambda,n) \sim \frac{4\lambda}{\bar{X}}\left(\frac{\bar{X}^2}{2\lambda}\right)^r \sum_{i=0}^{\infty} \frac{(2i+2r-3)!}{i!(r-2)!(-2G)^i}. \tag{47}$$

On using (47), or by applying the asymptotic series expansion (45) of J to (43), the expected values of the estimators of the cumulants become:

$$E(\bar{\kappa}_2 \mid \bar{X}; \lambda,n) \sim \lambda^{-1}\bar{X}^3(1 - 3G^{-1} + 15G^{-2} - 105G^{-3} + 945G^{-4} - \cdots) \tag{48.1}$$

$$E(\bar{\kappa}_3 \mid \bar{X}; \lambda,n) \sim 3\lambda^{-2}\bar{X}^5(1 - 10G^{-1} + 105G^{-2} - 1260G^{-3} + 17325G^{-4} - \cdots) \tag{48.2}$$

$$E(\bar{\kappa}_4 \mid \bar{X}; \lambda,n) \sim 15\lambda^{-3}\bar{X}^7(1 - 21G^{-1} + 378G^{-2} - 6930G^{-3} + 135135G^{-4} - \cdots) \tag{48.3}$$

(Tweedie [14]).

Experimental data involving the inverse Gaussian distribution exhibit a high correlation between changes in sample means and changes in sample variances in different sets of data. This fact can be accounted for theoretically by the relationship between the first and second cumulants ($\kappa_2 = \kappa_1^3/\lambda$, as can be seen from (6)). An estimate of the regression function of the sample variance S^2 (as defined in (24)) on the sample mean $\bar{X}$ has been found (Tweedie [14]). It is:

$$E(S^2 \mid \bar{X}; \lambda,n) = GJ\bar{X}^3/\lambda = Jn\bar{X}^2. \tag{49}$$

9. Related Distributions

Sometimes it may be convenient to use the reciprocal of a variate X having an inverse Gaussian distribution. Let this value be denoted by Y. (In terms of the moving particle (see Section 2) this corresponds to average speed.) The probability density function of Y is (see also Section 3)

$$p_Y(y) = \exp\left\{-\frac{\lambda y}{2} + \frac{\lambda}{\mu} - \frac{\lambda}{2\mu^2 y}\right\}[\lambda/2\pi y]^{\frac{1}{2}} \qquad (y > 0). \tag{50}$$

This distribution is called the *random walk distribution* (Wise [20], Wasan [18]).

The positive moments about zero for Y are the same as the negative moments about zero for the inverse Gaussian variate X and are given by (9) and (10).

The cumulant generating function of Y is

$$\Psi_Y(t; \phi,\lambda) = \phi(1 - (1 + 2t\lambda^{-1})^{\frac{1}{2}}) - \tfrac{1}{2}\ln(1 + 2t\lambda^{-1}). \tag{51}$$

The first two cumulants are

(52.1)
$$\kappa_1(Y) = \frac{1}{\mu} + \frac{1}{\lambda},$$

(52.2)
$$\kappa_2(Y) = \frac{1}{\lambda\mu} + \frac{2}{\lambda^2}.$$

It is apparent that the bias in the estimator of μ^{-1} is λ^{-1}. The expected mean square error in using Y as an estimator of $1/\mu$ is

(53)
$$E\left[\left(Y - \frac{1}{\mu}\right)^2\right] = (\phi + 3)/\lambda^2.$$

The mode of the density function of Y is located at

(54)
$$Y_{\text{mode}} = \frac{1}{\mu}\left[\left(1 + \frac{1}{4\phi^2}\right)^{\frac{1}{2}} - \frac{1}{2\phi}\right].$$

Wise [20] has shown that the density function has two points of inflection at values of y satisfying the equation

$$u^4 + 2u^3 + u = (\lambda/\mu)^2 + \tfrac{1}{4}$$

where

$$u = \tfrac{1}{2} + \tfrac{1}{2}(\lambda/\mu)\left(\frac{y}{\mu} - \frac{\mu}{y}\right).$$

Figure 2 is a plot of the density function of Y for different values of λ, reproduced from Tweedie [13].

Holla [3] has described a (discrete) compound Poisson distribution:

$$\text{Poisson } (\theta) \underset{\theta}{\wedge} \text{ Inverse Gaussian.}$$

10. Applications and Tables

Inverse Gaussian distributions have been used to explain the motion of particles influenced by Brownian movement (Wasan [18]). In particular they have been used to study the movement of particles in a colloidal suspension under an electric field (Tweedie [11]).

It has been suggested (Tweedie [13]) that because of the statistical independence between $\bar{X}$ and $1/\hat{\lambda}$ that an analogue of the analysis of variance for nested classifications can be performed. This analogue uses the tables of χ^2 and F developed for the analysis of variance where the measure of dispersion is given by (20.2).

Tables of the cumulative distribution function corresponding to (4.1) have been constructed by Wasan and Roy [19]. They give values of $\Pr[X \leq x]$ to four decimal places for $\mu = 5$ and

$$\lambda = 0.25(0.25)1.00(1)10,16(8)32.$$

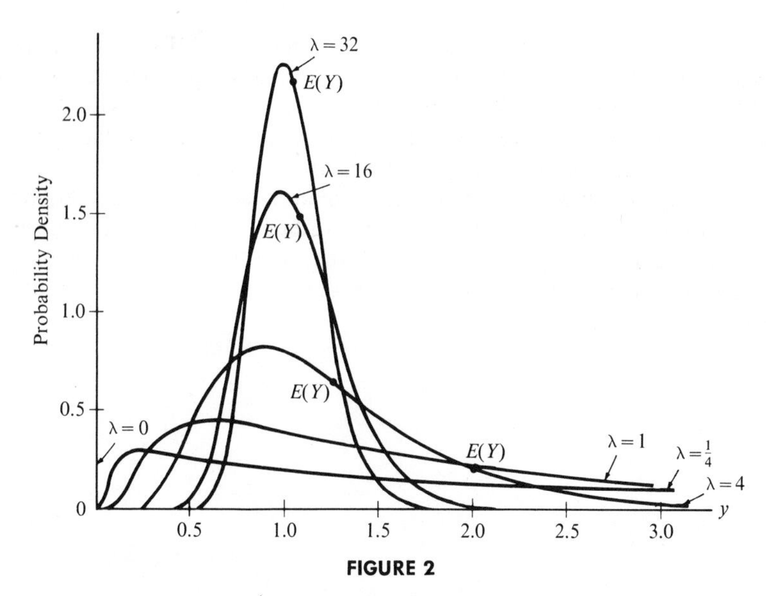

FIGURE 2

Density Functions of Reciprocals of Inverse Gaussian Variables
($\mu = 1$)

[*Note that* $\lim_{\lambda \to \infty} E(Y) = 1$]

For $\lambda < 0.25$ they suggest using a reciprocal gamma approximation; for $\lambda > 32$, a normal approximation. (See Section 5.) If it is desired to evaluate $\Pr[X \leq x]$ for a value of μ different from 5, then one can evaluate

$$\Pr[Y = \tfrac{1}{5}\mu X \leq \tfrac{1}{5}\mu x]$$

since Y has distribution (4.1) with μ equal to 5, and λ equal to $5 \times (\text{original } \lambda)/\mu$.

The intervals of x in the tables vary between 0.005 and 0.5. They were chosen so that the differences between successive tabulated values of $\Pr[X \leq x]$ rarely exceed 0.005.

Tables of the standard Wald probability density function (3) and cumulative distribution function (16) have been published by Krapivin [5]. They contain values of each function to six significant figures for

$$\phi = 0.01(0.01)0.1(0.1)4.0(0.2)5.0(0.5)10$$

and for various values of x. The coverage of x decreases with the increase in the values of ϕ. For example, when $\phi = 0.01$ one has

$$x = 0(0.0001)0.01(0.01)1(0.05)6(0.1)31(0.5)81(1)231(25)1181$$

and when $\phi = 2.0$ $x = 0(0.01)2(0.05)7(0.2)12.0$. (The tables also contain the modal values to six significant figures for

$$\phi = 0.0001(0.0001)0.0100(0.01)1.00(0.1)6.0(1)55{,}60(10)550,$$
$$800(100)5700{,}60000(10000)550000.$$

The main, and the most familiar, applications of Wald distributions are in sequential analysis (e.g. Wald [16], Bartlett [1]). However, some authors have used this distribution in various physical applications connected with diffusion processes with boundary conditions ([1]). It appears in the calculation of the distribution of time of first hitting the boundary in a random walk.

In Russian literature on electronics and radiotechnique the Wald distribution is often used, and several studies on this distribution have been published in these journals in recent years (e.g. Basharinov *et al.* [2], Sigangirov [10]). The extensive tables just described were also published by the Soviet Institute of Radiotechnique and Electronics.

REFERENCES

[1] Bartlett, M. S. (1956). *Introduction to Stochastic Processes*, London: Cambridge University Press.

[2] Basharinov, A. E., Fleishman, B. S. and Samochina, M. A. (1959). Binary accumulating systems with two threshold analyzers, *Radiotekhnika Élektronika*, **4,** 1419–1426 (In Russian)

[3] Holla, M. S. (1966). On a Poisson-inverse Gaussian distribution, *Metrika*, **11,** 115–121.

[4] Khatri, C. G. (1962). A characterization of the inverse Gaussian distribution, *Annals of Mathematical Statistics*, **33,** 800–803.

[5] Krapivin, V. F. (1965). *Tables of Wald's Distribution*, Moscow: Nauka. (In Russian)

[6] Patel, R. C. (1965). Estimates of parameters of truncated inverse Gaussian distribution, *Annals of the Institute of Statistical Mathematics, Tokyo*, **17,** 29–33.

[7] Roy, L. K. and Wasan, M. T. (1968). The first passage time distribution of Brownian motion with positive drift, *Mathematical Biosciences*, **3,** 191–204.

[8] Schrödinger, E. (1915). Zur Theorie der Fall — und Steigversuche an Teilchenn mit Brownscher Bewegung, *Physikalische Zeitschrift*, **16,** 289–295.

[9] Shuster, J. (1968). On the inverse Gaussian distribution function, *Journal of the American Statistical Association*, **63,** 1514–1516.

[10] Sigangirov, K. Sh. (1962). Representation of Wald's distribution by means of a normal distribution, *Radiotekhnika Élektronika*, **7,** 164–166 (In Russian)

[11] Tweedie, M. C. K. (1947). Functions of a statistical variate with given means, with special reference to Laplacian distributions, *Proceedings of the Cambridge Philosophical Society*, **43,** 41–49.

[12] Tweedie, M. C. K. (1956). Some statistical properties of inverse Gaussian distributions, *Virginia Journal of Science*, **7** (*New Series*), 160–165.

[13] Tweedie, M. C. K. (1957). Statistical properties of inverse Gaussian distributions, I, *Annals of Mathematical Statistics*, **28,** 362–377.

[14] Tweedie, M. C. K. (1957). *Ibid*, II, *Annals of Mathematical Statistics*, **28,** 696–705.

[15] National Bureau of Standards (1942). *Tables of Probability Functions*, **2.** Washington, D.C.: U.S. Government Printing Office.

[16] Wald, A. (1947). *Sequential Analysis*, New York: John Wiley & Sons, Inc.

[17] Wasan, M. T. (1968). On an inverse Gaussian process, *Skandinavisk Aktuerietidskrift*, 69–96.

[18] Wasan, M. T. (1968). *First Passage Time Distribution of Brownian Motion*, Monograph, Department of Mathematics, Queen's University, Kingston, Ontario.

[19] Wasan, M. T. and Roy, L. K. (1967). Tables of inverse Gaussian probabilities (Abstract) *Annals of Mathematical Statistics*, **38,** 299.

[20] Wise, M. E. (1966). Tracer dilution curves in cardiology and random walk and lognormal distributions, *Acta Physiologica Pharmacologica Neerlandica*, **14,** 175–204.

16

Cauchy Distribution

1. Definition and Properties

The special form of the Pearson Type VII distribution, with probability density function

$$(1) \qquad (\pi\lambda)^{-1}[1 + \{(x - \theta)/\lambda\}^2]^{-1} \qquad (\lambda > 0)$$

is called the *Cauchy distribution* ([5] equation (25), page 206).

The cumulative distribution function is

$$(2) \qquad \tfrac{1}{2} + \pi^{-1} \tan^{-1}[(x - \theta)/\lambda].$$

The parameters θ and λ are location and scale parameters respectively. The distribution is symmetrical about $x = \theta$. The median is θ; the upper and lower quartiles are $\theta \pm \lambda$. The probability density function has points of inflexion at $\theta \pm \lambda/\sqrt{3}$. It may be noted that the values of the cumulative distribution at the points of inflexion are 0.273 and 0.727, as compared with the corresponding values 0.159 and 0.841 for the normal distribution.

The probability density function of $Y = A + BX$ is of the same form as that of X, with θ replaced by $A + B\theta$ and λ replaced by $|B|\lambda$. The distribution does not possess finite moments of order greater than or equal to 1, and so does not possess a finite expected value or standard deviation. (However, θ and λ are location and scale parameters, respectively, and may be regarded as being analogous to mean and standard deviation.)

The most notable difference between the normal and Cauchy distributions is in the longer and flatter tails of the latter. These differences are illustrated in Tables 1 and 2.

TABLE 1

Comparison of Cauchy and Normal Distributions

x	$\Pr[X \leq x]$		x	$\Pr[X \leq x]$	
	Cauchy	Normal		Cauchy	Normal
0.0	0.5000	0.5000	2.2	0.8642	0.9311
0.2	0.5628	0.5537	2.4	0.8743	0.9472
0.4	0.6211	0.6063	2.6	0.8831	0.9602
0.6	0.6720	0.6571	2.8	0.8908	0.9705
0.8	0.7148	0.7053	3.0	0.8976	0.9785
1.0	0.7500	0.7500	3.2	0.9036	0.9845
1.2	0.7789	0.7919	3.4	0.9089	0.9891
1.4	0.8026	0.8275	3.6	0.9138	0.9924
1.6	0.8222	0.8597	3.8	0.9181	0.9948
1.8	0.8386	0.8876	4.0	0.9220	0.9965
2.0	0.8524	0.9113			

The Cauchy distribution is in standard form (5); the normal distribution has expected value zero and standard deviation $(0.67445)^{-1} = 1.4827$. The two distributions have the same median ($x = 0$) and upper and lower quartiles ($x = \pm 1$). As both distributions are symmetrical about $x = 0$, there is no need to tabulate negative values of x.

TABLE 2

Percentage Points of Standard Cauchy Distribution

$\Pr[X \leq x]$	x
0.5	0.0000
0.6	0.3249
0.7	0.7265
0.75	1.0000
0.8	1.3764
0.85	1.9626
0.9	3.0777
0.95	6.3138
0.975	12.7062

The characteristic function corresponding to (1) is

$$E(e^{itX}) = \exp\,[it\theta - |t|\lambda]. \tag{3}$$

If $X_1, X_2, \ldots, X_n$ are independent random variables with probability density functions

$$p_{X_j}(x) = (\pi\lambda_j)^{-1}[1 + \{(x - \theta_j)/\lambda_j^2\}]^{-1} \qquad (\lambda_j > 0; j = 1,2,\ldots,n) \tag{4}$$

then the characteristic function of $S_n = \sum_{j=1}^{n} X_j$ is

$$\exp\left[it \sum_{j=1}^{n} \theta_j - |t| \sum_{j=1}^{n} \lambda_j\right].$$

Hence S_n has a Cauchy probability density function, as in (1), with $\lambda = \sum_{j=1}^{n} \lambda_j$; $\theta = \sum_{j=1}^{n} \theta_j$. More generally, $\sum_{j=1}^{n} a_j X_j$ has a Cauchy probability density function, as in (1), with $\lambda = \sum_{j=1}^{n} |a_j|\lambda_j$; $\theta = \sum_{j=1}^{n} a_j\theta_j$. In particular, putting

$$\begin{aligned} \lambda_1 &= \lambda_2 = \cdots = \lambda_n = \lambda \\ \theta_1 &= \theta_2 = \cdots = \theta_n = \theta \\ a_1 &= a_2 = \cdots = a_n = n^{-1} \end{aligned}$$

the arithmetic mean $n^{-1} \sum_{j=1}^{n} X_j$ of n independent random variables each having probability density function (1) is seen to have the *same* Cauchy distribution as each of the X_j's.

Thus the Cauchy distribution is a 'stable' distribution and is also 'infinitely divisible'. Puglisi [24] has shown that a variable with a Cauchy distribution can be represented as the sum of two independent random variables, each having infinitely divisible, but *not* stable, distributions.

Springer and Thompson [30] have obtained explicit formulas for the probability density function of the product of n (≤ 10) independent and identically distributed Cauchy variables (each with $\theta = 0$, $\lambda = 1$).

There is no standardized form of the Cauchy distribution, as it is not possible to standardize without using (finite) values of mean and standard deviation, which do not exist in this case. However, a *standard form* is obtained by putting $\theta = 0$, $\lambda = 1$. The standard probability density function is

$$\pi^{-1}(1 + x^2)^{-1} \tag{5}$$

and the standard cumulative distribution function is

$$\tfrac{1}{2} + \pi^{-1}\tan^{-1}x. \tag{6}$$

(Note that this is the *t-distribution* with 1 degree of freedom — see Chapter 27.) Values of (6) are shown in Table 1. Percentage points (values X_α such that

$\Pr[X \leq X_\alpha] = \alpha$) are shown in Table 2.

If X_1 and X_2 are independent, and each has a standard Cauchy distribution, then $(b_1X_1 + b_2X_2)$ has a Cauchy distribution, and is in fact, distributed as $(|b_1| + |b_2|)X_1$. Pitman and Williams [23] have extended this result, obtaining (in a form suggested by Williams [32]):

"Suppose that the sets of numbers $\{w_j\}$ and $\{a_j\}$ are such that $w_j \geq 0$, $\sum_{j=1}^{\infty} w_j = 1$, and $\{a_j\}$ possesses no limit point. Then, if X has a standard Cauchy distribution, so does

$$\sum_{j=1}^{\infty} w_j(1 + a_jX)(a_j - X)^{-1}."$$

2. Order Statistics

If $X_1, X_2, \ldots, X_n$ are independent random variables each with probability density function (1), and $X_1' \leq X_2' \leq \cdots \leq X_n'$ are the corresponding order statistics, then

$$p_{X_r'}(x) = \frac{n!}{(r-1)!(n-r)!}\left(\frac{1}{2} + \frac{1}{\pi}\tan^{-1}\{(x-\theta)/\lambda\}\right)^{r-1} \times \left(\frac{1}{2} - \frac{1}{\pi}\tan^{-1}\{(x-\theta)/\lambda\}\right)^{n-r}(\pi\lambda)^{-1}(1 + \{(x-\theta)/\lambda\}^{-2})^{-1}. \tag{7}$$

If n is large then

$$\operatorname{var}(X_r') \doteqdot n^{-1}\lambda^2\pi^2(r/n)(1 - r/n)\operatorname{cosec}^4(\pi r/n) \tag{8.1}$$

$$\operatorname{cov}(X_r', X_s') \doteqdot n^{-1}\lambda^2\pi^2(r/n)(1 - s/n)\operatorname{cosec}^2(\pi r/n)\operatorname{cosec}^2(\pi s/n) \quad (r < s). \tag{8.2}$$

Note that the expected values of X_n' and X_1', and the variances of X_{n-1}' and X_2', are infinite.

The asymptotic distribution of the largest value, X_n' (as n increases) can be obtained from the general formulas of Fréchet [10], Fisher and Tippett [8] and von Mises [21]. The Cauchy distribution belongs to the class for which

$$\lim_{x\to\infty} [1 - F(x)]x^k = A > 0 \text{ with } k = 2.$$

(See Chapter 21, Section 3.)

The asymptotic distribution of the extremal quotient (ratio of greatest X to absolute value of least X—that is $X_n'/|X_1'|$) has been studied by Gumbel and Keaney [12].

Tables of expected values, variances, and covariances of order statistics for $n = 5(1)16(2)20$ are given in Barnett [2].

3. Estimation of Parameters

It is not possible to use the method of moments, to estimate θ and λ by equating sample and population first and second moments. (It may be remembered, incidentally, that it was shown in Section 1 of this chapter that the arithmetic mean, $n^{-1} \sum_{j=1}^{n} X_j = \bar{X}$, has the same distribution as any one of the X's. Thus $\bar{X}$ is no more informative than any single one of the X_j's, although one might feel it could be a useful estimator of θ.) It is possible to derive methods of estimation using moments of fractional order, but, in fact, these are not used.

The simple form of the cumulative distribution function (2) makes it possible to obtain simple estimators by equating population percentage points (quantiles) and sample estimators thereof.

The $100p\%$ quantile, X_p, of the distribution satisfies the equation

$$\Pr[X \leq X_p] = p;$$

a convenient estimator from the sample is the rth order statistic, X'_r, with $r = (n + 1)p$. We will denote this estimator by $\hat{X}_p$. The value of X_p depends on θ and λ.

If p_1 and p_2 are distinct numbers between 0 and 1 then the equations

$$\hat{X}_{p_j} = X_{p_j} \qquad (j = 1,2)$$

lead to the estimators

$$\tilde{\lambda} = (\hat{X}_{p_1} - \hat{X}_{p_2})(\cot \pi p_2 - \cot \pi p_1)^{-1} \tag{9.1}$$

$$\tilde{\theta} = (\hat{X}_{p_1} \cot \pi p_2 - \hat{X}_{p_2} \cot \pi p_1)(\cot \pi p_2 - \cot \pi p_1)^{-1}. \tag{9.2}$$

(Note that $\tan [\pi(p - \frac{1}{2})] = -\cot \pi p$.)

In the symmetrical case with $p_1 = p > \frac{1}{2} > 1 - p = p_2$,

$$\tilde{\lambda} = \tfrac{1}{2}(\hat{X}_p - \hat{X}_{1-p}) \tan [\pi(1 - p)] \tag{10.1}$$

$$\tilde{\theta} = \tfrac{1}{2}(\hat{X}_p + \hat{X}_{1-p}). \tag{10.2}$$

If the available observations can be represented by observed values of n independent random variables, $X_1, X_2, \ldots, X_n$, each with probability density function of form (1), and $\hat{X}_p$, $\hat{X}_{1-p}$ are appropriate order statistics (so that $E(\hat{X}_p) \doteqdot X_p$, $E(\hat{X}_{1-p}) \doteqdot X_{1-p}$), then

$$n \operatorname{var}(\tilde{\lambda}) \doteqdot \lambda^2[2\pi^2(1 - p)(2p - 1)] \operatorname{cosec}^2 2\pi p \tag{11.1}$$

$$n \operatorname{var}(\tilde{\theta}) \doteqdot \lambda^2[\tfrac{1}{2}\pi^2(1 - p)] \operatorname{cosec}^4 \pi p. \tag{11.2}$$

The estimators $\tilde{\lambda}$ and $\tilde{\theta}$ are uncorrelated; $\tilde{\theta}$ is an unbiased estimator of θ. The approximate variance (by (11.2)) of $\tilde{\theta}$ is minimized by taking $p = 55.65\%$; the corresponding value of the right-hand side of (11.2) is $2.33\lambda^2$. The Cramér-Rao lower bound for the variance of an unbiased estimator of θ is $2\lambda^2/n$, so the asymptotic efficiency of $\tilde{\theta}$ is 86%. (The quoted lower bound applies whether λ is known or not.)

Putting $r = n/2$ in (8.1), we see that the variance of the median is approximately $\frac{1}{4}\pi^2\lambda^2/n = 2.47\lambda^2/n$. The median is an unbiased estimator of θ with asymptotic efficiency 81%.

Rothenberg *et al.* [27] obtained an approximation to the variance of the symmetrical censored arithmetic mean, $\tilde{\theta}'$ say, where the censoring causes omission of the $\frac{1}{2}pn$ (approx.) lowest and $\frac{1}{2}pn$ (approx.) highest observed values in a random sample of size n. The formula they obtained is

$$(12) \qquad n\,\mathrm{var}(\tilde{\theta}') \doteqdot \lambda^2[p(1-p)^{-2}\cot^2(\tfrac{1}{2}\pi p) + 2\pi^{-1}(1-p)^{-2}\cot(\tfrac{1}{2}\pi p) - (1-p)^{-1}].$$

The statistic $\tilde{\theta}'$ is an unbiased estimator of θ. By taking $p \doteqdot 0.76$ the right-hand side of (12) attains its minimum value $2.28\lambda^2$. The asymptotic efficiency of the estimator $\tilde{\theta}'$ is then 88%. See also Bloch [4].

This 'optimum' $\tilde{\theta}'$ uses the values of only the central 24% of the observations, while the 'best' pair of order statistics to use in $\tilde{\theta}$ are separated by only the central 11% of observations. For the normal distribution (see Chapter 13, Section 6) considerably larger central groups of observations give 'optimum' results. This reflects the greater variability in the 'tails' of the Cauchy distribution.

Among estimators of form $\sum_{j=1}^{n} \alpha_j X'_j$ (i.e. linear functions of order statistics) the following values of α_j give asymptotically optimum estimators:

$$\text{For } \theta: \qquad \alpha_j = \frac{\sin 4\pi(j(n+1)^{-1} - \frac{1}{2})}{\tan \pi(j(n+1)^{-1} - \frac{1}{2})}$$

$$\text{For } \lambda: \qquad \alpha_j = \frac{8 \tan \pi(j(n+1)^{-1} - \frac{1}{2})}{\sec^4 \pi(j(n+1)^{-1} - \frac{1}{2})}.$$

(See Chernoff *et al.* [6].)

The Cramér-Rao lower bound of $2\lambda^2$ is the asymptotic value of $n\,\mathrm{var}(\hat{\theta})$ where $\hat{\theta}$ is the maximum likelihood estimator of θ (whether λ is known or not, since $\lim_{n\to\infty} n\,\mathrm{cov}(\hat{\theta},\hat{\lambda}) = 0$). If λ is unknown, the maximum likelihood estimators $\hat{\theta}$, $\hat{\lambda}$ are found by solving the equations

$$(13.1) \qquad \sum_{j=1}^{n} [1 + \{(X_j - \hat{\theta})/\hat{\lambda}\}^2]^{-1} = \tfrac{1}{2}n$$

$$(13.2) \qquad \sum_{j=1}^{n} X_j[1 + \{(X_j - \hat{\theta})/\hat{\lambda}\}^2]^{-1} = \tfrac{1}{2}n\hat{\theta}.$$

(Note that, on dividing (13.2) by (13.1) we see that $\hat{\theta}$ is a weighted mean of the X_j's.)

If the value of λ is known, then $\hat{\theta}$ satisfies the equation

$$(14) \qquad \sum_{j=1}^{n} (X_j - \hat{\theta})[1 + \{(X_j - \hat{\theta})/\lambda\}^2]^{-1} = 0.$$

There is no explicit form for the solution of this equation. It can be solved in a straightforward manner by trial and error. Alternatively, starting with a reasonably good unbiased estimator (e.g., $\tilde{\theta}$, $\tilde{\theta}'$, or the median) T, the standard iterative procedure may be used. It should be noted, however, that Barnett [1] has found that Equations (13) often have multiple roots, and the iterative method cannot be relied upon to give the root required. But Weiss [32] has shown that with T taken as the median, one iteration gives a statistic with the same asymptotic properties as a maximum likelihood estimator.

It is interesting to note that, even when λ is known, there is no minimum variance unbiased estimator of θ (see, e.g., Kendall and Stuart [13]).

4. Genesis and Application

The standard Cauchy distribution with $\theta = 0$, $\lambda = 1$, as in (5), is the distribution of central t with one degree of freedom (Chapter 27). It is thus the distribution of the ratio U/V, where U and V are independent unit normal variables. The common distribution of U and V need not be normal. For example, the ratio U/V has a Cauchy distribution if

$$p_U(u) = \sqrt{2}\,\pi^{-1}(1 + u^4)^{-1};\; p_V(v) = \sqrt{2}\,\pi^{-1}(1 + v^4)^{-1} \tag{15}$$

(Laha [16]). Other examples have been given by Laha [15], Mauldon [17], Fox [9], Steck [31] and Kotlarski [14]. Note that, since U and V have identical distributions, V/U and U/V must have identical distributions. Hence, if X has the probability density function (5), so does X^{-1}.

Since, provided b_1 and b_2 are not both zero,

$$\begin{aligned}(b_1(V/U) + b_2)^{-1}\\ &= b_1(b_1^2 + b_2^2)^{-1}[(b_1U - b_2V)/(b_2U + b_1V)] + b_2(b_1^2 + b_2^2)^{-1}\end{aligned}$$

it follows that if X has a standard Cauchy distribution so does $(b_1X + b_2)^{-1}$. This result was obtained in this way by Savage [28]; and by direct calculation by Menon [19]. (Note that if U and V are independent unit normal variables, so are $(b_1U - b_2V)(b_1^2 + b_2^2)^{-\frac{1}{2}}$ and $(b_2U + b_1V)(b_1^2 + b_2^2)^{-\frac{1}{2}}$ (see Chapter 13, Section 3).)

The Cauchy distribution is obtained as the limiting distribution of $n^{-1}\sum_{j=1}^{n} X_j^{-1}$ as $n \to \infty$, where $X_1, X_2, \ldots$ are independent identically distributed random variables with common density function $p_X(x)$ satisfying the conditions (E. J. G. Pitman):

(i) $p_X(0) > 0$;
(ii) $p_X(x)$ continuous at $x = 0$;
(iii) $p_X(x)$ possesses left-hand and right-hand derivatives at $x = 0$.

Since the reciprocal of a Cauchy variable also has a Cauchy distribution, it follows that the limiting distribution of the harmonic mean, under these conditions, is Cauchy.

The Cauchy distribution also arises in describing the distribution of the point of intersection P, of a fixed straight line, with another, variable straight line, randomly oriented in two dimensions through a fixed point A. The distance OP of the point of intersection from the pivot (O) of the perpendicular from A to the fixed line has a Cauchy distribution with $\theta = 0$.

The situation is represented diagrammatically in Figure 1. The angle $\angle OAP$ has a uniform (rectangular) distribution (Chapter 25) between $-\pi/2$ (corresponding to $OP = -\infty$) and $\pi/2$ (corresponding to $OP = +\infty$).

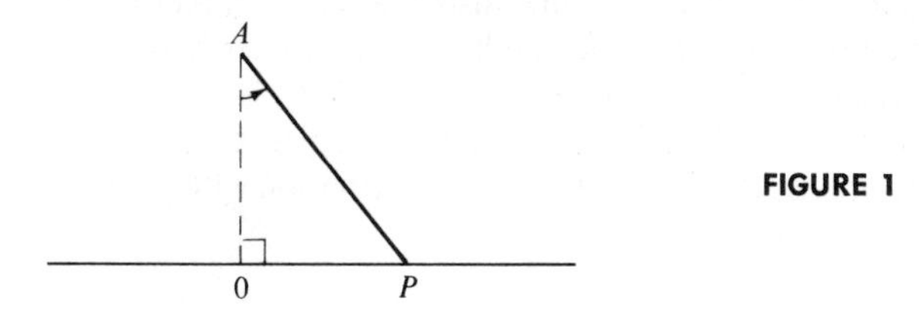

FIGURE 1

On the basis of this kind of model, the Cauchy distribution may be used to describe the distribution of points of impact of particles from a point-source (A) with a fixed straight line.

It may be noted that if the space is of $(s + 1)$, instead of two, dimensions, then the distance, r say, of the point of intersection with a fixed hyperplane from the foot of the perpendicular from A to the hyperplane is distributed as a multiple of central t with s degrees of freedom. (Chapter 27.)

Spitzer [29] has shown that if $(R(t),\theta(t))$ are the polar coordinates of a random point under standard Brownian motion at time t (with $R(0) > 0$) then the distribution of $\{\Delta\theta(t)\}(\frac{1}{2}\log t)^{-1}$ tends to a Cauchy distribution as t tends to infinity, where $\{\Delta\theta(t)\}$ denotes the total *algebraic* angle turned through up to time t (so that $\{\Delta\theta(t)\} \equiv [\theta(t) - \theta(0)] \bmod 2\pi$). Berman [3] established that if τ is the first time that $R(\tau) = r_1$, then $\{\Delta\theta(\tau)\}|\log(r_1/R(0))|^{-1}$ has a Cauchy distribution with $\theta = 0$, $\lambda = |\log(r_1/R(0))|$.

5. Characterization

Menon [18] [19], gave several characterizations of the Cauchy distribution. Among them is the following (from [19]):

"If $X_1, X_2, \ldots, X_n$ are independent, identically distributed random variables, then the common distribution is a Cauchy distribution if and only if, for *any* set of real numbers $\{b_j \neq 0, a_j\}$ $(j = 1,2,\ldots,n)$ there exist real numbers B $(\neq 0)$ and A such that $\sum_{j=1}^{n} (a_j + b_jX_j)^{-1}$ has the same distribution as $A(B + X_1)^{-1}$."

Obretenov [22] has given the following characterization:

"If X and Y are independent and have the same distribution, then a necessary and sufficient set of conditions for this to be a Cauchy distribution is (i) the characteristic function $\phi(t) = E(e^{itX}) = E(e^{itY})$ has a finite non-zero right-hand derivative at $t = 0$, i.e., $\phi_t'(0) = \lambda \neq 0$, and (ii) for any pair of positive real numbers a, b there is a positive real number c (depending on a and b) such that $aX + bY$ has the same distribution as cX (or cY)."

(Note that the normal distribution satisfies condition (ii), but not condition (i), when a and b are chosen so that $E(aX + bY) = 0$.)

A further remarkable characterization, due to Williams [33], is:

"If $(1 + aX)(a - X)^{-1}$ has the same distribution as X, and $\pi^{-1}\tan^{-1} a$ is not a rational number, then X has a standard Cauchy distribution."

A related characterization, also due to Williams [33], is:

"If $(\theta^2 + \lambda^2)(2\theta - X)^{-1}$ has the same distribution as X and $\pi^{-1}\tan^{-1}(\theta/\lambda)$ is not a rational number, then X has a Cauchy distribution with parameters θ, λ."

6. Related Distributions

We have already noted, in Section 1, that the Cauchy distribution is a central t distribution (Chapter 27) with one degree of freedom. It is related to other distributions in the same way as the t distributions.

Rider [25] has studied the properties of a system that he has named the *generalized Cauchy distributions.* The probability density function of a random variable X with such a distribution is of the form:

$$(16) \qquad p_X(x) = \frac{k\Gamma(h)}{2\lambda\Gamma(k^{-1})\Gamma(h - k^{-1})} \frac{1}{[1 + |(x - \theta)/\lambda|^k]^h} \qquad (\lambda, k, h, > 0).$$

For $k = 2$ and $\theta = 0$, X is distributed as $\lambda(2h - 1)^{-\frac{1}{2}}$ times a central t variable with $(2h - 1)$ degrees of freedom, and for $k = 2$, $h = 1$, X is distributed according to (1).

Symmetrically truncated Cauchy distributions, with density function (in standard form)

$$(17) \qquad [2(1 + x^2)\tan^{-1}\lambda]^{-1} \qquad (-\lambda \leq x \leq \lambda)$$

have been discussed in Derman [7]. The distribution is symmetrical about zero and has variance

$$(\lambda - \tan^{-1}\lambda)/(\tan^{-1}\lambda)$$

so the variance of the arithmetic mean of n independent variables having this distribution is

$$(18.1) \qquad n^{-1}(\lambda - \tan^{-1}\lambda)/(\tan^{-1}\lambda),$$

while, for large n, the variance of the median is approximately

$$n^{-1}(\tan^{-1}\lambda)^2. \tag{18.2}$$

(18.1) is larger than (18.2) for $\lambda > 3.41$, so that the median remains asymptotically more efficient, as an estimator of the center of a symmetrically truncated Cauchy distribution, provided not more than $\frac{1}{2} - \pi^{-1}\tan^{-1}(3.41) = 0.0908$ of the distribution is truncated at either end.

The standard *half-Cauchy* distribution has density function

$$2\pi^{-1}(1+x^2)^{-1} \qquad (0 < x). \tag{19}$$

The analogy with the half-normal distribution (Chapter 13, Section 7.1) is clear. Mijnheer [20] has carried out sampling experiments to investigate methods of estimating parameters of the general half-Cauchy distribution in which extreme observations are rejected.

The *folded Cauchy* distribution (obtained by folding the distribution (1) about $x = 0$) has density function

$$\begin{aligned}(\pi\lambda)^{-1}[\{1 + [(x-\theta)/\lambda]^2\}^{-1} + \{1 + [(x+\theta)/\lambda)^2\}^{-1}\\ = 2(\pi\lambda)^{-1}[\{1 + [(x-\theta)/\lambda]^2\}\{1 + [(x+\theta)/\lambda]^2\}]^{-1}\\ \times [1 + (x^2+\theta^2)/\lambda^2] \qquad (x > 0).\end{aligned} \tag{20}$$

For $\theta \le \lambda/\sqrt{3}$, the mode is at $x = 0$; for $\theta > \lambda/\sqrt{3}$ it is at

$$x = \theta[1 + (\lambda/\theta)^2]^{\frac{1}{2}}[2\{1 + (\lambda/\theta)^2\}^{-\frac{1}{2}} - 1]^{\frac{1}{2}}. \tag{21}$$

REFERENCES

[1] Barnett, V. D. (1966). Evaluation of the maximum likelihood estimator when the likelihood equation has multiple roots, *Biometrika*, **53,** 151–165.

[2] Barnett, V. D. (1966). Order statistics estimators of the location of the Cauchy distribution, *Journal of the American Statistical Association*, **61,** 1205–1217. (Correction, **63,** 383–385)

[3] Berman, S. M. (1967). An occupation time theorem for the angular component of plane Brownian motion, *Annals of Mathematical Statistics*, **38,** 25–31.

[4] Bloch, D. (1966). A note on the estimation of the location parameter of the Cauchy distribution, *Journal of the American Statistical Association*, **61,** 852–855.

[5] Cauchy, A. L. (1853). Sur les résultats moyens d'observations de même nature, et sur les résultats les plus probables, *Comptes Rendus de l'Académie des Sciences, Paris*, **37,** 198–206.

[6] Chernoff, H., Gastwirth, J. L. and Johns, M. V. (1967). Asymptotic distribution of linear combinations of functions of order statistics with applications to estimation, *Annals of Mathematical Statistics*, **38,** 52–72.

[7] Derman, C. (1964). Some notes on the Cauchy distribution, *National Bureau of Standards Technical Note*, Nos. 3–6.

[8] Fisher, R. A. and Tippett, L. H. C. (1928). Limiting forms of the frequency distributions of the smallest and the largest member of a sample, *Proceedings of the Cambridge Philosophical Society*, **24,** 180–190.

[9] Fox, C. (1965). A family of distributions with the same ratio property as normal distribution, *Canadian Mathematical Bulletin*, **8,** 631–636.

[10] Fréchet, M. (1927). Sur la loi de probabilité de l'écart maximum, *Annales de la Société Polonaise de Mathématique, Cracow*, **6,** 93–116.

[11] Fulvio, A. (1965). Sulla decomposizione de una variabile casuale seguente la legge di Cauchy, *Bolletino della Unione Matematica Italiana*, **20,** 177–180.

[12] Gumbel, E. J. and Keeney, R. D. (1950). The extremal quotient, *Annals of Mathematical Statistics*, **21,** 523–537.

[13] Kendall, M. G. and Stuart, A. (1961). *The Advanced Theory of Statistics*, **2.** London: Griffin.

[14] Kotlarski, I. (1960). On random variables whose quotient follows the Cauchy law, *Colloquium Mathematicum*, **7,** 277–284.

[15] Laha, R. G. (1959). On the law of Cauchy and Gauss, *Annals of Mathematical Statistics*, **30,** 1165–1174.

[16] Laha, R. G. (1959). On a class of distribution functions where the quotient follows the Cauchy law, *Transations of the American Mathematical Society*, **93,** 205–215.

[17] Mauldon, J. H. (1956). Characterizing properties of statistical distributions, *Quarterly Journal of Mathematics, Oxford*, **7,** 155–160.

[18] Menon, M. V. (1962). A characterization of the Cauchy distribution, *Annals of Mathematical Statistics*, **33,** 1267–1271.

[19] Menon, M. V. (1966). Another characteristic property of the Cauchy distribution, *Annals of Mathematical Statistics*, **37,** 289–294.

[20] Mijnheer, J. L. (1968). Steekproeven uit de halve Cauchy verdeling, *Statistica Neerlandica*, **22,** 97–101.

[21] Mises, R. von (1936). La distribution de la plus grande de n valeurs, *Revue Mathématique de l'Union Interbalkanique*, **1,** 1–20.

[22] Obretenov, A. (1961). A property characterizing the Cauchy distribution, *Fiziko-Matematichesko Spisaniye, Bülgarska Akademiya na Naukite*, **4,** (37); 40–43. (In Bulgarian)

[23] Pitman, E. J. G. and Williams, E. J. (1967). Cauchy-distributed functions of Cauchy variates, *Annals of Mathematical Statistics*, **38,** 916–918.

[24] Puglisi, Anna (1966). Sulla decomposizione della legge di probabilità di Cauchy, *Bolletino della Unione Matematica Italiana*, **21,** 12–18.

[25] Rider, P. R. (1957). Generalized Cauchy distributions, *Annals of the Institute of Statistical Mathematics, Tokyo*, **9,** 215–223.

[26] Rider, P. R. (1960). Variance of the median of samples from a Cauchy distribution, *Journal of the American Statistical Association*, **55,** 322–323.

[27] Rothenberg, T. J., Fisher, F. M. and Tilanus, C. B. (1964). A note on estimation from a Cauchy sample, *Journal of the American Statistical Association*, **59,** 460–463.

[28] Savage, L. J. (1966). *A geometrical approach to the special stable distributions*, Technical Report No. 1, Department of Statistics, Yale University.

[29] Spitzer, F. (1958). Some theorems concerning 2-dimensional Brownian motion, *Transactions of the American Mathematical Society*, **87,** 187–197.

[30] Springer, M. D. and Thompson, W. E. (1966). The distribution of products of independent random variables, *SIAM Journal of Applied Mathematics*, **14,** 511–526.

[31] Steck, G. P. (1958). A uniqueness property not enjoyed by the normal distribution, *Annals of Mathematical Statistics*, **29,** 604–606.

[32] Weiss, L. (1966). The relative maxima of the likelihood function. **II,** *Skardinavisk Aktuarietidskrift*, 119–121.

[33] Williams, E. J. (1969). Cauchy-distributed functions and a characterization of the Cauchy distribution, *Annals of Mathematical Statistics*, **40,** 1083–1085.

17

Gamma Distribution (Including Chi-Square)

1. Definition

A random variable X has a gamma distribution if its probability density function is of form

$$(1) \qquad p_X(x) = \frac{(x-\gamma)^{\alpha-1}\exp[-(x-\gamma)/\beta]}{\beta^\alpha\Gamma(\alpha)} \qquad (\alpha > 0, \beta > 0; x > \gamma).$$

This distribution is Type III of Pearson's system (Chapter 12, Section 4). It depends on three parameters α, β and γ.

The *standard form* of the distribution is obtained by putting $\beta = 1$ and $\gamma = 0$, giving

$$(2) \qquad p_X(x) = \frac{x^{\alpha-1}e^{-x}}{\Gamma(\alpha)} \qquad (x \geq 0).$$

If $\alpha = 1$, this is an *exponential distribution* (see Chapter 18). If α is a positive integer, it is an *Erlang distribution.*

The distributions of $Y = -X$, namely

$$(1)' \qquad p_Y(y) = \frac{(-y-\gamma)^{\alpha-1}\exp[(y+\gamma)/\beta]}{\beta^\alpha\Gamma(\alpha)} \qquad (y \leq -\gamma)$$

and

$$(2)' \qquad p_Y(y) = \frac{(-y)^{\alpha-1}e^y}{\Gamma(\alpha)} \qquad (y \leq 0)$$

are also gamma distributions, but such distributions rarely need to be considered, and we will not discuss them further here.

The probability integral of distribution (2) is

$$\Pr[X \leq x] = [\Gamma(\alpha)]^{-1} \int_0^x t^{\alpha-1} e^{-t}\, dt. \tag{3}$$

This is an *incomplete gamma function ratio.* The quantity

$$\Gamma_x(\alpha) = \int_0^x t^{\alpha-1} e^{-t}\, dt \tag{4}$$

is sometimes called an *incomplete gamma function*, but this name is also quite commonly applied to the ratio (3) (Chapter 1, Section 3).

This ratio depends on x and α, and it would be natural to use a notation representing it as a function of these variables. However, Pearson [96], found it more convenient to use $u = x\alpha^{-\frac{1}{2}}$ in place of x for tabulation purposes, and *defined* the incomplete gamma function as

$$I(u,\alpha-1) = \frac{1}{\Gamma(\alpha)} \int_0^{u\sqrt{\alpha}} t^{\alpha-1} e^{-t}\, dt. \tag{5}$$

The main importance of the (standard) gamma distribution in statistical theory is the fact that if $U_1, U_2, \ldots, U_\nu$ are independent unit normal variables, the distribution of $\sum_{j=1}^{\nu} U_j^2$ is of form (1) with $\alpha = \frac{1}{2}\nu$; $\beta = 2$; $\gamma = 0$. This particular form of gamma distribution is called a *chi-square distribution with* ν *degrees of freedom.* The corresponding random variable is often denoted by χ_ν^2, and we will follow this practice. It is clear that $\frac{1}{2} \sum_{j=1}^{\nu} U_j^2$ has a standard gamma distribution with $\alpha = \frac{1}{2}\nu$. Expressed symbolically:

$$p_{\chi_\nu^2}(x^2) = \{2^{\frac{1}{2}\nu}\Gamma(\tfrac{1}{2}\nu)\}^{-1}(x^2)^{\frac{1}{2}\nu-1} \exp(-\tfrac{1}{2}x^2) \qquad (x^2 \geq 0). \tag{6}$$

Although in the above definition, ν must be an integer, the distribution (6) is also called a "χ^2 distribution with ν degrees of freedom" if ν is *any* positive number.

2. Moments and Other Properties

The moment generating function of the standard gamma distribution (2) is

$$E[e^{tX}] = \{\Gamma(\alpha)\}^{-1} \int_0^\infty x^{\alpha-1} \exp[-(1-t)x]\, dx = (1-t)^{-\alpha} \qquad (t < 1). \tag{7}$$

The characteristic function is $(1 - it)^{-\alpha}$.

Since distributions of form (1) can be obtained from those of form (2) by the linear transformation $X = (X' - \gamma)/\beta$, there is no difficulty in deriving formulas for moments, generating functions, etc. for (1) from those for (2). Thus the moment generating function of χ_ν^2 is

$$E[e^{t\chi_\nu^2}] = (1-2t)^{-\frac{1}{2}\nu} \qquad (t < \tfrac{1}{2}) \tag{8}$$

and the characteristic function is $(1 - 2it)^{-\frac{1}{2}\nu}$.

Similarly, from the formula for the rth moment about zero of distribution (2):

$$\mu'_r = \{\Gamma(\alpha)\}^{-1}\int_0^\infty x^{\alpha+r-1}e^{-x}\,dx = \Gamma(\alpha + r)/\Gamma(\alpha), \tag{9}$$

the formula

$$\mu'_r(\chi^2_\nu) = 2^r\Gamma(\tfrac{1}{2}\nu + r)/\Gamma(\tfrac{1}{2}\nu) = \nu(\nu + 2)\cdots(\nu + 2\overline{r - 1}) \tag{10}$$

can be derived.

From (9) (and (10)) cumulants can be obtained. These are very simple. From (9):

$$\kappa_r = (r - 1)!\alpha, \tag{11}$$

while, for χ^2_ν from (10):

$$\kappa_r(\chi^2_\nu) = 2^{r-1}(r - 1)!\nu. \tag{12}$$

Hence for distribution (2)

$$E(X) = \text{var}(X) = \alpha;\ \mu_3 = 2\alpha;\ \mu_4 = 3\alpha^2 + 6\alpha \tag{13}$$

so that

$$\alpha_3 = \sqrt{\beta_1} = 2\alpha^{-\frac{1}{2}};\ \alpha_4 = \beta_2 = 3 + 6\alpha^{-1}. \tag{14}$$

Also, for the distribution of χ^2_ν:

$$E(\chi^2_\nu) = \nu;\ \text{var}(\chi^2_\nu) = 2\nu; \tag{15}$$

$$\alpha_3(\chi^2_\nu) = \sqrt{\beta_1(\chi^2_\nu)} = \sqrt{8/\nu};\ \alpha_4(\chi^2_\nu) = \beta_2(\chi^2_\nu) = 3 + 12\nu^{-1}.$$

The mean deviation of distribution (2) is

$$2\alpha^\alpha e^{-\alpha}/\Gamma(\alpha). \tag{16}$$

The mean deviation of χ^2_ν is:

$$\nu^{\frac{1}{2}\nu}e^{-\frac{1}{2}\nu}/\{2^{\frac{1}{2}\nu-1}\Gamma(\tfrac{1}{2}\nu)\}. \tag{17}$$

The coefficient of variation is $\alpha^{-\frac{1}{2}}$.

The standard distribution (2) has a single mode at $x = \alpha - 1$ if $\alpha \geq 1$. (Distribution (1) has a mode at $x = \gamma + \beta(\alpha - 1)$.) If $\alpha < 1$, $p_X(x)$ tends to infinity as x tends to zero; if $\alpha = 1$ (the standard exponential distribution), $\lim_{x\to 0} p_X(x) = 1$.

There are points of inflexion, equidistant from the mode, at

$$x = \alpha - 1 \pm \sqrt{\alpha - 1} \tag{18}$$

(provided the values are real and positive).

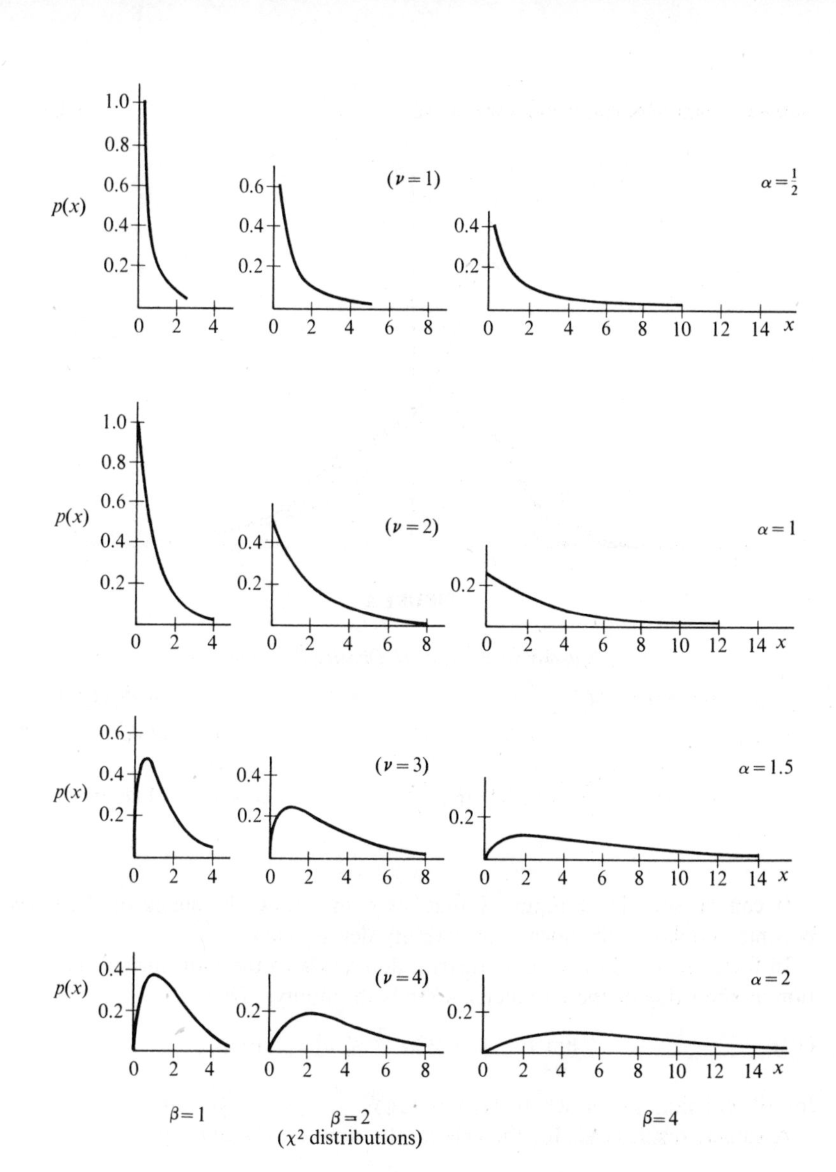

FIGURE 1

Gamma Density Functions

Some typical gamma probability density functions are shown in Figures 1 and 2. The center column in Figure 1 shows χ^2 distributions for $\nu = 1, 2, 3, 4$ ($\alpha = \frac{1}{2}, 1, \frac{3}{2}, 2$); the columns to either side also have $\gamma = 0$ and correspond to the same values of α (in any one row) but with $\beta = 1$ (left side) and 4 (right side) in place of $\beta = 2$ for the χ^2 distribution.

Figure 2 shows three different gamma distributions (1) each having the same expected value (zero) and standard deviation (unity).

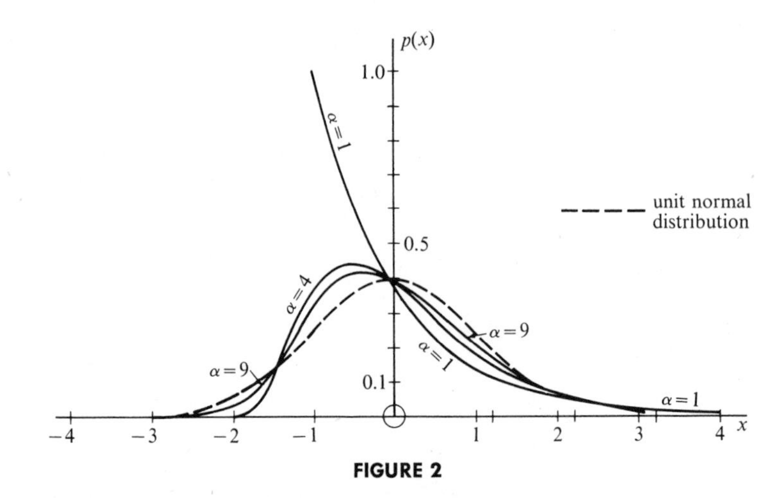

FIGURE 2

Standardized Type III Density Functions

$\alpha = 1;\ p(x) = \exp[-(x+1)]$	$(x > -1)$	Mode at -1
$\alpha = 4;\ p(x) = \frac{8}{3}(x+2)^3 \exp[-2(x+2)]$	$(x > -2)$	Mode at $-\frac{1}{2}$
$\alpha = 9;\ p(x) = \frac{2187}{4480}(x+3)^8 \exp[-3(x+3)]$	$(x > -3)$	Mode at $-\frac{1}{3}$

It can be seen from Figure 1 that, as α increases, the shape of the curve becomes similar to the normal probability density curve.

In fact, the standard gamma distribution tends to the unit normal distribution as the value of the parameter α tends to infinity. That is:

$$\lim_{\alpha\to\infty} \Pr[(X-\alpha)\alpha^{-\frac{1}{2}} \le u] = \Phi(u) \tag{19}$$

for all real values of u, where $\Phi(u) = (2\pi)^{-\frac{1}{2}}\int_{-\infty}^{u} \exp(-\frac{1}{2}t^2)\,dt$.

A similar result holds for the general distribution (1), namely:

$$\lim_{\alpha\to\infty} \Pr\left[\left\{\frac{(X-\gamma)}{\beta} - \alpha\right\}\alpha^{-\frac{1}{2}} \le u\right] = \Phi(u). \tag{20}$$

For χ_ν^2:

$$\lim_{\nu\to\infty} \Pr[(\chi_\nu^2 - \nu)(2\nu)^{-\frac{1}{2}} \le u] = \Phi(u). \tag{21}$$

It can be checked from (14) and (15) that $\alpha_3 \to 0$, $\alpha_4 \to 3$ (the values for the normal distribution) as α, ν, respectively, tend to infinity.

One of the most important properties of the distribution is the *reproductive property:* If X_1, X_2 are independent random variables each having a distribution of form (1), possibly with different values α', α'' of α, but with *common values of* β *and* γ, then $(X_1 + X_2)$ also has a distribution of this form, with the same value of β and γ, and with $\alpha = \alpha' + \alpha''$. In particular, if X_j is distributed as $\chi^2_{\nu_j}$ $(j = 1,2)$, then $(X_1 + X_2)$ is distributed as $\chi^2_{\nu_1+\nu_2}$ (whether or not ν_1,ν_2 are integers).

3. Genesis, Applications and Some History

Lancaster [78] quotes from Laplace [79] in which the latter obtains a gamma distribution as the posterior distribution of the 'precision constant' $(h = \frac{1}{2}\sigma^{-2})$, (Chapter 13, Section 1) given the values of n independent normal variables with zero mean and standard deviation σ (assuming a 'uniform' prior distribution for h). Quoting further from [78], Bienaymé [12] in 1838 obtained the (continuous) χ^2 distribution as the limiting distribution of the (discrete) random variable $\sum_{j=1}^{k} (N_i - np_i)^2(np_i)^{-1}$ when $(N_1,\ldots,N_k)$ have a joint multinomial distribution with parameters $n, p_1, p_2, \ldots, p_k$ (Chapter 11, Section 2).

The gamma distribution appears naturally in the theory associated with normally distributed random variables, as the distribution of the sum of squares of independent unit normal variables. If $U_1, U_2, \ldots, U_\nu$ are independent unit normal variables, then $\sum_{j=1}^{\nu} U_j^2$ has a χ^2 distribution with ν degrees of freedom (as has already been mentioned in Section 1).

Lancaster [78] has pointed out that the special case of this result, that if $V_1, V_2, \ldots, V_k$ are independent random variables each distributed as χ^2 with 2 degrees of freedom (i.e. exponentially) then their sum is distributed as χ^2 with $2k$ degrees of freedom, was demonstrated by Ellis [33] in 1844. The general result was proved by Bienaymé [13] in 1852. The result was also established, using a different method, by Helmert [53] in 1875. In the following year the same author [54] [55] proved that if $X_1, X_2, \ldots, X_n$ are independent identically normally distributed random variables, their common standard deviation being σ, then $\sum_{j=1}^{n} (X_j - \bar{X})^2$ (where $\bar{X} = n^{-1} \sum_{j=1}^{n} X_j$) is distributed as σ^2 times a variable having a χ^2_ν distribution, and further that this random variable and $\bar{X}$ are mutually independent. Kruskal [74] has suggested that the joint distribution of $\sum_{j=1}^{n} (X_j - \bar{X})^2$ and $\bar{X}$ be called *Helmert's distribution;* this suggestion is supported by Lancaster [78].

The gamma distribution appeared again in 1900 (Pearson [95]) as the approximate distribution (again a χ^2 distribution) for the 'chi-square statistics' used for various tests in contingency tables. (The *exact* distribution of this statistic is, of course, discrete (Chapter 11, Section 2).)

The use of the gamma distribution to approximate the distribution of quadratic forms (particularly positive definite quadratic forms) in multinormally distributed variables is well-established and widespread. One of the earliest examples was its use, in 1938, to approximate the distribution of the denominator in a test criterion for difference between expected values of two normal populations with possibly different variances (Welch [123]). It has been used in this way many times since. Some examples are discussed in Chapter 27 of this book. The use of gamma distributions to represent distributions of range and quasi-ranges in random samples from a normal population has been discussed in Chapter 13. In most applications the two parameter form ($\gamma = 0$)

$$p_X(x) = \frac{x^{\alpha-1}e^{-x/\beta}}{\beta^\alpha\Gamma(\alpha)} \qquad (x > 0; \alpha > 0, \beta > 0) \tag{22}$$

is used (this is equivalent to approximating by the distribution of $\frac{1}{2}\beta\chi^2_{2\alpha}$). However, the three-parameter form has also been used with good effect (see, for example, Pearson [93]).

The gamma distribution may be used in place of the normal distribution as 'parent' distribution in expansions of Gram-Charlier type (Chapter 12, Section 4.2). Series with Laguerre polynomial multipliers rather than Hermite polynomial multipliers are obtained here (Chapter 1, Section 3). Formulas for use with such expansions and their properties have been described by Khamis [65]. These *Laguerre series* have been used by Barton [6] and Tiku [116] [117] to approximate the distributions of 'smooth test' (for goodness-of-fit) statistics and non-central F (Chapter 30) respectively.

In applied work, gamma distributions give useful representations of many physical situations. They have been used to make realistic adjustments to exponential distributions in representing lifetimes in 'life-testing' situations. Of recent years, Weibull distributions have been more popular for this purpose, but this may not be permanent. The fact that a sum of independent exponentially distributed random variables has a gamma distribution (a special case of the 'reproductive' property mentioned in Section 1) leads to the appearance of gamma distribution in the theory of random counters and other topics associated with random processes in time, in particular in meteorological precipitation processes (Kotz and Neumann [73], Das [29]). Some other applications from diverse fields are indicated in papers [1], [8], [14], [44], [58], [68], [84], [109], and [124].

Gamma distributions share with lognormal distributions (Chapter 14) the ability to mimic closely a normal distribution (by choosing α large enough) while representing an essentially positive random variable (by choosing $\gamma \geq 0$).

4. Tables and Nomograms

One of the earliest tables of probability integrals of gamma distributions was published in 1902. It contains values of $\Pr[\chi^2_\nu > x]$ to six decimal places for $\nu = 2(1)29$ and $x = 1(1)30(10)70$, and was computed by Elderton [32].

In 1922, there appeared a comprehensive *Tables of the Incomplete* Γ-*Function*, edited by Pearson [96]. This contains values of $I(u,p)$ (defined as in (5)) to seven decimal places for $p = -1(0.05)0(0.1)5(0.2)50$ and u at intervals of 0.1. These are supplemented by a table of values of

$$\log I(u,p) - (p+1)\log u$$

for $p = -1(0.05)0(0.1)10$ and $u = 0.1(0.1)1.5$. This function was chosen to make interpolation easier, particularly for low values of p (Section 5).

Recently, Harter [47] has published tables of $I(u,p)$ to nine decimal places for $p = -0.5(0.5)74(1)164$ and u at intervals of 0.1.

It will be noted that [47] covers a greater range of values of p and has two extra decimal places, but [96] has p at finer intervals.

These are the most important direct tables of $I(u,p)$. Special values can be obtained from tables of the cumulative Poisson distribution using the formulas (Chapter 4, Section 11)

$$\Pr[\chi^2_{2\nu} \geq x] = e^{-x/2} \sum_{j=0}^{\nu-1} \{(\tfrac{1}{2}x)^j/j!\} \tag{23.1}$$

$$\Pr[\chi^2_{2\nu-1} \geq x] = e^{-x/2} \sum_{j=0}^{\nu-2} \{(\tfrac{1}{2}x)^{j+\frac{1}{2}}/\Gamma(j+\tfrac{3}{2})\} + 2\{1 - \Phi(\sqrt{x})\}. \tag{23.2}$$

Pearson [96] gave the more general formula for distribution (1):

$$\Pr[X \leq x] = e^{-y} \sum_{j=0}^{\infty} \{y^{\alpha+j}/\Gamma(\alpha + j + 1)\} \tag{24}$$

for $y = (x - \gamma)/\beta > 0$.

Salvosa [105] [106] has published tables of the probability integral, probability density function and its first six derivatives for distribution (1) with β and γ so chosen that X is standardized (that is $\beta = \frac{1}{2}\alpha_3; \gamma = -2/\alpha_3$). Values are given to six decimal places for $\alpha_3 (= 2\alpha^{-\frac{1}{2}}) = 0.1(0.1)1.1$ at intervals of 0.01 for x.

Cohen *et al.* [27] have calculated tables of the probability integral to nine decimal places for $\alpha_3 = 0.1(0.1)2.0(0.2)3.0(0.5)6.0$ at intervals of 0.01 for x.

Khamis and Rudert [66] have published extensive tables of the probability integral of (1), with $\gamma = 0$, $\beta = 2$, i.e.

$$\Pr[\chi^2_{2\alpha} \leq x] = \frac{1}{2^\alpha \Gamma(\alpha)} \int_0^x t^{\alpha-1} e^{-\frac{1}{2}t}\, dt \qquad (= I(\tfrac{1}{2}x\alpha^{-\frac{1}{2}}, \alpha - 1))$$

to ten decimal places for

$$\alpha = 0.05(0.05)10(0.1)20(0.25)\,70;$$
$$x = 0.0001(0.0001)0.0010(0.001)0.010(0.01)1.00(0.05)6.00(0.1)$$
$$16.0(0.5)66.0(1)166(2)250.$$

The varied intervals for x have been chosen to make interpolation easier. (Note that Pearson [96] used equal intervals for $u = \frac{1}{2}x\alpha^{-\frac{1}{2}}$ to this end.)

There are a number of tables of percentile points of the standard distributions (2). Thompson [115] (see also Pearson and Hartley [94]) gives tables of values $\chi^2_{\nu,\epsilon}$ such that

$$\Pr[\chi^2_\nu < \chi^2_{\nu,\epsilon}] = \epsilon \tag{25}$$

to six significant figures for

$$\nu = 1(1)30(10)100$$

$\epsilon = 0.005, 0.01, 0.025, 0.05, 0.1, 0.25, 0.5, 0.75, 0.9, 0.95, 0.975, 0.99, 0.995$ (also for $\epsilon = 0.999$ in [94]).

The tables of Vanderbeck and Cooke [119] use the same values of ϵ, with the addition of $\epsilon = 0.80$, and give $\chi^2_{\nu,\epsilon}$ to four decimal places or significant figures, whichever is the more accurate (except for $\nu = 1$), for $\nu = 1(1)300$. For $\nu > 30$ they used an approximation (Section 5).

Hald and Sinkbaek [45] have published tables of $\chi^2_{\nu,\epsilon}$ to three decimal places, or three significant figures (if more accurate), for $\nu = 1(1)100$ and

$$\epsilon = 0.0005, 0.001, 0.005, 0.01, 0.025, 0.05, 0.1(0.1)0.9, 0.95, 0.975, 0.99, 0.995, 0.999, 0.9995.$$

Harter [47] has published tables of $\chi^2_{\nu,\epsilon}$ to six significant figures for the values $\nu = 1(1)150(2)330$ and the above values of ϵ, plus $\epsilon = 0.0001$ and 0.9999. The portion of the tables for which $\nu = 1(1)100$ is also published in [46]. (The correct values for $\chi^2_{60;0.6}$, $\chi^2_{74;0.8}$ and $\chi^2_{80;0.4}$ are 62.1348, 83.9965 and 76.1879 respectively — as pointed out in Bol'shev's review of these tables.)

The less easily accessible tables of Vanderbeck and Cooke [119] give values of $\Pr[\chi^2_\nu > x]$ for $x = 0.(0.1)3.2$ with $\nu = 0.05(0.05)0.2(0.1)6.0$ and for $x = 3.2(0.2)7.0(0.5)10.0(1)35$ with $\nu = 0.1(0.1)0.4(0.2)6.0$. The inclusion of non-integer values of ν is a valuable feature of these tables. To facilitate interpolation when x is small, tables of

$$(\tfrac{1}{2}x)^{-\frac{1}{2}\nu}\Pr[\chi^2_\nu \le x]$$

are also given, for $x = 0.05(0.05)0.2(0.1)1.0$ with $\nu = 0.05(0.05)0.2(0.1)6.0$. For large values of x, a further table gives values of $\Pr[\chi^2_\nu > x]$ for $\sqrt{2x} - \sqrt{2\nu} = -4.0(0.1)4.8$ with $\sqrt{2/\nu} = 0.02(0.02)0.22(0.01)0.25$.

Thom [114] has given tables, to four decimal places, of

(i) $\Gamma_x(\alpha)/\Gamma(\alpha)$ for $\alpha = 0.5(0.5)15.0(1)36$
and $x = 0.0001, 0.001, 0.004(0.002)0.020(0.02)0.80(0.1)2.0(0.2)3.0(0.5)$— (the tabulation is continued for increasing x until the value of the tabulated function exceeds 0.9900),

and

(ii) values of x satisfying the equation

$$\Gamma_x(\alpha)/\Gamma(\alpha) = \epsilon$$

for $\alpha = 0.5(0.5)15.0(1)36$
and $\epsilon = 0.01, 0.05(0.05)0.95, 0.99$.

Harter [50] has published further tables of Type III distributions, giving the 0.01, 0.05, 0.1, 0.5, 2, 2.5, 4, 5, 10(10)90, 95, 96, 97.5, 98, 99, 99.5, 99.9, 99.95 and 99.99 percent points to 5 decimal places for $\sqrt{\beta_1} = 0.0(0.1)4.8(0.2)9.0$.

Russell and Lal [104] have constructed tables of $\Pr[\chi^2_\nu > x]$ to 5 decimal places for $\nu = 1(1)50$; $x = 0.001(0.001)0.01(0.01)0.1(0.1)10.0$.

A nomogram by Boyd [19], reproduced in Figure 3, connects values of $P = \Pr[\chi_\nu^2 \geq \chi_0^2]$ with those of ν and χ_0^2 by means of a straight-edge. Stammberger [113] gives another nomogram for the χ_ν^2 distribution.

In the Soviet Union the first comprehensive tables of the incomplete Gamma function and the χ^2 distribution were compiled by Slutskii [110]. (These tables are not easily accessible in the West.) They have the special feature of an auxiliary table for the compilation of the incomplete Gamma function in the region $p > 50$, which is not covered by Pearson [96].

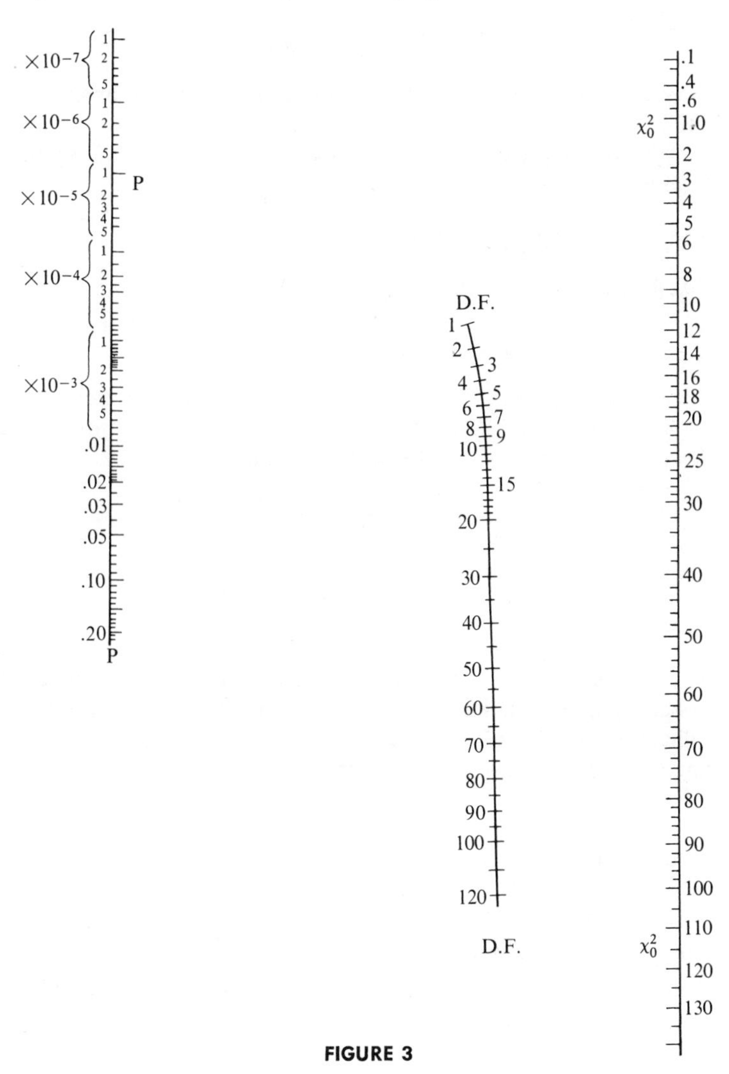

FIGURE 3

Boyd's χ^2 Nomogram

5. Approximations

The best known approximations for probability integrals of gamma distributions have been developed in connection with the χ^2 distribution. We will express these approximations in terms of the χ^2 distribution. Modifications to apply to general gamma distributions are direct, using the linear transformation $y = (x - \gamma)/\beta$.

It has been mentioned in Section 2, that as $\nu \to \infty$, the standardized χ_ν^2 distribution tends to a unit normal distribution. The simple approximation

$$\text{(26)} \qquad \Pr[\chi_\nu^2 < x] \doteq \Phi((x - \nu)(2\nu)^{-\frac{1}{2}})$$

is not very accurate, however, unless ν is rather large. Better approximations are obtained by utilizing the asymptotic normality of various functions of χ_ν^2, even though only approximate standardization is effected. Among the best known are Fisher's approximation [35]:

$$\text{(27)} \qquad \Pr[\chi_\nu^2 < x] \doteq \Phi(\sqrt{2x} - \sqrt{2\nu - 1})$$

and the Wilson-Hilferty [128] approximation:

$$\text{(28)} \qquad \Pr[\chi_\nu^2 < x] \doteq \Phi(\{(x/\nu)^{\frac{1}{3}} - 1 + \tfrac{2}{9}\nu^{-1}\}\sqrt{9\nu/2}).$$

Of these two approximations the second is definitely the more accurate. Both are much better than (26). Note that if (27) and (28) are used to obtain the following approximations to the percentile points $\chi_{\nu,\epsilon}^2$:

$$\text{(27)}' \qquad \chi_{\nu,\epsilon}^2 = \tfrac{1}{2}(U_\epsilon + \sqrt{2\nu - 1})^2$$

and

$$\text{(28)}' \qquad \chi_{\nu,\epsilon}^2 = \nu\left(U_\epsilon\sqrt{\frac{2}{9\nu}} + 1 - \frac{2}{9\nu}\right)^3$$

then the addition of $\frac{1}{6}(U_\epsilon^2 - 1)$ to (27)′ will make it very nearly equal to the (usually better) approximation (28)′, unless ν is small or ϵ is near to zero or 1. This is illustrated by the values in Table 1.

Putting $\nu = 2\alpha$ (28)′ gives the formula

$$2\alpha(\tfrac{1}{3}U_\epsilon\alpha^{-\frac{1}{2}} + 1 - \tfrac{1}{9}\alpha^{-1})^3$$

corresponding to

$$\text{(28)}'' \qquad \alpha(\tfrac{1}{3}U_\epsilon\alpha^{-\frac{1}{2}} + 1 - \tfrac{1}{9}\alpha^{-1})^3$$

for the $100\epsilon\%$ percentile of distribution (2). Pogurova [102] has given a table of corrections to add to formula (28)″ which gives values of percentile points correct to six significant figures for $\alpha \geq 25$ and ϵ between 0.0001 and 0.9999. Her table is reproduced here as Table 2.

Vanderbeck and Cooke [119] used the Cornish-Fisher expansion

TABLE 1

Comparison of Approximations to χ^2 Percentile Points

ϵ	ν	$\chi^2_{\nu,\epsilon}$	Approximation (27)′	Approximation (28)′	Difference (28)′ − (27)′	$\frac{1}{6}(U_\nu^2 - 1)$
0.01	5	0.5543	0.2269	0.5031	0.2762	
	10	2.5582	2.0656	2.5122	0.4466	0.7353
	25	11.5240	10.9215	11.4927	0.5712	
0.05	5	1.1455	0.9182	1.1282	0.3100	
	10	3.9403	3.6830	3.9315	0.2485	0.2843
	25	14.6114	14.3388	14.6086	0.2698	
0.10	5	1.6103	1.4765	1.6098	0.1333	
	10	4.8652	4.7350	4.8695	0.1345	0.1071
	25	16.4734	16.3503	16.4788	0.1285	
0.50	5	4.3515	4.5000	4.3625	−0.1375	
	10	9.3418	9.5000	9.3480	−0.1520	−0.1667
	25	24.3366	24.5000	24.3392	−0.1608	
0.90	5	9.2364	9.1658	9.2078	0.0420	
	10	15.9872	15.9073	15.9677	0.0604	0.1071
	25	34.3816	34.2920	34.3701	0.0781	
0.95	5	11.0705	10.7873	11.0439	0.2666	
	10	18.3070	18:0225	18.2918	0.2693	0.2843
	25	37.6525	37.3667	37.6452	0.2785	
0.99	5	15.0862	14.1850	14.4599	0.2749	
	10	23.2093	22.3463	23.2393	0.8930	0.7353
	25	44.3141	43.4904	44.3375	0.8471	

$$\chi^2_{\nu,\epsilon} \doteqdot \nu + \sqrt{2}\, U_\epsilon \nu^{\frac{1}{2}} + \tfrac{2}{3}(U_\epsilon^2 - 1) + \frac{1}{9\sqrt{2}}(U_\epsilon^3 - 7U_\epsilon)\nu^{-\frac{1}{2}} - \tfrac{1}{405}(106U_\epsilon^4 - 486U_\epsilon^2 + 168)\nu^{-1} + \frac{1}{4860\sqrt{2}}(9U_\epsilon^5 + 256U_\epsilon^3 - 433U_\epsilon)\nu^{-\frac{3}{2}}. \tag{29}$$

They provide a table comparing exact values with those given by (29) for $\nu = 10(10)100$,

$\epsilon = 0.005, 0.010, 0.025, 0.05, 0.1, 0.25, 0.5, 0.75, 0.9, 0.95, 0.975, 0.99, 0.995.$

For $\nu \geq 60$ the error in $\chi^2_{\nu,\epsilon}$ is only occasionally greater than 0.0001. Even for $\nu = 30$ this value was exceeded only for ϵ outside the range $0.1 - 0.9$.

An expression for $\Pr[\chi_\nu^2 \leq x]$ as a power series in x can be obtained in the

TABLE 2

Corrections ($\times 10^4$) *to add to* (28)″ $(t = (2\alpha)^{-1})$

ϵ \ t	0.20	0.19	0.18	0.17	0.16	0.15	0.14	0.13	0.12	0.11	0.10	0.09	0.08	0.07	0.06	0.05	0.04	0.03	0.02	0.01
0.0001	577	558	537	516	403	469	444	419	392	364	335	305	274	243	210	177	143	108	73	36
0.0005	397	382	367	350	333	316	298	279	260	241	221	201	180	159	137	115	92	70	47	23
0.0010	324	311	298	284	270	255	240	225	209	193	177	160	143	126	109	91	73	55	37	18
0.0050	169	161	153	145	137	128	120	112	103	95	86	78	69	60	52	43	34	26	17	8
0.0100	110	105	99	93	88	82	76	71	65	59	54	48	43	37	32	26	21	15	10	5
0.0200	59	55	52	48	45	42	39	35	32	29	26	23	20	17	15	12	9	7	4	2
0.0500	4	3	2	1	1	0	0	−0	−0	−0	−1	−1	−1	−1	−1	−1	−1	−1	−0	−0
0.1000	−23	−22	−21	−21	−20	−19	−18	−17	−16	−15	−14	−12	−11	−10	−8	−7	−6	−4	−3	−1
0.2000	−33	−32	−30	−29	−27	−26	−24	−22	−21	−19	−17	−15	−14	−12	−10	−8	−7	−5	−3	−1
0.3000	−29	−27	−26	−24	−23	−21	−20	−18	−17	−15	−14	−12	−11	−9	−8	−6	−5	−4	−2	−1
0.4000	−19	−18	−16	−15	−14	−13	−12	−11	−10	−9	−8	−7	−6	−5	−4	−3	−2	−2	−1	−0
0.5000	−6	−6	−5	−4	−4	−3	−3	−2	−2	−2	−1	−1	−1	−0	−0	−0	−0	−0	−0	−0
0.6000	7	7	7	7	6	6	6	6	5	5	5	4	4	4	3	3	2	1	1	0
0.7000	21	20	20	19	18	17	16	15	14	13	12	11	9	8	7	6	5	3	2	1
0.8000	34	33	31	29	28	26	24	23	21	19	17	16	14	12	10	8	7	5	3	1
0.9000	38	36	34	32	30	27	20	23	21	19	17	15	14	12	10	8	6	5	3	1
0.9500	21	19	18	16	15	13	12	11	9	8	7	6	5	4	3	3	2	1	1	0
0.9800	−30	−29	−29	−28	−27	−26	−24	−23	−22	−20	−19	−17	−16	−14	−12	−10	−8	−6	−4	−2
0.9900	−89	−86	−82	−78	−74	−70	−66	−61	−57	−53	−48	−44	−39	−34	−30	−25	−20	−15	−10	−5
0.9950	−164	−157	−149	−141	−133	−125	−117	−109	−101	−93	−85	−76	−68	−60	−51	−43	−34	−25	−17	−8
0.9990	−394	−374	−354	−334	−314	−294	−274	−253	−233	−213	−194	−174	−154	−134	−115	−95	−76	−56	−37	−18
0.9995	−516	−489	−462	−435	−409	−382	−355	−329	−302	−276	−250	−224	−198	−173	−147	−122	−97	−72	−48	−23
0.9999	−847	−801	−756	−710	−665	−620	−576	−532	−488	−445	−402	−359	−317	−276	−234	−194	−154	−114	−75	−37

following way:

$$(30)\qquad \Pr[\chi_\nu^2 \le x] = 2^{-\frac{1}{2}\nu}\{\Gamma(\tfrac{1}{2}\nu)\}^{-1}\int_0^x t^{\frac{1}{2}\nu-1}e^{-\frac{1}{2}t}\,dt$$

$$= 2^{-\frac{1}{2}\nu}\{\Gamma(\tfrac{1}{2}\nu)\}^{-1}\int_0^x \sum_{j=0}^{\infty}(-1)^j(2^j j!)^{-1}t^{\frac{1}{2}\nu+j-1}\,dt$$

$$= \frac{2(\frac{1}{2}x)^{\frac{1}{2}\nu}}{\Gamma(\frac{1}{2}\nu)}\sum_{j=0}^{\infty}(-1)^j\frac{x^j}{(\nu+2j)2^j j!}.$$

(Note that the series converges for all $x > 0$, and, for sufficiently large m, the true value lies between $\sum_{j=0}^{m}$ and $\sum_{j=0}^{m+1}$.)

The series may be used as a basis for approximately evaluating $\Pr[\chi_\nu^2 \le x]$ when x is small (say $x < \nu$).

Note that (30) can also be written in terms of Pearson's incomplete gamma function

$$(30)'\qquad I(u,p) = \frac{(\frac{1}{2}u\sqrt{p+1})^{p+1}}{\Gamma(p+1)}\sum_{j=0}^{\infty}(-1)^j\frac{(u\sqrt{p+1})^j}{(p+j+1)2^{j+1}j!}.$$

For u small, $u^{-(p+1)}I(u,p)$ is very approximately, a linear function of u. It is for this reason that the values

$$\log I(u,p) - (p+1)\log u$$

tabled by Pearson [96] lead to relatively easy interpolation (see Section 4).

Pearson and Hartley [94] have suggested an iterative procedure based on (30) for calculating values of $\chi_{\nu,\epsilon}^2$. Since $\Pr[\chi_\nu^2 < \chi_{\nu,\epsilon}^2] = \epsilon$, (30) can be written in the form

$$(30)''\qquad \chi_{\nu,\epsilon}^2 = 2\{\epsilon\Gamma(\tfrac{1}{2}\nu+1)\}^{2/\nu}\left\{\sum_{j=0}^{\infty}(-1)^j\frac{(\chi_{\nu,\epsilon}^2)^j\nu}{(\nu+2j)2^j j!}\right\}^{-2/\nu}.$$

Inserting a trial value of $\chi_{\nu,\epsilon}^2$ on the right hand side of (30)″, a new value is obtained. It is suggested that initially only the first term of the series be used, and that one additional term be included at each successive iteration.

Harter [47] used this, among other methods, in calculating his tables.

Gray *et al.* [40] have obtained the relatively simple approximation:

$$\frac{1}{x^{\alpha-1}e^{-x}}\int_x^{\infty}t^{\alpha-1}e^{-t}\,dt \doteqdot \frac{x}{x-\alpha+1}\left[1-\frac{\alpha-1}{(x-\alpha+1)^2+2x}\right]$$

which gives good results when x is sufficiently large. Gray *et al.* combine this formula with Stirling's approximation (Equation (31) of Chapter 1) to yield:

$$\Pr[\chi_\nu^2 > x] \doteqdot \frac{\nu^{\frac{3}{2}}e^{\frac{1}{2}\nu}}{(\nu+\frac{1}{6})\sqrt{\pi}}\left(\frac{x}{\nu}\right)^{\frac{1}{2}\nu}e^{-\frac{1}{2}x}\left[1-\frac{2(\nu-2)}{(x-\nu+2)^2+4\nu}\right].$$

This gives an accuracy of 3 decimal places when $\Pr[\chi_\nu^2 > x]$ is of order 0.1, even for ν as small as 2.

Wallace [121] has obtained definite bounds for the probability integral of the χ^2 distribution and for the corresponding equivalent unit normal deviate. He has shown that (for $x > n$).

$$d_\nu[1 - \Phi(w(x))] > \Pr[\chi_\nu^2 > x] > d_\nu e^{(9\nu)^{-1}}[1 - \Phi(w_2(x))] \tag{31}$$

where

$$d_\nu = (\tfrac{1}{2}\nu)^{\frac{1}{2}(\nu-1)}e^{-\frac{1}{2}\nu}\sqrt{2\pi}/\Gamma(\tfrac{1}{2}\nu)$$
$$w(x) = [x - \nu - \nu\log(x/\nu)]^{\frac{1}{2}}$$
$$w_2(x) = w(x) + \tfrac{1}{3}(2/\nu)^{\frac{1}{2}}.$$

Note that (from Stirling's formula) d_ν is very nearly equal to 1. Defining the equivalent normal deviate $u(x)$ by the equation

$$\Pr[\chi_\nu^2 > x] = 1 - \Phi(u(x)),$$

Wallace also showed that

$$w(x) \leq u(x) \leq w_2(x) + [w_2(x)]^{-1}\cdot max\,(0, d_\nu^{-1}e^{-(9\nu)^{-1}} - 1) \tag{32}$$

(the lower limit being valid for $\nu > 0.37$ and the *max* can be replaced by $0.6\nu^{\frac{1}{2}}$).

While (32) is less useful than the Wilson-Hilferty approximation for most of the distribution, it appears that (32), and especially the upper limit, give more useful approximations in the extreme upper tail.

Cornish **[a]*** reported higher accuracy for Hill's expansion **[b]**† of $u(x)$ in terms of $w(x)$ and $a_\nu = (2/\nu)^{\frac{1}{2}}$,

$$u(x) \doteqdot w + \tfrac{1}{3}a_\nu - \tfrac{1}{36}wa_\nu^2 - \tfrac{1}{1620}(w^2 - 13)a_\nu^3 + \tfrac{7}{38880}(6w^3 + 17w)a_\nu^4 + \cdots, \tag{33.1}$$

which gives $\Phi(u(x))$ to 5 decimal places for $\nu > 20$ or 8 decimal places with terms to a_ν^8. Moreover, Hill's expansion of Cornish-Fisher type:

(33.2)

$$\chi_{\nu,\epsilon}^2 = (\nu - \tfrac{2}{3})\exp\{U_\epsilon c_\nu - \tfrac{1}{6}U_\epsilon^2 + \tfrac{1}{36}(U_\epsilon^3 - U_\epsilon)c_\nu^{-1} - \tfrac{1}{1620}(6U_\epsilon^4 - 31U_\epsilon^2 - 32)c_\nu^{-2} + \tfrac{1}{38880}(9U_\epsilon^5 - 308U_\epsilon^3 - 481U_\epsilon)c_\nu^{-3}\},$$

where $c_\nu = [\frac{1}{2}\nu - \frac{1}{3}]^{\frac{1}{2}}$, is much more accurate than (29) for ϵ near to zero or 1.

If Y has the standard rectangular distribution (Chapter 25, Section 1)

$$p_Y(y) = 1 \qquad (0 < y < 1) \tag{33}$$

then $(-2\log Y)$ is distributed as χ^2 with 2 degrees of freedom. If Y_1,

*[a] Cornish, E. A. Fisher Memorial Lecture (1969), *37th I.S.I. Session, London.*

†[b] Hill, G. W. "Progress results on asymptotic approximations for Student's t and chi-squared" Personal communication (1969).

$Y_2, \ldots, Y_s$ each have distribution (33) and are independent then $\sum_{j=1}^{s} (-2 \log Y_j)$ is distributed as χ^2_{2s}; that is, it has a gamma distribution with $\alpha = s$, $\beta = 2$, $\gamma = 0$.

Using this relation, it is possible to generate gamma distributed variables from tables of random numbers. Extension to cases when α is not an integer can be effected by methods of the kind described by Bánkövi [4] (Chapter 24, Section 2).

If X has distribution (2), then the moment generating function of $\log X$ is

$$E[e^{t \log X}] = E[X^t] = \Gamma(\alpha + t)/\Gamma(\alpha).$$

Hence the rth cumulant of $\log X$ is

$$\kappa_r(\log X) = \psi^{(r-1)}(\alpha).$$

Note that for α large

$$\beta_1(\log X) \doteqdot \alpha^{-1}; \quad \beta_2(\log X) \doteqdot 3 + 2\alpha^{-1}$$

which may be compared with

$$\beta_1(X) = 4\alpha^{-1}; \quad \beta_2(X) = 3 + 6\alpha^{-1}.$$

The distribution of $\log X$ is more nearly normal than the distribution of X. Although this approximation is not generally used directly, it is often very useful when approximating the distributions of functions of independent gamma variables.

For example, suppose $X_1, X_2, \ldots, X_k$ are independent variables each distributed as χ^2 with ν degrees of freedom. Then the distribution of

$$R = \frac{\max(X_1, X_2, \ldots, X_k)}{\min(X_1, X_2, \ldots, X_k)}$$

may be approximated by noting that

$$\log R = \max(\log X_1, \ldots, \log X_k) - \min(\log X_1, \ldots, \log X_k)$$

is approximately distributed as the range of k independent normal variables each having the same expected value, and standard deviation

$$\sqrt{\psi^{(1)}(\tfrac{1}{2}\nu)} \doteqdot \sqrt{\frac{2}{\nu - 1}}.$$

6. Characterizations

If X_1 and X_2 are independent standard gamma random variables (i.e. having distributions of form (2), possibly with different values of α;α_1,α_2 say) then the random variables

$$(X_1 + X_2) \text{ and } X_1/(X_1 + X_2)$$

are mutually independent. (Their distributions are, respectively, a standard gamma with $\alpha = \alpha_1 + \alpha_2$, and a standard beta (Chapter 24) with parameters α_1, α_2.)

Lukacs [82] showed that this property characterizes the gamma distribution, in that if X_1 and X_2 are independent positive random variables, and $(X_1 + X_2)$ and $X_1/(X_1 + X_2)$ are mutually independent, then X_1 and X_2 must each have gamma distributions of form (1) with $\gamma = 0$, common β, but possibly different values of α.

If it be assumed that X_1 and X_2 have finite second moments and identical distributions it is sufficient to require that the regression function

$$E[(a_{11}X_1^2 + 2a_{12}X_1X_2 + a_{22}X_2^2)/(X_1 + X_2)^2 \mid (X_1 + X_2)]. \qquad (a_{11} + a_{22} \neq 2a_{12})$$

be independent of $(X_1 + X_2)$ to ensure that the common distribution is a gamma distribution (with $\gamma = 0$) (Laha [76]).

On the other hand, the distribution of X_1/X_2 is not sufficient to establish that each $|X_j|$ has a gamma distribution. If

$$p_{X_j}(x) = \{\Gamma(\alpha_j)\}^{-1}x^{\alpha_j-1}e^{-x} \qquad (x > 0; \quad \alpha_j > 0, j = 1,2)$$

then the probability density function of $G = X_1/X_2$ is

$$p_G(x) = [B(\alpha_1,\alpha_2)]^{-1}x^{\alpha_1-1}(1 + x)^{-(\alpha_1+\alpha_2)} \qquad (x > 0; \quad G > 0), \tag{34}$$

which is a Pearson Type VI distribution (see Chapter 12 and also Chapter 26). However it is possible for X_1 and X_2 to be independent, identically distributed positive random variables, and for $G = X_1/X_2$ to have distribution (34), without each X_j having a gamma distribution (Laha [77], Mauldon [85], Kotlarski [69] [70]). On the other hand Kotlarski [72] showed that the joint distribution of ratios X_1/X_3, X_2/X_3 (in a similar situation) does characterize the distribution (up to a constant multiplier).

It follows that any result depending *only* on the distribution (34) of the ratio X_1/X_2 cannot characterize the distribution of each X_j. In particular, it can be shown that if X_1 and X_2 are independent and identically distributed as in (1), then

$$\sqrt{\tfrac{1}{2}a}(\sqrt{X_1/X_2} - \sqrt{X_2/X_1})$$

has a t_{2a} distribution (as defined in Chapter 27), but this property is not sufficient to establish the form of the common distribution of X_1 and X_2 (given they are positive, independent and identically distributed).

However, if X_3 is a third random variable (with the same properties relative to X_1 and X_2) then the joint distribution of

$$\sqrt{\tfrac{1}{2}a}(\sqrt{X_1/X_2} - \sqrt{X_2/X_1}) \quad \text{and} \quad \sqrt{\tfrac{1}{2}a}(\sqrt{X_1/X_3} - \sqrt{X_3/X_1})$$

is sufficient to establish that common distribution is a gamma distribution with $\gamma = 0$. (Kotlarski [71].)

Khatri and Rao [67] have obtained the following characterizations of the gamma distribution, based on constancy of various regression functions:

(i) If $X_1, X_2, \ldots, X_n$ $(n \geq 3)$ are independent positive random variables and

$$Y_1 = \sum_{i=1}^{n} b_{1i}X_i \qquad (b_{1i} \neq 0,\ i = 1,2,\ldots,n)$$

$$Y_j = \prod_{i=1}^{n} X_i^{b_{ji}} \qquad (j = 2,\ldots,n)$$

with the $(n-1) \times n$ matrix (b_{ji}) $(j = 2,\ldots,n;\ \ i = 1,2,\ldots,n)$ non-singular, then the constancy of

$$E[Y_1 \mid Y_2,\ldots,Y_n]$$

ensures that the X's must have a common gamma distribution (unless they have zero variances).

Putting $b_{11} = b_{12} = \cdots = b_{1n} = 1$, and $b_{j,j-1} = -1;\ b_{j,j} = 1$, with all other b's zero, the condition becomes the constancy of

$$E\left[\sum_{j=1}^{n} X_j \,\middle|\, X_2/X_1, X_3/X_2, \ldots, X_n/X_1\right].$$

(ii) In the conditions of (i), if $E[X_j^{-1}] \neq 0$ $(j = 1,2,\ldots,n)$ and

$$Z_1 = \sum_{i=1}^{n} b_{1i}X_i^{-1}$$

$$Z_j = \sum_{i=1}^{n} b_{ji}X_j \qquad (j = 2,\ldots,n)$$

with the b's satisfying the same conditions as in (i), then the constancy of

$$E[Z_1 \mid Z_2, Z_3, \ldots, Z_n]$$

ensures that each X_j has a gamma distribution (not necessarily the *same* for all j)— unless they have zero variances. Choosing special values of b's as in (i) we obtain the condition that $E\left[\sum_j X_j^{-1} \,\middle|\, X_2 - X_1, \ldots, X_n - X_1\right]$ should be constant.

(iii) Under the same conditions as in (i), if $E[X_1 \log X_1]$ is finite then the constancy of

$$E\left[\sum_{j=1}^{n} a_j X_j \,\middle|\, \prod_{i=1}^{n} X_i^{b_i}\right]$$

with $\sum_{j=1}^{n} a_j b_j = 0$; $|b_n| > \max(|b_1|, |b_2|, \ldots, |b_{n-1}|)$ and $a_j b_j / a_n b_n < 0$ for all $j = 1, 2, \ldots, n-1$ ensures that X_1 has a gamma distribution (unless it has zero variance).

As a special case, putting $a_1 = a_2 = \cdots = a_n = 1$; $b_n = n - 1$; $b_1 = b_2 = \cdots = b_{n-1} = -1$, we obtain the condition as the constancy of

$$E\left[\sum_{j=1}^{n} X_j \,\middle|\, X_n^{n-1}\left\{\prod_{j=1}^{n-1} X_j\right\}^{-1}\right].$$

(iv) If $X_1, \ldots, X_n$ are independent, positive, *and identically distributed* random variables then if $E[X_i^{-1}] \neq 0$ $(i = 1,2,\ldots,n)$ and

$$E\left[\sum_{j=1}^{n} a_j X_j^{-1} \,\middle|\, \sum_{j=1}^{n} b_j X_j\right]$$

is constant, with the same conditions on the a's and b's as in (iii), the common distribution of the X's is a gamma distribution (unless it has a zero variance).

Giving the a's and b's the same special values as in (iii) we obtain the condition of constancy of

$$E\left[\sum_{j=1}^{n} X_j^{-1} \,\middle|\, X_n - \bar{X}\right]$$

where $\bar{X} = n^{-1} \sum_{j=1}^{n} X_j$.

Khatri and Rao [67] have obtained a number of further conditions characterizing gamma distributions.

If N is a random variable defined by

$$N = 0 \quad \text{if} \quad X_1 > x$$

and

$$N = n \quad \text{if} \quad X_1 + \cdots + X_n \leq x < X_1 + \cdots + X_{n+1}$$

where $X_1, X_2, \ldots, X_n$ are independent random variables each having distribution (1) with α an integer and with $\gamma = 0$ (*Erlang* distribution) then N has the generalized Poisson distribution (Chapter 4, Section 10)

$$\Pr[N \leq n] = e^{-x/\beta} \sum_{j=0}^{(n+1)\alpha-1} \{(x/\beta)^j/j!\}\,.$$

Nabeya [88] showed that for $\alpha = 1$, the converse is true, i.e. this is a characterization of the common distribution of $X_1, \ldots, X_n$ (*exponential*, in this case) given that they are positive, independent, identically distributed and continuous. Goodman [39] extended the result to apply to any positive integer value of $\alpha \leq 2$, thus providing a characterization of a gamma distribution.

7. Estimation

7.1 *Three Parameters Unknown*

We will first consider estimation for the three parameter distribution (1), (although in many cases it is possible to assume γ is zero, and estimate only α and β in (1)).

Given values of n independent random variables $X_1, X_2, \ldots, X_n$, each distributed as in (1), the equations satisfied by the maximum likelihood estimators $\hat{\alpha}, \hat{\beta}, \hat{\gamma}$ of α, β, γ respectively are

$$\sum_{j=1}^{n} \log (X_j - \hat{\gamma}) - n \log \hat{\beta} - n\psi(\hat{\alpha}) = 0 \tag{34.1$'$}$$

$$\sum_{j=1}^{n} (X_j - \hat{\gamma}) - n\hat{\alpha}\hat{\beta} = 0 \tag{34.2$'$}$$

$$-\sum_{j=1}^{n} (X_j - \hat{\gamma})^{-1} + n\{\hat{\beta}(\hat{\alpha} - 1)\}^{-1} = 0. \tag{34.3$'$}$$

From (34.3)$'$ it can be seen that if $\hat{\alpha}$ is less than 1, then some X_j's must be less than $\hat{\gamma}$. This is anomalous, since for $x < \gamma$ the probability density function (1) is zero. It is also clear that equations (34)$'$ will give rather unstable results if $\hat{\alpha}$ is near to 1, even though it exceeds 1. It is best, therefore, not to use these equations unless it is expected that $\hat{\alpha}$ is at least 2.5, say.

It is possible to solve equations (34)$'$ by iterative methods. A convenient (but not the only) method is to use (34.1)$'$ to determine a new value for $\hat{\beta}$, given $\hat{\alpha}$ and $\hat{\gamma}$; then (34.2)$'$ for a new $\hat{\gamma}$, given $\hat{\alpha}$ and $\hat{\beta}$, and (34.3)$'$ for a new $\hat{\alpha}$, given $\hat{\beta}$ and $\hat{\gamma}$.

The asymptotic variance-covariance matrix of $\sqrt{n}\,\hat{\alpha}$, $\sqrt{n}\,\hat{\beta}$ and $\sqrt{n}\,\hat{\gamma}$ is the inverse of the matrix

$$\begin{pmatrix} \psi'(\alpha) & \beta^{-1} & \beta^{-1}(\alpha-1)^{-1} \\ \beta^{-1} & \alpha\beta^{-2} & \beta^{-2} \\ \beta^{-1}(\alpha-1)^{-1} & \beta^{-2} & \beta^{-2}(\alpha-2)^{-1} \end{pmatrix}$$

The determinant of this matrix is

$$\beta^4\left[\frac{2\psi'(\alpha)}{\alpha-2} - \frac{2\alpha-3}{(\alpha-1)^2(\alpha-2)}\right].$$

Hence

$$\text{var}(\hat{\alpha}) \doteqdot 2n^{-1}[2\psi'(\alpha) - (2\alpha-3)(\alpha-1)^{-2}]^{-1} \tag{35.1}$$

$$\text{var}(\hat{\beta}) \doteqdot n^{-1}\beta^2[(\alpha-1)^2\psi'(\alpha) - \alpha + 2][2(\alpha-1)^2\psi'(\alpha) - 2\alpha + 3]^{-1} \tag{35.2}$$

$$\text{var}(\hat{\gamma}) \doteqdot n^{-1}\beta^2(\alpha-2)(\alpha\psi'(\alpha) - 1)[2\psi'(\alpha) - (2\alpha-3)(\alpha-1)^{-2}]^{-1}. \tag{35.3}$$

Using the approximation

$$\psi'(\alpha) \doteqdot \alpha^{-1} + \tfrac{1}{2}\alpha^{-2} + \tfrac{1}{6}\alpha^{-3}, \tag{36}$$

we obtain the simple formulas

$$\text{var}(\hat{\alpha}) \doteqdot 6n^{-1}\alpha^3 \tag{35.1$'$}$$

$$\text{var}(\hat{\beta}) \doteqdot 3n^{-1}\beta^2\alpha \tag{35.2$'$}$$

$$\text{var}(\hat{\gamma}) \doteqdot \tfrac{3}{2}n^{-1}\beta^2\alpha^3, \tag{35.3$'$}$$

giving the orders of magnitude of the variances when α is large.

Fisher [35] obtained the more precise approximation:

$$\text{var}(\hat{\alpha}) \doteqdot 6n^{-1}[(\alpha - 1)^3 + \tfrac{1}{5}(\alpha - 1)] \tag{37}$$

by using more terms in the expansion (36).

If the method of moments is used to estimate α, β, and γ, the following simple formulas are obtained:

$$\tilde{\gamma} + \tilde{\alpha}\tilde{\beta} = \overline{X} \tag{38.1}$$

$$\tilde{\alpha}\tilde{\beta}^2 = m_2 \tag{38.2}$$

$$2\tilde{\alpha}\tilde{\beta}^3 = m_3, \tag{38.3}$$

where

$$\overline{X} = n^{-1} \sum_{j=1}^{n} X_j; \quad m_2 = n^{-1} \sum_{j=1}^{n} (X_j - \overline{X})^2; \quad m_3 = n^{-1} \sum_{j=1}^{n} (X_j - \overline{X})^3$$

are the sample central, second and third moments. (Since this method would be used only when n is rather large, there is no need to attempt to make the estimators unbiased; it is not clear, also, whether this would improve the accuracy of estimation.) Note that (38.1) and (34.2)′ are identical.

From equations (38) the following formulas for the *moment estimators* $\tilde{\alpha}$, $\tilde{\beta}$, $\tilde{\gamma}$ are obtained:

$$\tilde{\alpha} = 4m_2^3/m_3^2 = 4/b_1 \qquad (\text{where } \sqrt{b_1} = m_3/m_2^{\frac{3}{2}}) \tag{39.1}$$

$$\tilde{\beta} = \tfrac{1}{2}m_3/m_2 \tag{39.2}$$

$$\tilde{\gamma} = \overline{X} - 2m_2^2/m_3. \tag{39.3}$$

Although these are simple formulas, the estimators are often, unfortunately, considerably less accurate than the maximum likelihood estimators $\hat{\alpha}$, $\hat{\beta}$, $\hat{\gamma}$.

It can be shown that if n and α are large

$$\text{var}(\tilde{\alpha}) \doteqdot 6\alpha(\alpha + 1)(\alpha + 5)n^{-1} \tag{40}$$

(Fisher [35]).

Comparing (37) and (40) it can be seen that the ratio of approximate values $\text{var}(\hat{\alpha})/\text{var}(\tilde{\alpha})$ is substantially less than 1 unless α is rather large. The ratio

$$\{(\alpha - 1)^3 + \tfrac{1}{5}(\alpha - 1)\}/\{\alpha(\alpha + 1)(\alpha + 5)\}$$

increases with α and reaches the value 0.8 at $\alpha = 39.1$.

On the other hand, we have already noted that when α is less than 2.5, the maximum likelihood estimators are of doubtful utility. It then becomes necessary to consider yet other methods of estimation.

When α is less than 1, the distribution is shaped like a reversed J, with the probability density function tending to infinity as x tends to γ (see Figure 1).

If n is large (as it usually is if a three-parameter distribution is being fitted) it is reasonable to estimate γ as the smallest observed value among $X_1, X_2, \ldots, X_n$, or a value slightly smaller than this. Estimation of α and β then proceeds as for the two parameter case, to be described later. Using the value of α so estimated, a new value for γ can be estimated, and so on.

7.2 *Two (or One) Parameters Unknown*

We now consider estimation when the value of one of the three parameters α, β and γ is known. The commonest situation is when the value of γ is known (usually it is zero). Occasionally α is known (at least approximately) but not β or γ. The third case, with β known but not α or γ, occurs rarely and we will not discuss it.

If γ is known, the maximum likelihood estimators of α and β might be denoted $\hat{\alpha}(\gamma)$, $\hat{\beta}(\gamma)$ to indicate their dependence on γ. We will, however, simply use $\hat{\alpha}$ and $\hat{\beta}$; no confusion between this use and that in Section 7.1 should arise.

If γ is known to be zero, the probability density function is of form (22). If $X_1, X_2, \ldots, X_n$ are independent random variables each having distribution (22), then equations for the maximum likelihood estimators $\hat{\alpha}$, $\hat{\beta}$ are

$$n^{-1} \sum_{j=1}^{n} \log X_j = \log \hat{\beta} + \psi(\hat{\alpha}), \tag{41.1}$$

$$\bar{X} = \hat{\alpha}\hat{\beta}. \tag{41.2}$$

From (41.2), $\hat{\beta} = \bar{X}/\hat{\alpha}$. Inserting this in (41.1) we obtain the following equation for $\hat{\alpha}$:

$$n^{-1} \sum_{j=1}^{n} \log X_j - \log \bar{X} = \psi(\hat{\alpha}) - \log \hat{\alpha}, \tag{41.3}$$

that is,

$$\log \left[\frac{\text{arithmetic mean } (X_1, X_2, \ldots, X_n)}{\text{geometric mean } (X_1, X_2, \ldots, X_n)} \right] = \log \hat{\alpha} - \psi(\hat{\alpha}).$$

The value of $\hat{\alpha}$ can be determined by inverse interpolation in a table of the function $[\log \alpha - \psi(\alpha)]$. Such a table has been published by Masuyama and Kuroiwa [84]. Chapman [23] has published a table giving the results of such inverse interpolation (i.e. values of $\hat{\alpha}$) corresponding to a few values of the ratio of arithmetic to geometric mean. (A more complete table is reported in [23] to be available from the Laboratory of Statistical Research, University of Washington.)

Greenwood and Durand [41] pointed out that the function $\alpha[\log \alpha - \psi(\alpha)]$ progresses much more smoothly than does $[\log \alpha - \psi(\alpha)]$ and so is more convenient for interpolation. They give a table of values of $\alpha[\log \alpha - \psi(\alpha)]$ *as a function of* $[\log \alpha - \psi(\alpha)]$ to eight decimal places for argument values 0.00(0.01)1.40; and to seven decimal places for argument values 1.4(0.2)18.0. This method eliminates the necessity of inverse interpolation and assures high accuracy using linear interpolation.

Wilk *et al.* [126] noted that the solution of equation (41.3) is very nearly a linear function of

$$H = \frac{\text{arithmetic mean}}{\text{arithmetic mean} - \text{geometric mean}},$$

except when $\hat{\alpha}$ is less than about 2. They give a table of solutions of equation (41.3) to 5 decimal places for

$$H = 1.000(0.001)1.010(0.002)1.030(0.005)1.080(0.01)1.16(0.02)1.40(0.05)$$
$$2.00(0.1)3.0(0.2)5.0(0.5)7.0(1)10(2)20(10)50.$$

For $H > 1.001$, linear interpolation gives four decimal place accuracy for $\hat{\alpha}$.

If $\hat{\alpha}$ is large enough, the approximation $\psi(\alpha) = \log(\alpha - \frac{1}{2})$ may be used. Then, from (41.3), we have

$$\frac{\text{arithmetic mean}}{\text{geometric mean}} \doteqdot \frac{\hat{\alpha}}{\hat{\alpha} - \frac{1}{2}},$$

that is

$$\hat{\alpha} \doteqdot \frac{\text{arithmetic mean}}{2(\text{arithmetic mean} - \text{geometric mean})} = \tfrac{1}{2}H.$$

For a better approximation $\frac{1}{12}(=0.08\dot{3})$ should be subtracted from the right hand side.

Thom [114] suggests the approximation $\hat{\alpha} \doteqdot \frac{1}{4}Y^{-1}(1 + \sqrt{1 + \frac{4}{3}Y})$ where

$$Y = \log \frac{\text{arithmetic mean}}{\text{geometric mean}}.$$

Thom further suggests adding the correction $[(\hat{\alpha} - 1)(24 - 96\hat{\alpha})^{-1} + 0.0092]$ if $\hat{\alpha} > 0.9$; and gives a table of corrections for $\hat{\alpha} < 0.9$. It is stated that with these corrections the value of $\hat{\alpha}$ should be correct to three decimal places.

Asymptotic formulas (as $n \to \infty$) for the variances of $\sqrt{n}\,\hat{\alpha}$ and $\sqrt{n}\,\hat{\beta}$, and the correlation between these statistics, are:

$$\text{(42)} \qquad \begin{cases} \text{var}(\sqrt{n}\,\hat{\alpha}) \doteqdot \alpha(\alpha\psi'(\alpha) - 1)^{-1} \\ \text{var}(\sqrt{n}\,\hat{\beta}) \doteqdot \beta^2\psi'(\alpha)(\alpha\psi'(\alpha) - 1)^{-1} \\ \text{corr}(\hat{\alpha},\hat{\beta}) \doteqdot -(\alpha\psi'(\alpha))^{-\frac{1}{2}}. \end{cases}$$

Masuyama and Kuroiwa [84] give tables with values of $\alpha(\alpha\psi'(\alpha) - 1)^{-1}$ and $\psi'(\alpha)(\alpha\psi'(\alpha) - 1)^{-1}$. If the approximation $\psi'(\alpha) \doteqdot (\alpha - \frac{1}{2})^{-1}$, useful for α large, is used, we have

$$\text{(43)} \qquad \begin{cases} \text{var}(\sqrt{n}\,\hat{\alpha}) \doteqdot 2\alpha(\alpha - \frac{1}{2}) \\ \text{var}(\sqrt{n}\,\hat{\beta}) \doteqdot \beta^2\alpha \\ \text{corr}(\hat{\alpha},\hat{\beta}) \doteqdot -\sqrt{1 - \frac{1}{2}\alpha^{-1}}. \end{cases}$$

Bowman and Shenton [17] have obtained expansions for the first few moments of $\hat{\alpha}$ and $\hat{\beta}$ up to terms in n^{-6}. As a simple good approximation to the expected value of $\hat{\alpha}$, they suggest (for $n \geq 4$; $\alpha \geq 1$)

$$\alpha + [3\alpha - \tfrac{2}{3} + \tfrac{1}{9}\alpha^{-1} + \tfrac{13}{405}\alpha^{-2}](n-3)^{-1}.$$

The error of this formula is stated to be less than 1.4%. An approximately unbiased estimator of α is

$$[(n-3)\hat{\alpha} + \tfrac{2}{3}]n^{-1}.$$

For $\frac{1}{2} \leq \alpha < 1$, the formula

$$E[\hat{\alpha}] \doteqdot \alpha + \frac{1.54777\alpha + 1.58102\alpha^2 - 0.67779\alpha^3}{n-3}$$

is suggested. The error of this formula is stated to be less than 4.3%.

These authors also investigated the approximate solutions suggested in [41], which are

(44.1)
$$\hat{\alpha} \doteqdot Y^{-1}(0.5000876 + 0.1648852Y - 0.0544274Y^2) \qquad (0 < Y \leq 0.5772)$$

(44.2)
$$\hat{\alpha} \doteqdot Y^{-1}(17.79728 + 11.968477Y + Y^2)^{-1}(8.898919 + 9.059950Y + 0.9775373Y^2) \qquad (0.5772 \leq Y \leq 17)$$

where

$$Y = \log \frac{\text{arithmetic mean}}{\text{geometric mean}}.$$

The error of (44.1) does not exceed 0.0088%, and that of (44.2) does not exceed 0.0054%.

If α is known but not β or γ, maximum likelihood estimators $\hat{\beta} = \hat{\beta}(\alpha)$, $\hat{\gamma} = \hat{\gamma}(\alpha)$ satisfy equations (34.2)′ and (34.3)′ with $\hat{\alpha}$ replaced by α. From (34.2)′:

$$\hat{\gamma} = \overline{X} - \alpha\hat{\beta}$$

and hence (34.3)′ can be written as an equation for β:

$$(\alpha - 1)\hat{\beta} = \left[n^{-1}\sum_{j=1}^{n}(X_j - \overline{X} + \alpha\hat{\beta})^{-1}\right]^{-1}. \tag{45}$$

Alternatively, using the first two sample moments, we have for moment estimators $\tilde{\beta} = \tilde{\beta}(\alpha)$ and $\tilde{\gamma} = \tilde{\gamma}(\alpha)$

$$\tilde{\gamma} = \overline{X} - \alpha\tilde{\beta}$$

$$\alpha\tilde{\beta}^2 = m_2 \qquad \text{(c.f. (38.1) and (38.2))},$$

whence

$$\tilde{\beta} = \sqrt{m_2/\alpha}; \qquad \tilde{\gamma} = \overline{X} - \sqrt{\alpha m_2}. \tag{46}$$

For this case (α known) for n large

$$\text{(47)} \qquad \begin{cases} \operatorname{var}(\hat{\beta}) \doteqdot \frac{1}{2}\beta^2 n^{-1} \\ \operatorname{var}(\tilde{\gamma}) \doteqdot \frac{1}{2}\beta^2\alpha(\alpha - 2)n^{-1} \\ \operatorname{corr}(\hat{\beta},\hat{\gamma}) \doteqdot -\sqrt{1 - 2\alpha^{-1}}, \end{cases}$$

while

$$\text{(48)} \qquad \begin{cases} \operatorname{var}(\tilde{\beta}) \doteqdot \frac{1}{2}\beta^2(1 + 3\alpha^{-1})n^{-1} \\ \operatorname{var}(\hat{\gamma}) \doteqdot \frac{1}{2}\beta^2\alpha(\alpha + 3)n^{-1} \\ \operatorname{corr}(\tilde{\beta},\hat{\gamma}) \doteqdot -(\alpha + 1)/(\alpha + 3). \end{cases}$$

The advantage of the maximum likelihood estimators is not so great in this case as when all three parameters have to be estimated.

If both γ and α are known, and it desired to estimate the scale parameter β, then both maximum likelihood and moment methods lead to the same estimator:

$$\hat{\beta} = \frac{\bar{X} - \gamma}{\alpha}.$$

Zubrzycki [130] has considered the case when β is known to exceed some positive number β_0. He has shown that, with a loss function $(\beta^* - \beta)^2/\beta^2$, where β^* denotes an estimator of β, then given a single observed value of X, estimators

$$\beta^* = (\alpha + 1)^{-1} X + b$$

with

$$\beta_0(\alpha + 1)^{-1} \le b \le 2\beta_0(\alpha + 1)^{-1}$$

have minimax risk (equal to $(\alpha + 1)^{-1}$) and are admissible in the class of estimators linear in X.

7.3 *Estimation Using Order Statistics*

A considerable amount of work has been done in evaluating the lower moments of order statistics $X'_1 \le X'_2 \le \cdots \le X'_n$ corresponding to sets of independent random variables $X_1, \ldots, X_n$ having a common standard gamma distribution of form (2). (Since γ and β are purely location and scale parameters, the results are easily extended to the general form (1).) Gupta ([43], also [42]) has given tables of the first four moments of all order statistics to six significant figures for sample sizes $n = 1(1)10$ and $\alpha = 1(1)5$; also of the least value (X'_1) for $n = 11(1)15$ and $\alpha = 1(1)5$. Breiter and Krishnaiah [20] have given values of these moments to five significant figures for all order statistics for $n = 1(1)16$ and $\alpha = 0.5(1)5.5(0.5)10.5$. Note that the two tables together cover $\alpha = 0.5(0.5)10.5$ for $n = 1(1)10$.

(Moments of order statistics of the exponential distribution ($\alpha = 1$) are discussed in more detail in Chapter 18, Section 6.)

We now take note that Kabe [62] has obtained a convenient formula for obtaining the characteristic function of any linear function $\sum_{j=1}^{n} a_j X'_j$ of the order statistics. The characteristic function is

$$E\left[\exp\left(it \sum_{j=1}^{n} a_j X'_j\right)\right] = n![\Gamma(\alpha)]^{-n} \int_0^{\infty} \cdots \int_0^{x_3} \int^{x_2} \left\{\prod_{j=1}^{n} x_j\right\}^{\alpha-1}$$
$$\times \exp\left\{-\sum_{j=1}^{n} (1 - ia_j t)x_j\right\} dx_1\, dx_2 \ldots dx_n.$$

Applying the transformation

$$x_r = \prod_{j=r}^{n} w_j$$

(so that $0 < w_j < 1$ for $j = 1,2,\ldots,(n-1)$, and $w_n > 0$) we obtain the formula

(49)

$$n![\Gamma(\alpha)]^{-n} \int_0^1 \int_0^1 \cdots \int_0^1 \int_0^{\infty} \left\{\prod_{j=1}^{n} w_j^{j\alpha-1}\right\} \exp\{-w_n D(w)]\, dw_n\, dw_{n-1} \ldots dw_1$$
$$= \frac{n!\Gamma(n\alpha)}{[\Gamma(\alpha)]^n} \int_0^1 \int_0^1 \cdots \int_0^1 \left\{\prod_{j=1}^{n-1} w_j^{j\alpha-1}\right\} [D(w)]^{-n\alpha}\, dw_{n-1} \ldots dw_1$$

where

$$D(w) = (1 - ia_n t) + w_{n-1}(1 - ia_{n-1}t) + w_{n-2}w_{n-1}(1 - ia_{n-2}t) + \cdots$$
$$+ w_1 w_2 \ldots w_{n-1}(1 - ia_1 t).$$

The multiple integral can be expanded as a series of beta functions. Although we will not use it here directly, equation (49) is very convenient as a starting point for studying the distributions of linear functions of order statistics from gamma distributions.

The distribution of X'_r, the rth smallest among n independent random variables each having distribution (2), has probability density function

(50)

$$p_{X'_r}(x) = \frac{n!}{(r-1)!(n-r)!} [\Gamma_x(\alpha)/\Gamma(\alpha)]^{r-1} [1 - \Gamma_x(\alpha)/\Gamma(\alpha)]^{n-r} x^{\alpha-1} e^{-x}/\Gamma(\alpha)$$
$$(x > 0).$$

In general this expression does not lend itself to simple analytic treatment. However, if α is a positive integer,

$$\frac{\Gamma_y(\alpha)}{\Gamma(\alpha)} = \frac{1}{\Gamma(\alpha)} \int_0^y t^{\alpha-1} e^{-t}\, dt = 1 - e^{-y} \sum_{j=0}^{\alpha-1} \{y^j/j!\} \qquad \text{(cf. (23.1))}$$

and so (50) becomes

$$p_{X'_r}(x) = \frac{n!}{(r-1)!(n-r)!}\left[1 - e^{-x}\sum_{j=0}^{\alpha-1}\{x^j/j!\}\right]^{r-1} \times \left[\sum_{j=0}^{\alpha-1}\{x^j/j!\}\right]^{n-r} x^{\alpha-1}e^{-(n-r+1)x}/(\alpha-1)! \quad (x > 0). \tag{51}$$

In this case it is possible to express all moments (of integer order) of X'_r, and all product moments (of integer orders) of order statistics as finite sums of terms involving factorials, although these expressions will usually be cumbersome. Johnson [60] has obtained an approximation to the distribution of range $(X'_n - X'_1)$ for Type III random samples.

Using Gupta's tables [43] it is possible to construct best linear unbiased estimators of the parameter β if α and γ are known. Coefficients of such estimators have been given by Musson [87]. Coefficients for best linear unbiased estimators, not using all of the sample values have been given by:

(i) Karns [64] using only one order statistic;

(ii) Bruce [22] using the least M values out of n;

(iii) Hill [56] using only the least *number* of order statistics (from a complete or censored sample) to give a specified efficiency relative to the best linear unbiased estimator using all available order statistics;

(iv) Särndal [107] using the best k order statistics. (Särndal also considers estimation of β and γ, α being known.)

Returning now to situations where it is necessary to estimate all three parameters α, β and γ, we consider maximum likelihood estimation when the least r_1 and greatest r_2 of the X's have been censored. The maximum likelihood equations are (introducing $\hat{Z}_j = (X'_j - \hat{\gamma})/\hat{\beta}$ for convenience):

$$\sum_{j=r_1+1}^{n-r_2} \log \hat{Z}_j - n\psi'(\hat{\alpha}) + \frac{\Gamma'(\hat{\alpha}) - \Gamma'_{\hat{Z}_{n-r_2}}(\hat{\alpha})}{\Gamma(\hat{\alpha}) - \Gamma_{\hat{Z}_{n-r_2}}(\hat{\alpha})} r_2 + \frac{\Gamma'_{\hat{Z}_{r_1+1}}(\hat{\alpha})}{\Gamma_{\hat{Z}_{r_1+1}}(\hat{\alpha})} r_1 = 0; \tag{52.1}$$

(52.2)

$$-(n - r_1 - r_2)\hat{\alpha} + \sum_{j=r_1+1}^{n-r_2} \hat{Z}_j + \frac{\hat{Z}_{n-r_2}^{\hat{\alpha}} e^{-\hat{Z}_{n-r_2}}}{\Gamma(\hat{\alpha}) - \Gamma_{\hat{Z}_{n-r_2}}(\hat{\alpha})} r_2 - \frac{\hat{Z}_{r_1+1}^{\hat{\alpha}} e^{\hat{Z}-r_1+1}}{\Gamma_{\hat{Z}_{r_1+1}}(\hat{\alpha})} r_1 = 0$$

$$-(\hat{\alpha} - 1)\sum_{j=r_1+1}^{n-r_2} \hat{Z}_j^{-1} + (n - r_1 - r_2) + \frac{\hat{Z}_{n-r_2}^{\hat{\alpha}-1} e^{-\hat{Z}_{n-r_2}}}{\Gamma(\hat{\alpha}) - \Gamma_{\hat{Z}_{n-r_2}}(\hat{\alpha})} r_2 - \frac{\hat{Z}_{r_1+1}^{\hat{\alpha}-1} e^{-\hat{Z}_{r_1+1}}}{\Gamma_{\hat{Z}_{r_1+1}}(\hat{\alpha})} r_1 = 0. \tag{52.3}$$

The equations simplify if either $r_1 = 0$ or $r_2 = 0$. For the case $r_1 = 0$ (censoring from above) a method of solving the equations is given by Harter and Moore [51].

Estimation is also simplified if the value of γ is known. Without loss of generality it may be arranged (if γ is known) to make $\gamma = 0$ (by adding, if necessary, a suitable constant to each observed value). For this case, with data censored from above, Wilk *et al.* [126] have provided tables which considerably facilitate solution of the maximum likelihood equations. They express these equations in terms of

(53)
$$P = \left(\prod_{j=1}^{n-r_2} X'_j\right)^{1/(n-r_2)} \Big/ X'_{n-r_2} \quad \text{and}$$

$$S = \left(\sum_{j=1}^{n-r_2} X'_j\right) \Big/ ((n - r_2)X'_{n-r_2});$$

that is, the ratios of the geometric and arithmetic means of the available observed values to their maximum. The maximum likelihood equations for $\hat{\alpha}$ and $\hat{\beta}$ are

(54.1)
$$(n - r_2)\log P = n[\psi'(\hat{\alpha}) - \log (X'_{n-r_2}/\hat{\beta})] - r_2\partial\log J(\hat{\alpha})/\partial\hat{\alpha}$$

(54.2)
$$SX'_{n-r_2}/\hat{\beta} = \hat{\alpha} - \{r_2/(n - r_2)\}e^{-X'_{n-r_2}/\hat{\beta}}/J(\hat{\alpha})$$

where

$$J(\hat{\alpha}) = \int_1^\infty t^{\hat{\alpha}-1}e^{-X'_{n-r_2}t/\hat{\beta}}\, dt.$$

Note that r_2 and n enter the equations only in terms of the ratio r_2/n and X'_{n-r_2} and $\hat{\beta}$ only as the ratio $X'_{n-r_2}/\hat{\beta}$. Wilk *et al.* [126] provide tables giving $\hat{\alpha}$ and $\hat{\mu} = \hat{\alpha}\hat{\beta}/X'_{n-r_2}$ to three decimal places for

$$n/r_2 = 1.0,\ 1.1,\ 1.2(0.2)2.0,\ 2.3,\ 2.6,\ 3.0$$
$$P = 0.04(0.04)1.00 \quad \text{and} \quad S = 0.08(0.04)1.00.$$

The values for $n/r_2 = 1$ correspond, of course, to uncensored samples. A special table, which we have already mentioned in Section 7.2, is provided for this case.

Wilk *et al.* [127] discuss generalizations and modifications of these techniques for estimation of an unknown common scale parameter based on order statistics from a sample of gamma random variables with *known* shape parameters not necessarily all equal.

If α is known, it is possible to use "gamma probability paper," as described by Wilk *et al.* [125] to estimate β and γ graphically.

This entails plotting the observed order statistics against the corresponding expected values for the standard distribution (2) (which, of course, depends on α) or if these are not available, the values ξ_j satisfying the equations

$$j/(n + 1) = [\Gamma(\alpha)]^{-1} \int_0^{\xi_j} x^{\alpha-1}e^{-x}\, dx.$$

8. Related Distributions

The relationship between the chi-square and Poisson distributions has been mentioned several times; the latest is in Sections 4 and 5 of this chapter.

We also recall that if Y has the standard uniform (rectangular) distribution

$$p_Y(y) = 1 \qquad (0 \le y \le 1)$$

then $Z = -\log Y$ has the exponential distribution

$$p_Z(z) = e^{-z} \qquad (0 \le z),$$

which is a special form of gamma distribution. If $Y_1, Y_2, \ldots, Y_k$ are independent random variables each distributed as Y, and $Z_j = -\log Y_j$ $(j = 1,\ldots,k)$, then $Z_{(k)} = \sum_{j=1}^{k} Z_j$ has a gamma distribution with parameters $\alpha = k, \beta = 1, \gamma = 0$. ($2Z_{(k)}$ is distributed as χ^2_{2k}; see Section 5).

Relationships between gamma and beta distributions are described in Chapter 24 (see also Section 6 of this chapter).

Apart from noting these interesting relationships, we will devote this section to an account of four other classes of distributions which are related to gamma distributions:

(i) truncated gamma distributions,
(ii) compound gamma distributions,
(iii) transformed gamma distributions,
(iv) 'generalized' gamma distributions as defined by Stacy [111]; these are distributions of variables X such that $[(X - \gamma)/\beta]^c$ has a standard gamma distribution of form (2). (If the latter is an exponential distribution, then we have, of course, a Weibull distribution (Chapter 20).)

8.1 *Truncated Gamma Distributions*

The most common form of truncation of gamma distributions, when used in life-testing situations, is truncation from above. This is omission of values exceeding a fixed number τ, which is usually (though not always) known. If τ is not known, and the distribution before truncation is of the general form (1) there are four parameters $(\alpha,\beta,\gamma,\tau)$ to estimate, and technical problems become formidable. However, it is not difficult to construct fairly simple (but quite likely not very accurate) formulas for estimating these parameters.

Fortunately, it is often possible to assume that γ is zero in these situations (see, e.g., Parr and Webster [91] for examples), and we will restrict ourselves to this case. We will suppose that we have observations which can be regarded as observed values of independent random variables $X_1, X_2, \ldots, X_n$, each having the probability density function

$$\frac{x^{\alpha-1}e^{-x/\beta}}{\int_0^\tau t\alpha^{-1}e^{-t/\beta}dt} \qquad (0 \le x \le \tau). \tag{55}$$

Estimation of the parameters α and β has been discussed by Chapman [23], Cohen [25] [26], Das [29], Des Raj [30] and Iyer and Singh [57].

The moments of distribution (55) are conveniently expressed in terms of incomplete gamma functions

(56) $$\mu_r'(X) = \beta\Gamma_{\tau 1\beta}(\alpha + r)/\Gamma_{\tau 1\beta}(\alpha)$$

or, in terms of Pearson's incomplete gamma function

(56)′ $$\mu_r'(X) = \alpha^{(r)} \frac{I(\tau(\alpha + r)^{-\frac{1}{2}},\alpha + r - 1)}{I(\tau\alpha^{-\frac{1}{2}},\alpha - 1)}.$$

8.2 *Compound Gamma Distributions*

Starting from (1), compound gamma distributions can be constructed by assigning joint distributions to α, β and γ.

The great majority of such distributions which are used in applied work start from (2) (i.e., with $\gamma = 0$) and assign a distribution to one of α and β (usually β).

If β^{-1} be supposed, itself, to have a standard gamma distribution with

(57) $$p_{\beta^{-1}}(x) = \frac{b^\delta x^{\delta-1}e^{-xb}}{\Gamma(\delta)} \qquad (0 \le x),$$

the resulting compound distribution has probability density function

(58) $$p_X(x) = \frac{\Gamma(\alpha + \delta)}{\Gamma(\alpha)\Gamma(\delta)} x^{\alpha-1}(x + b)^{-(\alpha+\delta)} \qquad (0 \le x).$$

This belongs to Type VI of Pearson's system (Chapter 12, Section 4). This result can be expressed formally

$$\text{Gamma } (\alpha,\beta,0) \underset{\beta^{-1}}{\wedge} \text{Gamma } (\delta,b^{-1},0) \equiv \text{Type V}.$$

The non-central χ^2 distribution (Chapter 28) is a mixture of gamma distributions, each with β equal to 2, and α distributed as $(v + 2j)$ where j is a Poisson variable. Formally

$$\text{Gamma } (\alpha,2,0) \underset{\frac{1}{2}(\alpha-\nu)}{\wedge} \text{Poisson } (\tfrac{1}{2}\lambda) \equiv \chi_\nu'^2(\lambda)$$

or equivalently

$$\chi_\alpha^2 \underset{\frac{1}{2}(\alpha-\nu)}{\wedge} \text{Poisson } (\tfrac{1}{2}\lambda) \equiv \chi_\nu'^2(\lambda).$$

The Planck distributions (Chapter 33, Section 5) are also mixtures of gamma distributions. Bhattacharya [11] has considered the distribution

$$\text{Gamma } (\alpha,\beta,0) \underset{\beta}{\wedge} \text{Gamma } (\alpha',\beta',0).$$

This is considerably more complicated than when β^{-1} has a gamma distribution. The probability density function is

$$2[\beta'\Gamma(\alpha)\Gamma(\alpha')]^{-1}(x/\beta')^{\frac{1}{2}(\alpha+\alpha')-1}K_{\alpha'-\alpha}(2\sqrt{x/\beta'}) \qquad (0 < x)$$

where $K_{\alpha'-\alpha}(2\sqrt{x/\beta'})$ is the modified Bessel function of the third kind, of order $\alpha' - \alpha$.

8.3 *Transformed Gamma Distributions*

The various approximately normalizing transformations of the chi-square distribution described in Section 5 ($\sqrt{\chi_\nu^2}$, $(\chi_\nu^2)^{\frac{1}{3}}$, $\log \chi_\nu^2$) are particular cases of transformation of gamma distribution. Olshen [90], in 1937, published an account of a systematic investigation into the distribution of $\log X$ when X has a standard gamma distribution (2).

The moment generating function of $\log X$ is

$$E[e^{t \log X}] = E[X^t] = \Gamma(\alpha + t)/\Gamma(\alpha), \tag{59}$$

and the cumulant generating function is

$$\log \Gamma(\alpha + t) - \log \Gamma(\alpha).$$

Hence

$$\kappa_r(\log X) = \psi^{(r-1)}(\alpha). \tag{60}$$

Introducing the approximation

$$\psi^{(s)}(\alpha) \doteqdot (-1)^{s-1}(s-1)!(\alpha - \tfrac{1}{2})^{-s} \qquad (s \geq 1)$$

we see that the shape factors of $\log Y$ are approximately

$$\begin{aligned} \alpha_3(\log X) &\doteqdot -(\alpha - \tfrac{1}{2})^{-\frac{1}{2}} \\ \alpha_4(\log X) &\doteqdot 3 + 2(\alpha - \tfrac{1}{2})^{-1}. \end{aligned} \tag{61}$$

By comparison with (14) it can be seen that these are nearer to the 'normal values' 0 and 3 than are the moment ratios of the original distribution of X. Thus, the approximation "$\log X$ is normally distributed with expected value $\psi(\alpha)$ and variance $\psi'(\alpha)$" is likely to be fairly accurate for α sufficiently large. The accuracy of this approximation has been studied by Bartlett and Kendall [5].

If the approximation is accepted, it provides an approximation to the distribution of

$$R = \max(X_1, \ldots, X_n)/\min(X_1, \ldots, X_n)$$

where the X_j's are independent chi squared random variables with ν degrees of freedom. For

$$\begin{aligned} \log R &= \max(\log X_1, \ldots, \log X_n) - \min(\log X_1, \ldots, \log X_n) \\ &= \text{range}(\log X_1, \ldots, \log X_n) \end{aligned}$$

is appropriately distributed as

$$\sqrt{\psi'(\tfrac{1}{2}\nu)} \cdot (\text{range of } n \text{ independent unit normal variables}).$$

The distribution of $\sqrt{\chi_\nu^2}$ (χ_ν *distribution or chi-distribution with* ν *degrees of freedom*) has probability density function

(62) $$p_{\chi_\nu}(x) = [2^{\frac{1}{2}\nu-1}\Gamma(\tfrac{1}{2}\nu)]^{-1}x^{\nu-1}e^{-\frac{1}{2}x^2} \qquad (x \geq 0).$$

For the case $\nu = 2$, this is sometimes called the *Rayleigh* distribution (this terminology is especially favored by engineers). Archer [3] and Siddiqui [108] give a useful summary of properties. Moments of χ_ν are easily calculated from the formula

(63) $$\mu_r'(\chi_\nu) = \mu_{\frac{1}{2}r}'(\chi_\nu^2) = \Gamma(\tfrac{1}{2}(\nu + r))/\Gamma(\tfrac{1}{2}\nu).$$

Johnson and Welch [61] give formulas for computing the first six cumulants of χ_ν; they also give the expansion

(64) $$E[\chi_\nu] = \sqrt{\nu}\{1 - \tfrac{1}{4}\nu^{-1} + \tfrac{1}{32}\nu^{-2} + \tfrac{5}{128}\nu^{-3} - \tfrac{21}{2048}\nu^{-4} - \ldots\}.$$

Geldston [38] has studied the distribution of $k \log(1 + bx) = Y$, when X has a Rayleigh distribution. Mixtures of Rayleigh distributions are discussed in Siddiqui and Weiss [109] and in Krysicki [75].

8.4 *Generalized Gamma Distributions*

If it be supposed that $\{(Z - \gamma)/\beta\}^c = X$ (with $c > 0$) has the standard gamma distribution (2) then the probability density function of Z is

(65) $$p_Z(z) = \frac{c(z - \gamma)^{c\alpha-1}}{\beta^{c\alpha}\Gamma(\alpha)} \exp\left[-\left(\frac{z - \gamma}{\beta}\right)^c\right] \qquad (z \geq \gamma)\cdot$$

This was defined (with $\gamma = 0$) by Stacy [111] as the family of *generalized gamma distributions*. It includes Weibull distributions ($\alpha = 1$), half-normal distributions ($\alpha = \frac{1}{2}, c = 2, \gamma = 0$), and of course, ordinary gamma distributions ($c = 1$).

Since $\{(Z - \gamma)/\beta\}^c$ has a standard gamma distribution, it is clear that (from (9))

(66) $$\begin{aligned} E[\{(Z - \gamma)/\beta\}^r] &= E[\{(Z - \gamma)/\beta\}^{(r/c)c}] \\ &= \Gamma(\alpha + r/c)/\Gamma(\alpha). \end{aligned}$$

The moments of Z can be deduced from (66). We may note, in particular that if $c = 2$ then $(Z - \gamma)^2$ has a gamma distribution, so that if $c = 2$, $\gamma = 0$, $\beta = 2$ and $\alpha = \frac{1}{2}\nu$ then Z is distributed as χ_ν. If the values of c and γ are known, problems of estimation can be reduced to similar problems for ordinary gamma distributions by using the transformed variable $(Z - \gamma)^c$.

The two parameters c and α define the shape of the distribution (65). These distributions cover an area in the (β_1, β_2) plane. Note that, as for the Weibull distribution, there is a value $c = c(\alpha)$ for which $\beta_1 = 0$. For $c < c(\alpha)$, $\sqrt{\beta_1} > 0$; for $c > c(\alpha)$, $\sqrt{\beta_1} < 0$.

Maximum likelihood estimation of c, α and β (assuming γ to be known) has been described by Parr and Webster [91].

Stacy and Mihram [112] also assuming γ to be known (equal to zero), extended the definition of generalized gamma distributions to include negative (non-zero) values of c, by replacing the multiplier c in (65) by $|c|$. (Note that if c is negative, the rth moment of Z is infinite if $\alpha + r/c \leq 0$.)

Stacy and Mihram proposed a method of estimation based on the moments of $\log Z$. (Of course, if the known value of γ is not zero, $\log (Z - \gamma)$ would be used.)

The moment generating function of

$$T = \log (Z/\beta)$$

is

$$E[e^{Tt}] = E[(Z/\beta)^t] = \Gamma(\alpha + t/c)/\Gamma(\alpha).$$

Hence the cumulant generating function of T is

$$\log \Gamma(\alpha + t/c) - \log \Gamma(\alpha)$$

and

$$\kappa_r(T) = c^{-r}\psi^{(r-1)}(\alpha)$$

so that

$$\begin{cases} \kappa_1(\log Z) = c^{-1}\psi(\alpha) + \log \beta \\ \kappa_r(\log Z) = c^{-r}\psi^{(r-1)}(\alpha) \qquad (r \geq 2). \end{cases} \tag{67}$$

The equations

$$\begin{cases} \mu_1'(\log Z) = c^{-1}\psi(\alpha) + \log \beta \\ \mu_2(\log Z) = c^{-2}\psi'(\alpha) \\ \mu_3(\log Z) = c^{-3}\psi''(\alpha) \end{cases} \tag{68}$$

can be written in the form

$$\mu_3/\mu_2^{\frac{3}{2}} = \psi''(\alpha)/[\psi'(\alpha)]^{\frac{3}{2}} \tag{69.1}$$

$$c = \mu_2\psi''(\alpha)/[\mu_3\psi'(\alpha)] \tag{69.2}$$

$$\beta = \exp [\mu_1' - c^{-1}\psi(\alpha)]. \tag{69.3}$$

Replacing population moments in (69) by sample moments and solving successively for α, c and β gives the required moment estimators.

Harter [48] has studied maximum likelihood estimation of all four parameters α, β, γ and c. He gives tables of asymptotic variances and covariances and includes an extract from these tables in [49], which also contains the results of a number of applications of the method of maximum likelihood to complete samples of size 40, and also to such samples censored by exclusion of the greatest 10, 20, or 30 observed values. From the evidence of some sampling experiments, it appears that

(a) maximum likelihood estimates have a definite bias in samples of this size,
(b) estimation of α and c have high negative correlations.

Generalized gamma distributions were discussed as early as 1925 by Amoroso [2], who fitted such a distribution to an observed distribution of income rates. Between 1925 and 1962, however, there appeared to be little interest in this family of distributions. An interesting physical model generating generalized Gamma distributions is described by Lienhard and Meyer in [81].

The mixture of two generalized gamma distributions (each having the range of variation from zero to infinity) has been discussed by Wasilewski [122]. The distribution of the quotient (and more recently, of the product) of two independent generalized Gamma variables was derived by Malik [83] (see also [37]). Plucinska [100] used two generalized gamma distributions, one for negative and one for positive values of the argument, to construct a new class of distribution functions; mixtures of such distributions are used in [101]. She had previously [99] discussed the distribution obtained by reflecting a generalized gamma distribution with $\gamma = 0$ about the origin, giving the density function:

$$p_Z(z) = \frac{c|z|^{c\alpha-1}}{2\beta^{c\alpha}\Gamma(\alpha)} \exp[-(|z|/\beta)^c]. \tag{70}$$

Borghi [16] has discussed distributions obtained by a similar reflection of the standard gamma distribution. The properties of such reflected distributions are very easily obtained from those of their parent distributions. Harvey [52] has considered a similar but different form of distribution, with density function

$$K\cdot(\alpha + \beta|x - \xi|)^c e^{-b|x-\xi|} \qquad (\alpha,\beta,b,c > 0) \tag{71}$$

with

$$K = \left[2\int_0^\infty (\alpha + \beta t)^c e^{-bt}\, dt\right]^{-1} = \frac{b^{c+1}}{2\beta^c} e^{-\alpha b/\beta}\left[\int_{\alpha b/\beta}^\infty t^c e^{-t}\, dt\right]^{-1}.$$

An attractive feature of this class of distributions, as compared with those of Borghi (and also the latter of Plucinska's) is that the density is not zero, in general, at the point of symmetry.

If α, β, b and c are known, the maximum likelihood estimator of ξ is found as a solution of the equation.

$$\beta c \sum_{j=1}^n (\alpha + \beta|X_j - \hat{\xi}|)^{-1} \operatorname{sgn}(X_j - \hat{\xi}) = b \sum_{j=1}^n \operatorname{sgn}(X_j - \hat{\xi}). \tag{72}$$

(Harvey actually considers the special case $\alpha = c$, $\beta = b$.)

The 'modified normal distributions' of Box and Tiao [18] are reflections about the start of generalized gamma distributions with $c\alpha = 1$. They will be discussed in Section 2 of Chapter 33.

REFERENCES

[1] Alexander, G. N. (1962). The use of the gamma distribution in estimating regulated output from storages, *Transactions in Civil Engineering, Institute of Engineers, Australia*, **4,** 29–34.

[2] Amoroso, L. (1925). Ricerche intorno alla curva dei redditi, *Annali di Mathematica, Serie IV, Tomo II*, 123–159.

[3] Archer, C. O. (1967). *Some properties of Rayleigh distributed random variables and of their sums and products*, Technical Memo. TM–67–15, Naval Missile Center, Point Mugu, California.

[4] Bánkövi, G. (1964). A note on the generation of beta distributed and gamma distributed random variables, *Mathematical Proceedings of the Hungarian Academy of Science, Series A*, **9,** 555–562.

[5] Bartlett, M. S. and Kendall, D. G. (1946). The statistical analysis of variance heterogeneity and the logarithmic transformation, *Journal of the Royal Statistical Society, Series B*, **8,** 128–138.

[6] Barton, D. E. (1953). The probability distribution function of a sum of squares, *Trabajos de Estadística*, **4,** 199–207.

[7] Basu, A. P. (1964). Estimates of reliability for some distributions useful in life testing, *Technometrics*, **6,** 215–219.

[8] Beard, R. E. (1948). Some experiments in the use of the incomplete gamma function for the approximate calculation of actuarial functions, *Proceedings of the Centennial Assembly, Institute of Actuaries*, **2,** 89–107.

[9] Beard, R. E. (1947). Some notes on approximate product-moment integration, *Journal of the Institute of Actuaries*, **73,** 356–403. (Discussion 404–416.)

[10] Berndt, G. D. (1958). Power functions of the gamma distribution, *Annals of Mathematical Statistics*, **29,** 302–306.

[11] Bhattacharya, S. K. (1966). A modified Bessel function model in life testing, *Metrika*, **11,** 133–144.

[12] Bienaymé, I. J. (1838). Mémoire sur la probabilité des résultats moyens des observations; demonstration directe de la règle de Laplace, *Mémoires de l'Académie de Sciences de l'Institut de France, Paris, Series Étrangers*, **5,** 513–558

[13] Bienaymé, I. J. (1852). Mémoire sur la probabilité des erreurs d'aprés la méthode de moindres carrés, *Liouville's Journal de Mathématiques Pures et Appliquées*, **17,** 33–78.

[14] Birnbaum, Z. W. and Saunders, S. C. (1958). A statistical model for life-length of materials, *Journal of the American Statistical Association*, **53,** 151–160.

[15] Blischke, W. R., Glinski, A. M., Johns, M. V., Mundle, P. B. and Truelove, A. J. (1965). *On non-regular estimation, minimum variance bounds and the Pearson Type III distribution*, Aerospace Research Laboratory, Wright-Patterson Air Force Base, Ohio, Report ARL 65-177.

[16] Borghi, O. (1965). Sobre una distribución de frecuencias, *Trabajos de Estadística*, **16,** 171–192.

[17] Bowman, K. O. and Shenton, L. R. (1968). *Properties of estimators for the gamma distribution*, Report CTC–1, Union Carbide Corp., Oak Ridge, Tennessee.

[18] Box, G. E. P. and Tiao, G. C. (1962). A further look at robustness via Bayes's theorem, *Biometrika*, **49,** 419–432.

[19] Boyd, W. C. (1965). A nomogram for chi-square, *Journal of the American Statistical Association*, **60,** 344–346.

[20] Breiter, M. C. and Krishnaiah, P. R. (1967). *Tables for the moments of gamma order statistics*, Aerospace Research Laboratory, Wright-Patterson Air Force Base, Ohio, Report ARL 67-1066. (Also *Sankhyā*, *Series B*, **30,** 59–72.)

[21] Broeder, G. G. den (1955). On parameter estimation for truncated Pearson Type III distributions, *Annals of Mathematical Statistics*, **26,** 659–663.

[22] Bruce, R. A. (1964). *Estimation of the scale parameter of the gamma distribution by the use of M order statistics*, Unpublished thesis, Air Force Institute of Technology, Wright-Patterson Air Force Base, Ohio.

[23] Chapman, D. G. (1956). Estimating the parameters of a truncated gamma distribution, *Annals of Mathematical Statistics*, **27,** 498–506.

[24] Clark, C. (1951). Urban population densities, *Journal of the Royal Statistical Society*, *Series A*, **114,** 490–496.

[25] Cohen, A. C. (1950). Estimating parameters of Pearson Type III populations from truncated samples, *Journal of the American Statistical Association*, **45,** 411–423.

[26] Cohen, A. C. (1951). Estimation of parameters in truncated Pearson frequency distributions, *Annals of Mathematical Statistics*, **22,** 256–265.

[27] Cohen, A. C., Helm, F. R. and Sugg, M. (1969). *Tables of Areas of the Standardized Pearson Type III Density Function*, Report NASA CR-61266, NASA, Marshall Space Flight Center, Alabama.

[28] Czuber, E. (1891). *Theorie der Beobachtungsfehler*, Leipzig: Teubner.

[29] Das, S. C. (1955). The fitting of truncated Type III curves to daily rainfall data, *Australian Journal of Physics*, **7,** 298–304.

[30] Des Raj (1953). Estimation of the parameters of Type III populations from truncated samples, *Journal of the American Statistical Association*, **48,** 336–349.

[31] Dubey, S. D. (1966). *Compound gamma, beta and F distributions*, Dearborn, Michigan: Ford Motor Company.

[32] Elderton, W. P. (1902). Tables for testing the goodness of fit of theory to observation, *Biometrika*, **1,** 155–163.

[33] Ellis, R. L. (1844). On a question in the theory of probabilities, *Cambridge Mathematical Journal*, **4,** 127–132.

[34] Fisher, R. A. (1921). On the mathematical foundations of theoretical statistics, *Philosophical Transactions of the Royal Society of London*, *Series A*, **222,** 309–368.

[35] Fisher, R. A. (1922). On the interpretation of χ^2 from contigency tables and calculation of P., *Journal of the Royal Statistical Society*, *Series A*, **85,** 87–94.

[36] Fisher, R. A. and Cornish, E. A. (1960). The percentile points of distributions having known cumulants, *Technometrics*, **2,** 209–225.

[37] Garti, Y. and Consoli, T. (1954). Sur la densité de probabilité du produit de variables aléatoires de Pearson du Type III (pp. 301–309 of *Studies in Mathematics and Mechanics*, presented to R. von Mises, New York: Academic Press).

[38] Geldston, S. (1962). *Probability distribution at the output of a logarithmic receiver*, Research Report PIBMRI–1087–62, Microwave Research Institute, Polytechnic Institute of Brooklyn, New York.

[39] Goodman, L. A. (1952). On the Poisson-gamma distribution problem, *Annals of the Institute of Statistical Mathematics, Tokyo*, **3,** 123–125.

[40] Gray, H. L., Thompson, R. W. and McWilliams, G. V. (1969). A new approximation for the chi-square integral, *Mathematics of Computation*, **23,** 85–89.

[41] Greenwood, J. A. and Durand, D. (1960). Aids for fitting the gamma distribution by maximum likelihood, *Technometrics*, **2,** 55–65.

[42] Gupta, S. S. (1960). Order statistics from the gamma distribution, *Technometrics*, **2,** 243–262.

[43] Gupta, S. S. (1962). Gamma distribution, (pp. 431–450 in *Contributions to Order Statistics*, (Ed., A. E. Sarhan and B. G. Greenberg), New York: John Wiley & Sons, Inc.).

[44] Gupta, S. S. and Groll, P. A. (1961). Gamma distribution in acceptance sampling based on life tests, *Journal of the American Statistical Association*, **56,** 942–970.

[45] Hald, A. and Sinkbaek, S. A. (1950). A table of percentage points of the χ^2 distribution, *Skandinavisk Aktuarietidskrift*, **33,** 168–175.

[46] Harter, H. L. (1964). A new table of percentage points of the chi-square distribution, *Biometrika*, **51,** 231–239.

[47] Harter, H. L. (1964). *New Tables of the Incomplete Gamma Function Ratio and of Percentage Points of the Chi-square and Beta Distributions*, Washington, D.C.: U.S. Government Printing Office.

[48] Harter, H. L. (1966). *Asymptotic variances and covariances of maximum-likelihood estimators, from censored samples, of the parameters of a four-parameter generalized gamma population*, Report 66-0158, Aerospace Research Laboratories, Wright-Patterson Air Force Base, Ohio.

[49] Harter, H. L. (1967). Maximum likelihood estimation of the parameters of a four-parameter generalized gamma population from complete and censored samples, *Technometrics*, **9,** 159–165.

[50] Harter, H. L. (1969). A new table of percentage points of the Pearson Type III distribution, *Technometrics*, **11,** 177–187.

[51] Harter, H. L. and Moore, A. H. (1965). Maximum likelihood estimation of the parameters of gamma and Weibull populations from complete and from censored samples, *Technometrics*, **7,** 639–643.

[52] Harvey, H. (1967). *A Family of Averages in Statistics*, Morrisville, Pennsylvania: Annals Press.

[53] Helmert, F. R. (1875). Über die Berechnung des wahrscheinlichen Fehlers aus einer endlichen Anzahl wahrer Beobachtungsfehler, *Zeitschrift für angewandte Mathematik und Physik*, **20,** 300–303.

[54] Helmert, F. R. (1876). Die Genauigkeit der Formel von Peters zur Berechnung des wahrscheinlichen Beobachtungsfehlers directer Beobachtungen gleicher Genauigkeit, *Astronomische Nachrichten*, **88,** columns 113–120.

[55] Helmert, F. R. (1876). Über die Wahrscheinlichkeit der Potenzsummen der Beobachtungsfehler und über einige damit in Zusammenhänge stehende Fragen, *Zeitschrift für angewandte Mathematik und Physik*, **21,** 192–218.

[56] Hill, T. D. (1965). *Estimation of the scale parameter of the gamma distribution by the use of L order statistics*, Unpublished thesis, Air Force Institute of Technology, Wright-Patterson Air Force Base, Ohio.

[57] Iyer, P. V. K. and Singh, N. (1963). Estimation of the mean and the standard deviation of a Type III population from censored samples, *Journal of the Indian Statistical Association*, **1,** 161–166.

[58] Jackson, O. A. Y. (1969). Fitting a Gamma or log-normal distribution to fibre-diameter measurements of wool tops, *Applied Statistics*, **18,** 70–75.

[59] Jöhnk, M. D. (1964). Erzeugung von betaverteilten und gammaverteilten Zufallszahlen, *Metrika*, **8,** 5–15.

[60] Johnson, N. L. (1952). Approximations to the probability integral of the distribution of range, *Biometrika*, **39,** 417–419.

[61] Johnson, N. L. and Welch, B. L. (1939). On the calculation of the cumulants of the χ-distribution, *Biometrika*, **31,** 216–218.

[62] Kabe, D. G. (1966). Dirichlet's transformation and distributions of linear functions of ordered gamma variates, *Annals of the Institute of Statistical Mathematics, Tokyo*, **18,** 367–374.

[63] Kagan, A. M. and Ruhin, A. L. (1967). On the estimation theory of the scale parameter, *Teoriya Veroyatnostei i ee Primeneniya*, **12,** 735–741. (In Russian)

[64] Karns, R. C. (1963). *Scale parameter estimation of the gamma probability function based on one order statistic*, Unpublished thesis, Air Force Institute of Technology, Wright-Patterson Air Force Base, Ohio.

[65] Khamis, S. H. (1960). Incomplete gamma function expansions of statistical distribution functions, *Bulletin of the International Statistical Institute*, **37,** 385–396.

[66] Khamis, S. H. and Rudert, W. (1965). *Tables of the incomplete gamma function ratio: chi-square integral, Poisson distribution*, Darmstadt: Justus von Leibig.

[67] Khatri, C. G. and Rao, C. R. (1968). Some characterizations of the gamma distribution, *Sankhyā, Series A*, **30,** 157–166.

[68] Klinken, J. van (1961). A method for inquiring whether the Γ-distribution represents the frequency distribution of industrial accident costs, *Actuariële Studiën*, **3,** 83–92.

[69] Kotlarski, I. (1962). On pairs of independent variables whose quotients follow some known distribution, *Colloquium Mathematicum*, **9,** 151–162.

[70] Kotlarski, I. (1965). A certain problem connected with the gamma distribution, *Zeszyty Naukowe Politechniki Warszawskiej*, **6,** 21–28. (In Polish)

[71] Kotlarski, I (1966). On characterizing the chi-square distribution by the Student law, *Journal of the American Statistical Association*, **61,** 976–981.

[72] Kotlarski, I. (1967). On characterizing the gamma and the normal distribution, *Pacific Journal of Mathematics*, **20,** 69–76.

[73] Kotz, S. and Neumann, J. (1963). On distribution of precipitation amounts for the periods of increasing length, *Journal of Geophysical Research*, **68,** 3635–3641, (Addendum: **69,** 800–801).

[74] Kruskal, W. H. (1946). Helmert's distribution, *American Mathematical Monthly*, **53,** 435–438.

[75] Krysicki, W. (1963). Application de la méthode des moments a l'estimation des paramètres d'un mélange de deux distributions de Rayleigh, *Revue de Statistique Appliquée*, **11,** No. 4, 25–34.

[76] Laha, R. G. (1964). On a problem connected with beta and gamma distributions, *Transactions of the American Mathematical Society*, **113,** 287–298.

[77] Laha, R. G. (1954). On a characterization of the gamma distribution, *Annals of Mathematical Statistics*, **25,** 784–787.

[78] Lancaster, H. O. (1966). Forerunners of the Pearson χ^2, *Australian Journal of Statistics*, **8,** 117–126. (See also W. Kruskal's review in *Mathematical Reviews*, **36,** 916.)

[79] Laplace, P. S. (1836). *Théorie Analytique des Probabilités*, (Supplement to the third edition).

[80] Lavender, D. E. (1967). *On the distribution of the sum of independent doubly truncated gamma variables*, Report NASA CR-61184.

[81] Lienhard, J. H. and Meyer, P. L. (1967). A physical basis for the generalized gamma distribution, *Quarterly of Applied Mathematics*, **25,** 330–334.

[82] Lukacs, E. (1965). A characterization of the gamma distribution, *Annals of Mathematical Statistics*, **26,** 319–324.

[83] Malik, H. J. (1967). Exact distribution of the quotient of independent generalized gamma variables, *Canadian Mathematical Bulletin*, **10,** 463–466.

[84] Masuyama, M. and Kuroiwa, Y. (1952). Table for the likelihood solutions of gamma distribution and its medical applications, *Reports of Statistical Application Research*, (*JUSE*), **1,** 18–23.

[85] Mauldon, J. G. (1956). Characterizing properties of statistical distributions, *Quarterly Journal of Mathematics* (*Oxford, 2nd Series*) **27,** 155–160.

[86] Menon, M. V. (1966). Characterization theorems for some unimodal distributions, *Journal of the Royal Statistical Society, Series B*, **28,** 143–145.

[87] Musson, T. A. (1965). *Linear estimation of the location and scale parameters of the Weibull and gamma probability distributions by the use of order statistics*, Unpublished thesis, Air Force Institute of Technology, Wright-Patterson Air Force Base, Ohio.

[88] Nabeya, S. (1950). On a relation between exponential law and Poisson's law, *Annals of the Institute of Statistical Mathematics, Tokyo*, **2,** 13–16.

[89] Norris, N. (1966). *Maximum-likelihood estimators of the relative scale parameters of Type III and Type IV populations* (Presented at Institute of Mathematical Statistics meeting, August, 1966, New Brunswick, New Jersey).

[90] Olshen, A. C. (1937). Transformations of the Pearson Type III distributions, *Annals of Mathematical Statistics*, **8,** 176–200.

[91] Parr, B. van, and Webster, J. T. (1965). A method for discriminating between failure density functions used in reliability predictions, *Technometrics*, **7,** 1–10.

[92] Patil, G. P. and Seshadri, V. (1964). Characterization theorems for some univariate probability distributions, *Journal of the Royal Statistical Society, Series B*, **26,** 286–292.

[93] Pearson, E. S. (1959). Note on an approximation to the distribution of non-central χ^2, *Biometrika*, 46, 364.

[94] Pearson, E. S. and Hartley, H. O. (1954). *Biometrika Tables for Statisticians*, **1,** London: Cambridge University Press.

[95] Pearson, K. (1900). On a criterion that a given system of deviations from the probable in the case of a correlated system of variables is such that it can be reasonably supposed to have arisen from random sampling, *Philosophical Magazine*, 5*th Series*, **50,** 157–175.

[96] Pearson, K. (Ed.) (1922). *Tables of the Incomplete* Γ-*Function*, London: H.M. Stationery Office, (Since 1934: Cambridge University Press).

[97] Pearson, K. (1931). Historical note on the distribution of the standard deviation of samples of any size drawn from an indefinitely large normal parent population, *Biometrika*, **23,** 416–418.

[98] Pitman, E. J. G. (1939). Tests of hypotheses concerning location and scale parameters, *Biometrika*, **31,** 200–215.

[99] Plucinska, A. (1965). On certain problems connected with a division of a normal population into parts, *Zastosowania Matematyki*, **8,** 117–125. (In Polish)

[100] Plucinska, A. (1966). On a general form of the probability density function and its application to the investigation of the distribution of rheostat resistence, *Zastosowania Matematyki*, **9,** 9–19. (In Polish)

[101] Plucinska, A. (1967). The reliability of a compound system under consideration of the system elements prices, *Zastosowania Matematyki*, **9,** 123–134. (In Polish)

[102] Pogurova, V. I. (1965). On the calculation of quantiles of the Γ-distribution, *Teoriya Veroyatnostei i ee Primeneniya*, **10,** 746–749. (In Russian. English translation, pp. 677–680.)

[103] Roberts, C. and Geisser, S. (1966). A necessary and sufficient condition for the square of a random variable to be gamma, *Biometrika*, **53,** 275–277.

[104] Russell, W. and Lal, M. (1969). *Tables of Chi-square Probability Function* Department of Mathematics, Memorial University of Newfoundland, St. Johns. (Reviewed in *Mathematics of Computation*, **23,** 211–212.)

[105] Salvosa, L. R. (1929). *Generalizations of the Normal Curve of Error*, Ann Arbor, Michigan: Edwards Brothers, Inc.

[106] Salvosa, L. R. (1930). Tables of Pearson's Type III function, *Annals of Mathematical Statistics*, **1,** 191–198.

[107] Särndal, C.-E. (1964). Estimation of the parameters of the gamma distribution by sample quantiles, *Technometrics*, **6,** 405–414.

[108] Siddiqui, M. M. (1962). Some problems connected with Rayleigh distributions, *Journal of Research, National Bureau of Standards*, **66D,** 167–174.

[109] Siddiqui, M. M. and Weiss, G. H. (1963). Families of distributions for hourly median power and instantaneous power of received radio signals, *Journal of Research, National Bureau of Standards*, **67D,** 753–762.

[110] Slutskii, E. E. (1950). *Tablitsi dlya Vichisleniya Nepolnoi* Γ-*funktsii i Veroyatnosti* χ^2, (Tables for Computing the Incomplete Gamma function and χ^2 probabilities), Moscow: Akademia Nauk SSSR. (Ed. A. N. Kolmogorov.)

[111] Stacy, E. W. (1962). A generalization of the gamma distribution, *Annals of Mathematical Statistics*, **33,** 1187–1192.

[112] Stacy, E. W. and Mihram, G. A. (1965). Parameter estimation for a generalized gamma distribution, *Technometrics*, **7,** 349–358.

[113] Stammberger, A. (1967). Über einige Nomogramme zur Statistik, *Wissenshaftliche Zeitschrift der Humboldt-Universität Berlin, Mathematisch-Naturwissenschaftliche Reihe*, **16,** 1, 86–93.

[114] Thom, H. C. S. (1968). *Direct and Inverse Tables of the Gamma Distribution*, Silver Spring, Maryland; Environmental Data Service.

[115] Thompson, Catherine M. (1941). Tables of the percentage points of the χ^2-distribution, *Biometrika*, **32,** 187–191.

[116] Tiku, M. L. (1964). Approximating the general non-normal variance ratio sampling distribution, *Biometrika*, **51,** 83–95.

[117] Tiku, M. L. (1964). Laguerre series forms of noncentral χ^2 and F distributions, *Biometrika*, **52,** 415–427.

[118] Tricomi, F. G. (1950). Sulle funzione gamma incomplete, *Annali di Matematica Pura ed Applicata*, **31,** 263–279.

[119] Vanderbeck J. P. and Cooke, J. R. (1961). *Extended Table of Percentage Points of the Chi-Square Distribution*, U.S. Naval Ordnance Test Station, China Lake, California (Nauweps Report 7770).

[120] Wallace, D. L. (1958). Asymptotic approximations to distributions, *Annals of Mathematical Statistics*, **29,** 635–654.

[121] Wallace, D. L. (1959). Bounds on normal approximations to Student's and the chi-square distributions, *Annals of Mathematical Statistics*, **30,** 1121–1130.

[122] Wasilewski, M. J. (1967). Sur certaines propriétés de le distribution gamma généralisée, *Revue de Statistique Appliquée*, **15,** No. 1, 95–105.

[123] Welch, B. L. (1938). The significance of the difference between two means when the population variances are unequal, *Biometrika*, **29,** 350–361.

[124] Wilk, M. B. and Gnanadesikan, R. (1964). Graphical methods for internal comparisons in multiresponse experiments, *Annals of Mathematical Statistics*, **35,** 613–631.

[125] Wilk, M. B., Gnanadesikan, R. and Huyett, M. J. (1962). Probability plots for the gamma distribution, *Technometrics*, **4,** 1–20.

[126] Wilk, M. B., Gnanadesikan, R. and Huyett, M. J. (1962). Estimation of parameters of the gamma distribution using order statistics, *Biometrika*, **49,** 525–545.

[127] Wilk, M. B., Gnanadesikan, R. and Lauh, E. (1966). Scale parameter estimation from the order statistics of unequal gamma components, *Annals of Mathematical Statistics*, **37,** 152–176.

[128] Wilson, E. B. and Hilferty, M. M. (1931). The distribution of chi-square, *Proceedings of the National Academy of Sciences, Washington*, **17,** 684–688.

[129] Wishart, J. (1927). On the approximate quadrature of certain skew curves, with an account of the researches of Thomas Bayes, *Biometrika*, **19,** 1–38. (Correction: **19,** 442.)

[130] Zubrzycki, S. (1966). Explicit formulas for minimax admissible estimators in some cases of restrictions imposed on the parameter, *Zastosowania Matematyki*, **9,** 31–52.

18

Exponential Distribution

1. Definition and Introduction

The random variable X has an *exponential (or negative exponential) distribution* if it has a probability density function of form:

$$p_X(x) = \sigma^{-1} \exp [-(x - \theta)/\sigma] \qquad (x > \theta;\ \ \sigma > 0). \tag{1}$$

Figure 1 gives a graphical representation of this function, with $\theta > 0$.

This is a special case of the gamma distribution, the subject of Chapter 17. The exponential distribution has a separate chapter because of its considerable importance and widespread use in statistical procedures.

Very often it is reasonable to take $\theta = 0$. The special case of (1) so obtained is called the *one-parameter* exponential distribution. If $\theta = 0$ and $\sigma = 1$, the distribution is called the *standard* exponential distribution.

The mathematics associated with the exponential distribution is often of a simple nature. It is often possible to obtain explicit formulas in terms of elementary functions, without troublesome quadratures. For this reason, models constructed from exponential variables are sometimes used as an approximate representation of other models, more 'natural' for a particular application.

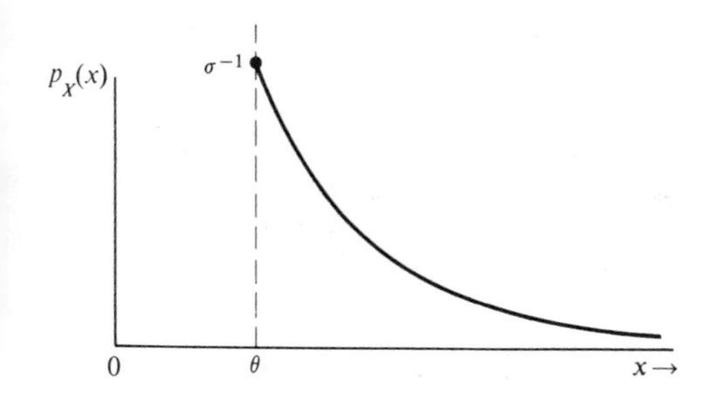

FIGURE 1

Exponential Density Function

2. Genesis

There are many situations in which one would expect an exponential distribution to give a useful description of observed variation. One of the most widely quoted is that of events recurring 'at random in time.' In particular, suppose that the future 'lifetime' of an individual has the same distribution, no matter how 'old' it is at present. This can be written formally (X representing lifetime)

$$\Pr[X \leq x_0 + x \mid X > x_0] = \Pr[X \leq x]$$

for all $x_0 > 0$ and all $x > 0$. X must be a continuous positive random variable. If it has a probability density function $p_X(x)$ then the conditional probability density function, given that X is greater than x_0, is

$$p_X(x)/[1 - F_X(x_0)] \qquad (x > x_0 > 0).$$

Since the conditional distribution of the future lifetime $(X - x_0)$ is the same as the (unconditional) distribution of X, we have, say,

$$\frac{p_X(x_0)}{1 - F_X(x_0)} = p_X(0) = p_0. \tag{2}$$

It follows that if $F_X(x_0) \neq 1$, $p_0 > 0$ and $F_X(x)$ satisfies the differential equation

$$\frac{dF_X(x)}{dx} = p_0\,[1 - F_X(x)]$$

whence $1-F_X(x) \propto e^{-p_0 x}$. Introducing the condition $\lim_{x \to 0} F_X(x) = 0$ we find that

$$1 - F_X(x) = e^{-p_0 x} \tag{3}$$

that is,

$$F_X(x) = 1 - e^{-p_0 x} = p_0 \int_0^x e^{-p_0 t}\, dt.$$

This shows that the probability density function of X is of form (1) with $\theta = 0$, $\sigma = p_0^{-1}$.

There are other situations in which exponential distributions appear to be the most natural. Many of these do, however, have as an essential feature the random recurrence (often in time) of an event.

In applying the Monte Carlo method it is often required to transform random variables from standard rectangular distribution to exponential random variables.

An ingenious method was suggested at an early date by von Neumann [52]. Let $\{X_i; i = 0,1,\ldots\}$ be a sequence of independent random variables from the standard rectangular distribution, and define a random variable N taking positive integer values through $\{X_i\}$ by the inequalities

$$X_1 < X_0, \quad \sum_{j=1}^{2} X_j < X_0, \ldots, \quad \sum_{j=1}^{N-1} X_j < X_0, \quad \sum_{j=1}^{N} X_j > X_0.$$

We 'accept' the sequence $\{X_i\}$ if N is odd, otherwise we 'reject' it and repeat the process until N turns out odd. Let T be the number of sequences rejected before an odd N appears ($T = 0,1,\ldots,$) and X_0 be the value of the first variable in the accepted sequence. Then $Y = T + X_0$ is an exponential random variable with the standard density e^{-x}.

A rather more convenient method was suggested by Marsaglia [48]. Let N be a non-negative random integer with the geometric distribution (Chapter 5, Section 2)

$$\Pr(N = n) = (1 - e^{-\lambda})e^{-n\lambda}, \qquad n = 0, 1, \ldots$$

and let M be a positive random integer with the zero-truncated Poisson distribution (Chapter 4, Section 10)

$$\Pr(M = m) = (1 - e^{-\lambda})^{-1}e^{-\lambda}\lambda^m/m! \qquad m = 1, 2, \ldots .$$

Finally let $\{X_i; i = 1,2,\ldots\}$ be a sequence of independent random variables each having a standard rectangular distribution (Chapter 25). Then

$$Y = \lambda\{N + \min(X_1,\ldots,X_M)\}$$

has the standard exponential distribution.

Sibuya [74] gave a statistical interpretation of the procedure, recommended that the value of the parameter λ be taken as 0.5 or log 2 and extended the technique to the chi-square distribution. Bánkövi [2] also investigated a similar technique.

A table of 10,000 exponential random numbers is given in Barnett [3].

3. Some Remarks on Recent History

Over the last 15 years the study of estimators based on samples from an exponential population has been closely associated with the study of order statistics. In 1952 Lloyd [47] described a method for obtaining the best linear unbiased estimators (B.L.U.E's) of the parameters of a distribution, using order statistics. In 1953 Epstein and Sobel published a paper [26] which presented the maximum likelihood estimator of the scale parameter σ, of the one-parameter exponential distribution in the case of censoring from the right. In 1954 the same authors [27] extended the foregoing analysis to the two-parameter exponential distribution. In that same year, Sarhan [69] employed the method derived by Lloyd to obtain the B.L.U.E.'s of σ and θ for the two-parameter exponential distribution in the case of no censoring. In the following year, 1955, Sarhan [70] extended his results to censoring. Sarhan noted that in the special case of the one-parameter exponential distribution his results agreed with those of Epstein and Sobel and therefore his estimator of σ was not only the best linear unbiased estimator but also the maximum likelihood estimator of σ. In 1960 Epstein [23] extended his own results to estimators of σ and θ for the one- and two-parameter exponential distributions in the cases of censoring from the right and/or left. For the two-parameter exponential distribution his maximum likelihood estimators coincided with the B.L.U.E.'s of Sarhan [70] but for the one-parameter exponential distribution there was agreement only in the case of censoring from the right. (References [16]–[28] reflect the numerous contributions of Epstein and Sobel and subsequently of Epstein in propagating the exponential distribution for industrial applications.)

In the light of the applicability of order statistics to the exponential distribution it became quite natural to attempt estimation of the parameter by use of the sample quasi-ranges. In 1959 Rider [64] derived the probability density function and the cumulants of the quasi-range of the standardized exponential distribution and in 1960 Fukuta [30] derived "best" linear estimators of σ and θ by two sample quasi-ranges. The next step would quite reasonably be that of determining those two order statistics which would supply the best linear unbiased estimator of σ and θ for the two-parameter distribution and this was, in fact, done numerically by Sarhan *et al.* [73] in 1963. They employed the method of Lloyd to obtain the best linear estimators of σ and θ based on the two order statistics $X'_{1;n}$ and $X'_{m;n}$, and then compared numerically the relative efficiencies of the estimators for various pairs of values $(1,m)$. In 1961, Harter [36], using a similar approach to that of Sarhan *et al.*, presented the best linear estimators of σ for the one-parameter distribution based on one and on two order statistics. Harter mentioned in this paper that he was not aware of any analytical process by which the optimum pair of order statistics $X'_{1;n}$ and $X'_{m;n}$ can be determined. In 1963 Siddiqui [76] presented an analytical method based on the Euler-Maclaurin formula, for obtaining the optimum pair of B.L.U.E. order statistics. Since 1963 a considerable number of additional, more refined, results have been obtained, some of which are presented in Section 6.

4. Moments and Generating Functions

The moment generating function of a random variable X with probability density function (1) is

$$E[e^{tX}] = (1 - \sigma t)^{-1}e^{t\theta} \qquad (4)$$
$$(= (1 - \sigma t)^{-1} \text{ if } \theta = 0).$$

The characteristic function is $(1 - i\sigma t)^{-1}e^{it\theta}$.

The central moment generating function is

$$E[e^{t(X-\theta-\sigma)}] = (1 - \sigma t)^{-1}e^{-t\sigma}.$$

The cumulant generating function is $\log E[e^{tX}] = t\theta - \log(1 - \sigma t)$. Hence the cumulants are

$$\kappa_1 = \theta + \sigma \qquad (= E(X)) \qquad (5)$$
$$\kappa_r = (r-1)!\sigma^r \qquad (r > 1).$$

Putting $r = 2, 3, 4$ we find

$$\text{var}(X) = \mu_2 = \sigma^2$$
$$\mu_3 = 2\sigma^3$$
$$\mu_4 = 9\sigma^4.$$

Note that if $\theta = 0$ and $\sigma = 1$ then $E(X) = 1 = \text{var}(X)$ and so the standard exponential distribution is also standardized.

The first two moment ratios are

$$\sqrt{\beta_1} = 2; \qquad \beta_2 = 9.$$

The mean deviation is

$$2\sigma \int_1^\infty (x - 1)e^{-x}\,dx = 2e^{-1}\sigma.$$

Note that

$$\frac{\text{mean deviation}}{\text{standard deviation}} = \frac{2}{e} = 0.736.$$

The median of the distribution is $\theta + \sigma \log_e 2$. The 'mode' of this distribution is at the lowest value, θ, of the range of variation.

The information generating function ($(u - 1)$-th frequency moment) is $\sigma^{1-u}u^{-1}$.

The entropy is $1+\log \sigma$.

5. Estimation

If $X_1, X_2, \ldots, X_n$ are independent random variables each having the probability density function (1), then the maximum likelihood estimators of θ and σ are

$$\hat{\theta} = \min (X_1, X_2, \ldots, X_n) \tag{6}$$

$$\hat{\sigma} = n^{-1} \sum_{i=1}^{n} (X_i - \hat{\theta}) = \overline{X} - \hat{\theta}.$$

If θ be known, the maximum likelihood estimator of σ is $(\overline{X} - \theta)$. If σ be known, $\hat{\theta}$ above is still the maximum likelihood estimator of θ.

The probability density function of $\hat{\theta}$ is

$$p_{\hat{\theta}}(t) = (n/\sigma) \exp [- n(t - \theta)/\sigma] \qquad (t > \theta) \tag{7}$$

which is of the same form as (1) with σ replaced by σ/n. The variance of $\hat{\theta}$ is therefore σ^2/n^2 and its expected value is $\theta + \sigma/n$. It is interesting to note that the variance is proportional to n^{-2} and not to n^{-1}.

The expected value of $\hat{\sigma} = (\overline{X} - \hat{\theta})$ is $\sigma(1 - n^{-1})$ and its variance is

$$\sigma^2[n^{-1} + n^{-2} - 2n^{-3}].$$

The expected value of $(\overline{X} - \theta)$ is σ and its variance is $\sigma^2 n^{-1}$.

A function of special interest in some applications is the value of the probability that X exceeds a value x_0. If $\theta = 0$ so that

$$p_X(x) = \sigma^{-1} \exp(-x/\sigma) \qquad (x > 0; \sigma > 0) \tag{8}$$

then the required value is $e^{-x_0/\sigma}$. Inserting the maximum likelihood estimator $\hat{\sigma} = n^{-1} \sum_{i=1}^{n} X_i$ in place of σ would give the estimator $\exp\left(-x_0 n \Big/ \sum_{i=1}^{n} X_i\right)$. This is the maximum likelihood estimator of $e^{-x_0/\sigma}$. It is biased, but a minimum variance unbiased estimator may be obtained by using the Blackwell-Rao theorem (as described in Chapter 1).

The statistic

$$T = \begin{cases} 1 & \text{if} \quad X_1 > x_0 \\ 0 & \text{if} \quad X_1 \le x_0 \end{cases}$$

is an unbiased estimator of $\exp(-x_0/\sigma)$. Since $\sum_{i=1}^{n} X_i$ is a complete sufficient statistic for σ (and so also for $\exp(-x_0/\sigma)$) the required minimum variance unbiased estimator is

$$E\left[T \,\middle|\, \sum_{i=1}^{n} X_i\right] = \Pr\left[X_1 > x_0 \,\middle|\, \sum_{i=1}^{n} X_i\right]. \tag{9}$$

The ratio $X_1 \Big/ \left[\sum_{i=1}^{n} X_i\right]$ has a beta distribution with parameters 1, $n-1$, *and is independent of* $\sum_{i=1}^{n} X_i$ (Chapter 17, Section 6).

Hence

$$E\left(T \,\middle|\, \sum_{i=1}^{n} X_i\right) = (n-1)\int_{x_0/\sum_1^n X_i}^{1} (1-z)^{n-2}\, dz \tag{10}$$

$$= \begin{cases} \left(1 - \dfrac{x_0}{\sum_{i=1}^{n} X_i}\right)^{n-1} & x_0 < \sum_{i=1}^{n} X_i \\ 0 & x_0 \ge \sum_{i=1}^{n} X_i, \end{cases}$$

which is the required minimum variance unbiased estimator. This formula was obtained by Pugh [59]. (Pugh claims this is the "best" estimator, but does not compare its mean square error with that of competing estimators.)

In a monograph [43] Kulldorff discussed a general theory of estimation based on grouped or partially grouped samples. "Grouped" means that, in disjoint intervals of distribution range, the individual sample values are not available but only the numbers of observed values which have fallen in the intervals.

The distribution of the observed numbers is a multinomial distribution (Chapter 11, Section 2) with probabilities which are functions of the parameter. If individual observations are available in *some* intervals the sample is partially grouped.

Kulldorff devoted a large part of his book to the estimation of the exponential distribution because of its simplicity. The cases studied included completely or partially grouped data, θ unknown, σ unknown, both θ and σ unknown, a finite number of intervals, an infinite number of intervals, intervals of equal length, and intervals of unequal length.

Here we describe only the maximum likelihood estimator of σ when θ is known, the intervals are not of equal length and their number is finite.

Let $0 = x_0 < x_1 < \cdots < x_{k-1} < \infty$ be the dividing points and $N_1, \ldots, N_k$, $\left(\sum_{i=1}^{k} N_i = n\right)$ the numbers of observed values in the respective intervals. Then the maximum likelihood estimator, $\hat{\sigma}$, is the unique solution of

$$\sum_{i=1}^{k-1} \frac{N_i(x_i - x_{i-1})}{e^{(x_i - x_{i-1})/\hat{\sigma}} - 1} - \sum_{i=2}^{k} N_i x_{i-1} = 0$$

(which exists if and only if $N_1 < n$ and $N_k < n$). For large n

$$n \operatorname{var}(\hat{\sigma}^{-1}) \doteqdot \left(\sum_{i=1}^{k-1} \frac{(x_i - x_{i-1})^2}{e^{x_i/\sigma} - e^{x_{i-1}/\sigma}}\right)^{-1}.$$

For a given k the dividing points which minimize the asymptotic variance are given by

$$x_i/\sigma = \sum_{j=k-i}^{k-1} \delta_j$$

where

$$\delta_1 = g^{-1}(2), \qquad \delta_i = g^{-1}(2 + \delta_{i-1} - g(\delta_{i-1}))$$

and

$$g(x) = x(1 - e^{-x})^{-1}.$$

For example,

$$\begin{aligned} k &= 2 \qquad x_1/\sigma = 1.5936 \\ k &= 3 \qquad x_1/\sigma = 1.0176, \quad x_2/\sigma = 2.6112. \end{aligned}$$

The simplicity of mathematical analysis for the exponential distribution permits us to construct convenient Bayesian estimators of parameters of (1). Some recent results in this area (for censored samples) are presented in Varde [81] who also compared their performance with more natural (at least in this case) and efficient maximum likelihood and minimal variance unbiased estimators.

The sampling distribution of the maximum likelihood estimator of parameter σ in (8), based on a "time censored" sample was derived by Bartholomew [4].

6. Estimators Using Selected Quantiles

The exponential distribution is frequently used for life-test data. It is supposed that lifetime can be represented by a random variable with probability density function

$$p_X(x) = \sigma^{-1} \exp(-x/\sigma) \qquad (x > 0;\ \sigma > 0) \tag{8$'$}$$

that is, as in (1) but with $\theta = 0$.

If a number (n) of items are observed with lifetimes commencing simultaneously, then, as each life concludes, observations of lifetime become available sequentially, starting with the shortest lifetime, X_1', of the n items, followed by the 2nd, 3rd . . . shortest lifetimes X_2', X_3', . . . respectively. Clearly it will be advantageous if useful inferences can be made at a relatively early stage, without waiting for completion of the longer lifetimes. This means that inference must be based on observed values of the r, say, shortest lifetimes, or in more general terms, the first *r order statistics*. On account of the practical importance of these analyses, statistical techniques have been worked out in considerable detail. Here we will describe only methods of estimation, but a considerable range of test procedures is also available.

The theory of order statistics from exponential distributions is quite simple, because the cumulative distribution functions are simple in form. The probability density function of the rth smallest value, X_r', among n independent random variables each having a standard exponential distribution (i.e. as in (1), with $\theta = 0$, $\sigma = 1$) is

$$p_{X_r'}(x) = \frac{n!}{(r-1)!(n-r)!} e^{-(n-r+1)x}(1 - e^{-x})^{r-1} \qquad (0 \le x).$$

The joint density function of X_1', X_2', . . . , X_n' is

$$p_{X_1',\ldots,X_n'}(x_1,\ldots,x_n) = n! \exp\left(-\sum_{j=1}^{n} x_j\right) \qquad (0 \le x_1 \le x_2 \le \cdots \le x_n).$$

Transforming to new variables

$$X_1';\ V_2 = X_2' - X_1';\ V_3 = X_3' - X_2';\ \ldots;\ V_n = X_n' - X_{n-1}'$$

we find

$$p_{X_1',V_2,\ldots,V_n}(x_1,v_2,\ldots,v_n)$$
$$= n! \exp[-nx_1 - (n-1)v_2 - (n-2)v_3 - \cdots - 2v_{n-1} - v_n]$$
$$(0 \le x_1; 0 \le v_j, j = 2,\ldots,n).$$

Hence we see that

(i) X_1', V_2, V_3, . . . , V_n are mutually independent
(ii) the distribution of X_1' is exponential (1) with $\theta = 0$, $\sigma = n^{-1}$
(iii) the distribution of V_j is exponential (1) with $\theta = 0$,

$$\sigma = (n-j+1)^{-1} \qquad (j = 2,\ldots,n).$$

Since $X'_j = X'_1 + V_2 + \cdots + V_j$ $(j \geq 2)$ it follows that all linear functions of the order statistics can be expressed as linear functions of the independent random variables $X'_1, V_2, \ldots, V_n$. This form of representation (suggested by Sukhatme [78] and Rényi [63]—see also Epstein and Sobel [27] for a similar result) is very helpful in solving distribution problems associated with methods described in the remainder of this section. A similar kind of representation can be applied to gamma distributions, though the results are not so simple.

It is necessary to distinguish between *censoring*, in which the order statistics that will be used (e.g. the r smallest values) are decided in advance, and *truncation* in which the range of values that will be used (e.g. all observations less than T) is decided in advance (regardless of how many observations fall within the specified limits). Truncation by omission of all observations less than a fixed value $T_0(> 0)$, has the effect that observed values may be represented by a random variable with probability density function

(11) $$\sigma^{-1} \exp [-(x - T_0)/\sigma] \qquad (x > T_0; \quad \sigma > 0)$$

which is again of form (1) (with T_0 (known) replacing θ), and so presents no special difficulties. However, if (as is more commonly the case) truncation is by omission of all values greater than $T_0(> 0)$ then the corresponding probability density function is

(12)
$$p_X(x) = [1 - \exp(-T_0/\sigma)]^{-1}\sigma^{-1} \exp(-x/\sigma) \qquad (0 < x < T_0; \quad \sigma > 0).$$

If m observations are obtained, they can be represented by independent random variables $X_1, X_2, \ldots, X_m$ each with distribution (12). The maximum likelihood equation for an estimator $\hat{\sigma}_{T_0}$ of σ is

(13) $$\hat{\sigma}_{T_0} = m^{-1} \sum_{j=1}^{m} x_j + T_0[\exp(T_0/\hat{\sigma}_{T_0}) - 1]^{-1}.$$

This equation may be solved by an iterative process. In this work the table of Barton *et al.* [5] is useful.

We will now restrict ourselves to censored samples only. We will give details only for the case when $\theta = 0$ (so that the probability density function is as in (8)) and when censoring results in omission of the largest $(n - r)$ values (i.e. observation of the r smallest values, where r is specified prior to obtaining the observations).

The joint probability density function of the $r(> 1)$ smallest observations in a random sample of size n is

(14)
$$p_{X'_1,\ldots,X'_r}(x_1,\ldots,x_r) = \frac{n!}{(n-r)!\sigma^r} \exp\left[-\left(\sum_{j=1}^{r-1} x_j + (n - r + 1)x_r\right)\Big/\sigma\right]$$

$$(0 \leq x_1 \leq x_2 \cdots \leq x_r).$$

The maximum likelihood estimator of σ is

(15) $$Y'_r = r^{-1}\left(\sum_{j=1}^{r-1} X'_j + (n - r + 1)X'_r\right).$$

This statistic is distributed as $(\frac{1}{2}\sigma/r) \times (\chi^2$ with $2r$ degrees of freedom). The expected value of Y'_r is therefore σ, and its variance is σ^2/r. The limits

$$2rY'_r/\chi^2_{2r,1-\frac{1}{2}\alpha} \quad \text{and} \quad 2rY'_r/\chi^2_{2r,\frac{1}{2}\alpha}$$

define a confidence interval for σ with confidence coefficient $100(1 - \alpha)\%$.

A wide variety of estimators of σ and θ based on order statistics is available. Many references at the end of this chapter contain discussions of such estimators ([4], [11], [14], [15], [17], [21], [28], [37], [51]) as well as other estimation procedures. Among the problems discussed are estimation:

(*a*) by linear functions of a limited number (usually not more than about five) of order statistics — this includes both the choice of coefficients in the linear function, and of the order statistics to be used,

(*b*) when only the r *largest* values are observed,

(*c*) when predetermined numbers of both the smallest and largest values are omitted,

(*d*) conversely when only these values are known — that is, the middle group of observations is omitted.

In all cases formulas have been obtained appropriate to estimation of σ, knowing θ; of θ, knowing σ; and of both θ and σ, neither being known. We now discuss some of the more useful of these formulas.

The variance-covariance matrix for the order statistics $X'_1 \leq X'_2 \leq \cdots \leq X'_n$ has elements

(16) $$\operatorname{var}(X'_r) = \sigma^2 \sum_{j=1}^{r} (n - j + 1)^{-2} = \operatorname{cov}(X'_r, X'_s) \qquad (r < s).$$

Also

(17) $$E(X'_r) = \theta + \sigma \sum_{j=1}^{r} (n - j + 1)^{-1}.$$

From these relationships it is straightforward to construct best linear unbiased estimators based on k selected order statistics $X'_{n_1}, X'_{n_2}, \ldots, X'_{n_k}$ with

$$n_1 < n_2 < \cdots < n_k.$$

It will be convenient to use the notation

(18) $$w_{mi} = \sum_{j=n_{i-1}}^{n_i - 1} (n - j)^{-m}$$

with $n_0 = 0$, and w_{10}/w_{20}, $w_{1,k+1}/w_{2,k+1}$ each defined to be zero.

If θ is known, the best linear unbiased estimator of σ is

$$\tilde{\sigma} = \left[\sum_{i=1}^{k}\left(\frac{w_{1i}}{w_{2i}} - \frac{w_{1,i+1}}{w_{2,i+1}}\right)X'_{n_i} - \frac{w_{11}}{w_{21}}\theta\right]\left[\sum_{i=1}^{k}\frac{w_{1i}^2}{w_{2i}}\right]^{-1}. \tag{19}$$

The variance of $\tilde{\sigma}$ is $\sigma^2\left(\sum_{i=1}^{k} w_{1i}^2 w_{2i}^{-1}\right)^{-1}$.

Some special cases are:

(*a*) $n_i = i$; $k = n$ (complete sample):

$$\tilde{\sigma} = n^{-1}\sum_{i=1}^{n} X_i; \quad \text{var}(\tilde{\sigma}) = \sigma^2/n.$$

(*b*) $n_i = r_1 + i$; $k = n - r_1 - r_2$ (censoring r_1 smallest and r_2 largest values):

$$\tilde{\sigma} = K^{-1}\left[\frac{\sum_{i=1}^{r_1+1}(n-i+1)^{-1}}{\sum_{i=1}^{r_1+1}(n-i+1)^{-2}}(X'_{r_1+1} - \theta) - (n - r_1)X'_{r_1+1} + (n - r_1 - k)X'_{r_1+k} + \sum_{i=r_1+1}^{r_1+k} X'_i\right] \tag{20}$$

with

$$K = \left\{\sum_{i=1}^{r_1+1}(n-i+1)^{-1}\right\}^2\left\{\sum_{i=1}^{r_1+1}(n-i+1)^{-2}\right\}^{-1} + k - 1.$$

(If $r_1 = 0$, this is the case of omission of the r_2 largest values only, already discussed. If $r_2 = 0$, it is the case of omission of the r_1 smallest values only.)

(*c*) $n_i = i$ for $i = 1, 2, \ldots, r_1$; $n_j = n - r_2 + j - r_1$ for $j = r_1 + 1, \ldots, r_1 + r_2$; $k = r_1 + r_2$.

This corresponds to censoring by omission of a central group of values, retaining only the r_1 smallest and r_2 largest values. In this case

$$\begin{aligned}\tilde{\sigma} = {} & [r_1 + r_2 - 1 + w'^2/w]^{-1} \\ & \times\left[\sum_{i=1}^{r_1-1} X'_i + (n - r_1 + 1 - w'/w)X'_{r_1} + (w'/w - r_2 + 1)X'_{r_2+1} + \sum_{i=r+2}^{n} X'_i - n\theta\right],\end{aligned} \tag{21}$$

where $w' = \sum_{i=r_1+1}^{n-r_2+1}(n-i+1)^{-1}$; $w = \sum_{i=r_1+1}^{n-r_2+1}(n-i+1)^{-2}$.

(*d*) $k = 1$ (estimation from a single order statistic).

In this case $\tilde{\sigma} = \left[\sum_{i=1}^{n_1}(n-i+1)^{-1}\right]^{-1}(X'_{n_1} - \theta)$ and

$$\text{(22)} \qquad \operatorname{var}(\tilde{\sigma}) = \sigma^2 \left[\sum_{i=1}^{n_1} (n - i + 1)^{-2} \right] \left[\sum_{i=1}^{n_1} (n - i + 1)^{-1} \right]^2.$$

The variance is a minimum (with respect to n_1) when n_1 is the nearest integer to $0.79681(n + 1) - 0.39841 + 1.16312(n + 1)^{-1}$ (Siddiqui [76]). Epstein [25] has noted that the efficiency of the unbiased estimator based on $(X_r' - \theta)$ is never less than 96% if $r/n \leq \frac{1}{2}$; or 90% if $r/n \leq \frac{2}{3}$.

(*e*) $k = 2$ (estimation from two order statistics X_{n_1}', X_{n_2}').

The variance of σ is a minimum when n_1 is the nearest integer to $0.6386(n + \frac{1}{2})$ and n_2 is the nearest integer to $0.9266(n + \frac{1}{2})$. (Siddiqui [76]).

For small samples Sarhan and Greenberg [71] give the following optimal choices:

Sample size (n):	2–4	5–7	8–11	12–15	16–18	19–21
n_1	$n - 1$	$n - 2$	$n - 3$	$n - 4$	$n - 6$	$n - 7$
n_2	n	n	n	n	$n - 1$	$n - 1$

If θ is not known, the optimal choices are different. Sarhan and Greenberg [71] give the following optimal choices for n_1 and n_2 (with $k = 2$):

Sample size (n):	2–6	7–10	11–15	16–20	21
n_1	1	1	1	1	1
n_2	n	$n - 1$	$n - 2$	$n - 3$	$n - 4$

The best (in fact, only) linear unbiased estimator of σ, using only X_{n_1}' and X_{n_2}' is

$$(X_{n_2}' - X_{n_1}') \left[\sum_{j=n_1}^{n_2-1} (n - j)^{-1} \right]^{-1}$$

and its variance is

$$\left[\sum_{j=n_1}^{n_2-1} (n - j)^{-2} \right] \left[\sum_{j=n_1}^{n_2-1} (n - j)^{-1} \right]^{-2} \sigma^2.$$

The optimal value of n_1 is always 1, whatever the values of n and n_2.

Kulldorff [41] considered the problem of choosing $n_1, n_2, \ldots, n_k$ from the whole sample to minimize $\operatorname{var}(\tilde{\sigma})$ or $\operatorname{var}(\tilde{\theta})$ for k fixed. He tabulated the optimal n's and coefficients for small values of k and n. Earlier Sarhan and Greenberg [71] treated the asymptotic (large n) case (for θ known) giving the optimal percentiles $n_1/n, n_2/n, \ldots, n_k/n$, for k fixed. These tables are reproduced by Ogawa [55]. Saleh and Ali [68] and Saleh [67] proved the uniqueness of the optimal selection and extended the results to censored cases.

The table of Zabransky *et al.* [82] is most exhaustive and covers uncensored and censored samples, finite and asymptotic cases for a wide range. Sibuya [75] gave the algorithms for computing their tables, and unified previous results in simpler form.

Epstein [25] has given a number of useful results. He points out that $(X'_r - \theta)$ is approximately distributed as $\frac{1}{2}(\sigma/r)$ (χ^2 with $2r$ degrees of freedom). Approximate confidence intervals for σ can be constructed on this basis. Epstein also gives formulas for the minimum variance unbiased estimators of θ and σ for the two-parameter distribution (1), based on $X'_1, X'_2, \ldots, X'_r$.

These are

$$\sigma^* = (r - 1)^{-1} \sum_{j=1}^{r-1} (n - j)(X'_{j+1} - X'_j)$$

$$\theta^* = X'_1 - n^{-1}\sigma^*;$$

σ^* is distributed as $\frac{1}{2}\sigma(r - 1)^{-1} \times$ (χ^2 with $2(r - 1)$ degrees of freedom).

$100(1 - \alpha)\%$ confidence limits for σ are

$$2(r - 1)\sigma^*/\chi^2_{2(r-1),1-\alpha/2} \qquad \text{and} \qquad 2(r - 1)\sigma^*/\chi^2_{2(r-1),\alpha/2}$$

(using the notation of Chapter 17, Section 4).

$100(1 - \alpha)\%$ confidence limits for θ are

$$X'_1 - F_{2,2(r-1),1-\alpha}\sigma^* n^{-1} \qquad \text{and} \qquad X'_1$$

(using the notation of Chapter 26, Section 1).

If only $X'_{r_1+1}, \ldots, X'_{n-r_2}$ are to be used, the minimum variance unbiased estimators are

$$\sigma^* = (n - r_1 - r_2 - 1)^{-1} \sum_{j=r_1+1}^{n-r_2-1} (n - j + 1)(X'_{j+1} - X'_j);$$

$$\theta^* = X'_{r_1+1} - \sigma^* \sum_{j=0}^{r_1} (n - j)^{-1};$$

σ^* is distributed as $\frac{1}{2}\sigma(n - r_1 - r_2 - 1)^{-1} \times$ (χ^2 with $2(n - r_1 - r_2 - 1)$ degrees of freedom) and

$$\text{var}(\theta^*) = \sigma^2\left[\sum_{j=0}^{r_1} (n - j)^{-2} + (n - r_1 - r_2 - 1)^{-1}\left\{\sum_{j=0}^{r_1} (n - j)^{-1}\right\}^2\right].$$

7. Characterizations

I. If X and Y are independent random variables with absolutely continuous distributions, and min (X,Y) and $(X - Y)$ are mutually independent random variables, then both X and Y have exponential distributions with a common location parameter (θ) but with possibly different scale parameters (σ).

(Proved by Ferguson [29]. A slight extension is due to Crawford [12].)

II. If $X_1, X_2, \ldots, X_n$ are independent, and each has the same absolutely continuous distribution, and $X'_1 \leq X'_2, \ldots, \leq X'_n$ are the corresponding ordered variables, then, if the conditional expectation $E[X'_{m+1} - X'_m \mid X'_m = w]$

is independent of w for fixed m, $1 \leq m < n$, the X's have exponential distributions. (Srivastava [77].)

Govindarajulu [33] has shown that the condition that X_1' and $(X_2' - X_1', X_3' - X_1', \ldots, X_n' - X_1')$ are independent characterizes an exponential distri, bution. He also states that independence of X_1' and $\sum_{j=2}^{n} (X_j' - X_1')$ characterizes exponential distributions.

Rossberg [66] has reported that the conditions that $\sum_{j=k}^{n} c_j X_j'$ (with $\sum_{j=k}^{n} c_j = 0$ and c_j's not all zero) and X_h' $(h \leq k)$ are independent ensure that the distribution is of exponential form. If, further,

$$\Pr[X_k' - X_{k-1}' \geq x] = (1 - F_X(x))^{n-k+1}$$

then the start of the distribution is at zero.

III. If X_1, X_2 have the same absolutely continuous distribution, then if the distribution of $X_1/(X_1 + X_2)$ is uniform over the range 0 to 1 (and is independent of $(X_1 + X_2)$), the common distribution of X_1 and X_2 is exponential with location parameter θ equal to zero. (Patil and Seshadri [57].)

IV. Guerrieri [35] has shown that the following properties also characterize the exponential distribution:

(*a*) The variance of the conditional distribution, given that the variable takes values exceeding x (greater than θ) does not depend on x.

(*b*) As for (*a*) 'mean deviation' replacing 'variance'.

(*c*) As for (*a*) "mean difference" (expected value of modulus of difference between two independent, identically distributed variables) replacing "variance".

V. The argument used in Section 2 shows that the conditions

(i) $\Pr[X \leq x_0 + x \mid X > x_0] = \Pr[X \leq x]$ for all $x_0 > 0$ and all $x > 0$,

(ii) X has a probability density function,

characterize distributions of form (8). Reinhardt [62] has extended this result, showing that (i) can be replaced by the weaker condition that

"$E[X - x_0 \mid X > x_0] > 0$ does not depend on x for any $x_0 > 0$"

and that (ii) can be omitted.

8. Applications

As has already been mentioned in Section 1, the exponential distribution is applied in a very wide variety of statistical procedures. Currently among the most prominent applications are those in the field of life-testing. The lifetime (or life characteristic, as it is often called) can often be usefully represented by an exponential random variable, with (usually) a relatively simple associated theory. Sometimes the representation is not adequate; in such cases a modification of the exponential distribution (very often a Weibull distribution (Chapter

20)) is used.

Another application is producing usable approximate solutions to difficult distributional problems. The approximation [(80) of Chapter 29] to the distribution of a quadratic form is a case in point. An ingenious application of the exponential distribution to approximate a sequential procedure is due to Ray [61]. He wished to calculate the distribution of the smallest n for which $\sum_{i=1}^{n} U_i^2 < K_n$, where $U_1, U_2, \ldots$ are independent unit normal variables and $K_1, K_2, \ldots$ are specified positive constants. By replacing this by the distribution of the smallest *even* n he obtained a problem in which the sums $\sum_{i=1}^{n} U_i^2$ are replaced by sums of independent exponential variables (actually χ^2's with two degrees of freedom each).

9. Related Distributions

There are many important distributions closely related to the exponential distribution.

If a variable $X(> \theta)$ is such that $Y = (X - \theta)^c$ has the exponential distribution

$$p_Y(y) = \sigma^{-1}e^{-y/\sigma} \qquad (y > 0) \tag{8$'$}$$

then X is said to have a *Weibull distribution* with shape parameter c. It is necessary that c should be greater than -1, as otherwise the integral of

$$p_X(x) = \sigma^{-1}c(x - \theta)^{c-1} \exp\left[-(x - \theta)^c/\sigma\right] \qquad (x > \theta)$$

between $x = \theta$ and $x = \theta' > \theta$ will be infinite. (See Chapter 20 for a detailed discussion of this distribution.) If c and θ are known (and often θ is known to be zero) the transformed variable $Y = (X - \theta)^c$ may be used, and the well-developed techniques associated with the exponential distribution become applicable. If c be not known, special techniques are needed; these are discussed in Chapter 20.

If $Y = e^{-X}$ has an exponential distribution of form (8)′ then X has a distribution of 'extreme value' form (Chapter 21).

Another distribution of theoretical, and some practical, importance is the *double* (or *bilateral*) exponential distribution. This is the distribution of a random variable X with probability density function

$$p_X(x) = (2\sigma)^{-1} \exp\left(-|x - \theta|/\sigma\right) \qquad (\sigma > 0). \tag{23}$$

It is also known as *Laplace's First Law of Error*. (*Laplace's Second Law of Error* is the normal distribution.) The exponential distribution might be regarded as a double exponential 'folded' about $x = \theta$. The double exponential distribution is the subject of Chapter 23.

The fact that the exponential distribution belongs to the class of gamma distributions has been mentioned in Section 1. Taking $\theta = 0$ and $\sigma = 2$ in (1), one obtains the distribution of χ^2 with 2 degrees of freedom. From this it can be seen that if $X_1, X_2, \ldots, X_n$ are independent random variables each with probability density function (1) then their arithmetic mean is distributed as $\theta + \frac{1}{2}n^{-1}\sigma \times$ (χ^2 with $2n$ degrees of freedom). We further note that if the quadratic form $\sum_{i,j=1}^{n} a_{ij}U_iU_j$ in independent unit normal variables $U_1, U_2, \ldots,$ U_n has a matrix A with eigenvalues $\lambda_1, \lambda_2, \ldots, \lambda_n$ then by an appropriate linear transformation of the U's

$$\sum_{i,j=1}^{n} a_{ij}U_iU_j = \sum_{j=1}^{n} \lambda_j U_j^{*2} \tag{24}$$

where $U_1^*, U_2^*, \ldots, U_n^*$ are independent unit normal variables. If the non-zero λ's are equal in pairs then $\sum_{j=1}^{n} \lambda_j U_j^{*2}$ can be expressed as a linear function of independent variables each distributed as χ^2 with two degrees of freedom, and so exponentially. The distribution theory of such a variable is much simpler than that of a general quadratic form in normal variables (Chapter 29).

In fact, if $X_1, \ldots, X_n$ are independent standard exponential variables, then the linear form

$$Y = \sum_{j=1}^{n} \lambda_j X_j \qquad \lambda_j \neq \lambda_k$$

has the probability density function

$$p_Y(y) = \sum_{j=1}^{n} \left(\prod_{k \neq j} (\lambda_j - \lambda_k)^{-1} \right) \lambda_j^{n-2} e^{-y/\lambda_j} \; (y > 0). \tag{25}$$

This is a special type of mixture of exponential distributions and called the *general gamma* or the *general Erlang* distribution. It is used in queueing theory, reliability theory and psychology. Some special patterns of λ's have been discussed by Likeš [45] [46], McGill and Gibbon [49] and others.

The relationship between the exponential and rectangular distributions is discussed in more detail in the chapter on the latter (Chapter 25). Here we note only the following property: If $X_1, X_2, \ldots, X_n$ each have probability density function (8′) then $Y_j = \sum_{i=1}^{j} X_i \Big/ \left[\sum_{i=1}^{n} X_i\right]$ for $j = 1, \ldots, (n-1)$ are distributed as the order statistics of a random sample of size $(n-1)$ from a rectangular distribution over the range 0 to 1.

The following relation between Poisson and exponential distributions is of importance in applications (see also Chapter 4, Sections 1 and 9): If $T_1, T_2, \ldots,$ are a succession of independent random variables each having the same density function

$$p_{T_j}(t) = \sigma^{-1} \exp(-t/\sigma) \qquad (t > 0; \sigma > 0)$$

and the random variable N is defined by the inequalities

$$T_1 + T_2 + \cdots + T_N = \leq \tau < T_1 + T_2 + \cdots + T_{N+1}$$

then N has a Poisson distribution with expected value τ/σ.

An exponential distribution truncated from below is still an exponential distribution, with the same scale parameter (see equation (2)).

An exponential distribution truncated by exclusion of values exceeding x_0 has the density function

$$p_X(x) = \sigma^{-1}\{1 - \exp[-(x_0 - \theta)/\sigma]\}^{-1} \exp[-(x - \theta)/\sigma] \quad (x_0 > x > \theta). \tag{26}$$

The expected value of this distribution is

$$\theta + \sigma\left[1 - \frac{x_0 - \theta}{\sigma}\left\{\exp\left[\frac{x_0 - \theta}{\sigma}\right] - 1\right\}^{-1}\right].$$

If $X_1, X_2, \ldots, X_n$ are independent random variables, each having the distribution (26) then the statistic $\sum_{j=1}^{n} X_j$ is sufficient for σ, if θ be known. Bain and Weeks [1] have shown that the density function of $Y = \sum_{j=1}^{n} X_j - n\theta$ is

$$p_Y(y) = [\sigma\{1 - \exp[-(x_0 - \theta)/\sigma]\}]^{-n} e^{-y/\sigma}[(n-1)!]^{-1} \times \sum_{j=0}^{j_0} (-1)^j \binom{n}{j} [y - j(x_0 - \theta)]^{n-1} \tag{27.1}$$

for $j_0(x_0 - \theta) < y < (j_0 + 1)(x_0 - \theta)$; $j_0 = 0,1,2,\ldots,(n-1)$.

The cumulative distribution function is

$$F_Y(y) = \{1 - \exp[-(x_0 - \theta)/\sigma]\}^{-n} \sum_{j=0}^{j_0} (-1)^j \binom{n}{j} e^{-j(x_0-\theta)/\sigma} \times \Pr[\chi^2_{2n} < 2(j_0 + d - j)(x_0 - \theta)/\sigma], \tag{27.2}$$

where $y = (x_0 - \theta)(j_0 + d)$, j_0 being an integer and $0 \leq d \leq 1$.

Deemer and Votaw [14] show that if θ be known to be zero, then the maximum likelihood estimator $\hat{\sigma}$ of σ is the solution of the equation

$$\overline{X}x_0^{-1} = \hat{\sigma}x_0^{-1} - (e^{x_0/\hat{\sigma}} - 1)^{-1}$$

provided $\overline{X} < \frac{1}{2}x_0$. If $\overline{X} > \frac{1}{2}x_0$ then $\hat{\sigma}$ is infinite — this may be taken to mean that a truncated exponential distribution is inappropriate. (Formally, it indicates that a rectangular distribution (Chapter 25) over the range 0 to x_0 should be used.)

In [14] there is a table of $\overline{X}/x_0$ to 4 decimal places as a function of $\hat{\sigma}^{-1}$ for

$\hat{\sigma}^{-1} = 0.01(0.01)0.89$. For n large

$$n \operatorname{var}(\hat{\sigma}^{-1}) \doteqdot [\sigma^2 - x_0^2 e^{-x_0/\sigma}(1 - e^{-x_0/\sigma})^{-2}]^{-1}.$$

10. Mixtures of Exponential Distributions

Recently mixed exponential distributions have come into prominence through their life testing applications. A batch of electronic tubes may consist of two sub-populations, each sub-population having its own characteristic mean life, these two sub-populations being in a proportion of p to $q = 1 - p$. (Each unit of the population may be regarded as if it contains a "tag" which indicates the sub-population to which it belongs and hence defines the way in which that particular item will fail.) Finally, the failure times for the ith sub-population, $(i = 1,2,)$ are assumed to be independently distributed with probability density function

$$p_T(t) = \sigma_i^{-1} \exp(-t/\sigma_i) \qquad (t \geq 0)$$

so that

$$\Pr[T \leq t_0] = 1 - \exp(-t_0/\sigma_i) \qquad (t_0 \geq 0).$$

If p is the proportion of units belonging to sub-population $i = 1$, then the cumulative distribution function for the population is

$$p[1 - \exp(-t/\sigma_1)] + (1 - p)[1 - \exp(-t/\sigma_2)] \qquad (t \geq 0) \tag{28.1}$$

and the density function is

$$p\sigma_1^{-1}e^{-t/\sigma_1} + (1 - p)\sigma_2^{-1}e^{-t/\sigma_2} \qquad (t \geq 0). \tag{28.2}$$

This distribution has also been applied (under the name of *Schuhl distribution*) to the distances between elements in traffic (Petigny [58]). Mixtures of exponential distributions have been found to represent some demographic distributions (Susara and Pathala [79]).

In 1939, Gumbel [34] gave a method for fitting a mixture of two exponential distributions, using the first two sample moments.

Cases where σ has a continuous distribution have been studied by Bhattacharya and Holla [9] and Bhattacharya [8]. In [8] it is shown that if σ has an exponential distribution (with parameters $0,\beta$) the probability density function of T is $2\beta^{-1}K_0(2(x/\beta)^{\frac{1}{2}})$ $(0 < x)$, where $K_0(\cdot)$ is a modified Bessel function of the third kind of zero order. If σ has a beta distribution (Chapter 24) over the range $(0,\theta)$ the probability density function can be expressed in terms of a Whittaker function. In [9] the special case where σ has a uniform distribution is considered.

We now consider methods of estimating the parameters p, σ_1, and σ_2 in (28.1)–(28.2).

Case A: Relative magnitude of sub-population parameters is not known.

Mendenhall and Hader [50] derived the following equations for the maximum likelihood estimators, expressed in terms of

$$\hat{p} = \frac{R_1}{n} + \hat{k}\,\frac{(n - R)}{n}, \tag{29.1}$$

$$\hat{\beta}_1 = \bar{x}_1 + \hat{k}\,\frac{(n - R)}{R_1}, \tag{29.2}$$

$$\hat{\beta}_2 = \bar{x}_2 + \frac{(1 - \hat{k})(n - R)}{R_2}, \tag{29.3}$$

where

$$\bar{x}_i = \frac{\sum_{j=1}^{R_i} X_{ij}}{R_i}, \qquad (i = 1,2)$$

and

$$\hat{k} = \frac{1}{1 + (\hat{q}/p)\exp[1/\hat{\beta}_1 - 1/\hat{\beta}_2]}. \tag{29.4}$$

Here:

n = the number of units on test;
t_0 = the time at which test is terminated;
R = the total number of failures during test;
R_1 = the number of failures from sub-population 1; R_2 the number of failures from sub-population 2 (note that $R_1 + R_2 = R$), and

$$X_{ij} = \frac{T_{ij}}{t_0},$$

T_{ij} being the time at which the jth failure from ith population occurs, and $\hat{\beta}_i$ is the estimator of $\beta_i = \sigma_i/t_0$ $(i = 1,2)$.

Now the estimators of β_1, β_2, and p must be obtained by solving the simultaneous equations (29). A method of solving these, approximately, is given by Mendenhall and Hader [50].

Case B: Relative magnitude of sub-population parameters is known.

In the special case $\beta_1 = \beta_2 = \beta$ our maximum likelihood estimators become

$$\hat{p} = \frac{r_1}{r}. \tag{29.1$'$}$$

$$\hat{\beta} = \frac{r_1\bar{x}_1 + r_2\bar{x}_2 + (n - r)}{r}, \tag{29.2$'$}$$

In general we have to use the method in Case A to obtain estimators, but if we have crossover, i.e.,

$$\hat{\beta}_1 > \hat{\beta}_2 \text{ while } \beta_1 \leq \beta_2$$

or

$$\hat{\beta}_1 < \hat{\beta}_2 \text{ while } \beta_1 \geq \beta_2$$

we use the above (Case B) estimators corresponding to the special case $\beta_1 = \beta_2 = \beta$.

Rider [65] uses the "method of moments," to obtain rather involved estimators of mixed exponential parameters.

Mixtures of generalized exponential distributions have also found application in "mine dust" analysis, (Tallis and Light [80]). In particular the problem which arises is that of estimating the parameters α, β, γ of the distribution with density function

$$p_X(x) = \gamma A e^{-\alpha\sqrt{x}} + (1 - \gamma) B e^{-\beta\sqrt{x}} \qquad (x > a)$$

where A, B, a are known constants.

The method of maximum likelihood is often regarded as the standard method of estimating parameters. In this case, however, we would obtain rather complicated equations, and methods using *half-moment* equations have been developed (Joffe [39]). These are less complicated, and, although they must be solved iteratively, they can be handled on a desk calculator. The three half-moment equations are

$$(30) \qquad m'_{\frac{1}{2}r} = \tilde{\gamma}\mu'_{\frac{1}{2}r,\tilde{\alpha}} + (1 - \tilde{\gamma})\mu'_{\frac{1}{2}r,\tilde{\beta}} \qquad (r = 1, 2, 3)$$

where

$$m'_{\frac{1}{2}r} = \frac{1}{n}\sum_{1}^{n} X_i^{\frac{1}{2}r} \qquad (n \text{ is the numbers of elements})$$

and

$$\mu'_{\frac{1}{2}r,\tilde{\alpha}} = \frac{\tilde{\alpha}^2 e^{\tilde{\alpha}\sqrt{a}}}{2(1 + \tilde{\alpha}\sqrt{a})}\int_a^\infty x^{\frac{1}{2}r} e^{-\alpha\sqrt{x}}\, dx.$$

Use may be made of the recurrence relationship

$$\int_a^\infty x^{\frac{1}{2}r} e^{-\alpha\sqrt{x}}\, dx = \frac{2a^{\frac{1}{2}(r+1)} e^{-\alpha\sqrt{a}}}{\alpha} + \frac{r+1}{\alpha}\int_a^\infty x^{\frac{1}{2}(r-1)} e^{-\alpha\sqrt{x}}\, dx.$$

One can express $\tilde{\beta}$ as a function of $\tilde{\alpha}$ and $\tilde{\gamma}$ and the three half-moments as follows

$$(31.1) \qquad \hat{\beta} = \frac{P - \tilde{\gamma}Q}{R - (\tilde{\gamma}/\tilde{\alpha})Q}$$

where

$$P = 4m'_1 - 3\sqrt{am'_{\frac{1}{2}}}$$

$$Q = 4\mu'_{1,\tilde{\alpha}} - 3\sqrt{a\mu'_{\frac{1}{2},\tilde{\alpha}}}$$

$$R = m'_{\frac{3}{2}} - \sqrt{am'_1}.$$

$\tilde{\gamma}$ may be written as a function of $\tilde{\alpha}$ and the three half-moments:

$$(31.2) \qquad \tilde{\gamma} = \frac{PU - SR}{Q\left(U - \dfrac{S}{\tilde{\alpha}}\right) - \tau\left(R - \dfrac{P}{\tilde{\alpha}}\right)}$$

where

$$U = m'_1 - \sqrt{am'_{\frac{1}{2}}}$$

$$S = 3m'_{\frac{1}{2}} - 2\sqrt{a}$$

$$\tau = 3\mu'_{\frac{1}{2},\tilde{\alpha}} - 2\sqrt{a}.$$

A solution of our equations can be obtained iteratively by choosing a value of $\tilde{\alpha}$, and consequent values from (31.2) and (31.1) of $\tilde{\gamma}$ and $\tilde{\beta}$.

11. A Generalization of the Exponential Distribution

Estimation of θ for the cumulative distribution function

$$1 - e^{-x/\theta}(1 - F(x)),$$

(where $F(x)$ is the cumulative distribution function of a non-negative random variable) was treated by Gercbah [31]). He suggested an estimator based on a random sample of size $n = km$. Divide the sample into m groups at random, with $X'_{1(j)}$ the smallest value in the jth group. Then under certain monotonicity conditions on $F(x)$

$$\hat{\theta} = km^{-1} \sum_{j=1}^{m} X'_{1(j)}$$

is a consistent estimator of θ as k and m tend to infinity.

Also:

$$E(\hat{\theta}) = \theta - \frac{f'(0)}{1 - F(0)} \frac{\theta^3}{k} + o\left(\frac{1}{k}\right) \tag{32.1}$$

$$V(\hat{\theta}) = \frac{\theta^2}{m} - \frac{4f'(0)}{1 - F(0)} \cdot \frac{\theta^4}{km} + o\left(\frac{1}{km}\right) \tag{32.2}$$

where $f(x) = F'(x)$.

By using normal approximations a confidence interval for θ with approximate confidence coefficient $(1 - \alpha)$ is obtained as

$$\hat{\theta} \pm \left(K_{\alpha/2}\theta/\sqrt{m} + \frac{f'(0)\theta^3}{(1 - F(0))k}\right).$$

An 'optimal' choice of (k,m) is obtained by minimizing the quantity in the parentheses.

REFERENCES

[1] Bain, L. T. and Weeks, D. L. (1964). A note on the truncated exponential distribution, *Annals of Mathematical Statistics*, **35,** 1366–1367.

[2] Bánkövi, G. (1964). A decomposition-rejection technique for generating exponential random variables, *Mathematical Proceedings of the Hungarian Academy of Science*, **9,** *Series A*, 573–581.

[3] Barnett, V. D. (1965). *Random Negative Exponential Deviates, Tracts for Computers*, XXVII, London: Cambridge University Press.

[4] Bartholomew, D. J. (1963). The sampling distribution of an estimate arising in life testing, *Technometrics*, **5,** 361–374.

[5] Barton, D. E., David, F. N. and Merrington, M. (1963). Table for the solution of the exponential equation, exp $(b) - b/(1 - p) = 1$, *Biometrika*, **50,** 169–176.

[6] Basu, A. P. (1965). On characterizing the exponential distribution by order statistics, *Annals of the Institute of Statistical Mathematics, Tokyo*, **17,** 93–96.

[7] Basu, A. P. (1965). On some tests of hypotheses relating to the exponential distribution when some outliers are present, *Journal of the American Statistical Association*, **60,** 548–559.

[8] Bhattacharya, S. K. (1966). A modified Bessel function model in life testing, *Metrika*, **11,** 133–144.

[9] Bhattacharya, S. K. and Holla, M. S. (1965). On a life test distribution with stochastic deviations in the mean, *Annals of the Institute of Statistical Mathematics, Tokyo*, **17,** 97–104.

[10] Carlson, P. G. (1958). Tests of hypothesis on the exponential lower limit, *Skandinavisk Aktuarietidskrift*, **41,** 47–54.

[11] Cohen, A. C. (1963). Progressively censored samples in life testing, *Technometrics*, **5,** 327–339.

[12] Crawford, G. B. (1966). Characterization of geometric and exponential distributions, *Annals of Mathematical Statistics*, **37,** 1790–1795.

[13] David, F. N. and Johnson, N. L. (1948). The probability integral transformation when parameters are estimated from the sample, *Biometrika*, **35,** 182–190.

[14] Deemer, W. L. and Votaw, D. F. (1955). Estimation of parameters of truncated or censored exponential distributions, *Annals of Mathematical Statistics*, **26,** 498–504.

[15] El-Sayyad, G. M. (1967). Estimation of the parameter of an exponential distribution, *Journal of the Royal Statistical Society*, *Series B*, **29,** 525–532.

[16] Epstein, B. (1954). Truncated life tests in the exponential case, *Annals of Mathematical Statistics*, **25,** 555–564.

[17] Epstein, B. (1957). Simple estimators of the parameters of exponential distributions when samples are censored, *Annals of the Institute of Statistical Mathematics, Tokyo*, **8,** 15–26.

[18] Epstein, B. (1958). Exponential distribution and its role in life testing, *Industrial Quality Control*, **15,** 4–9.

[19] Epstein, B. (1960). Tests for the validity of the assumption that the underlying distribution of life is exponential, Part I, *Technometrics*, **2,** 83–101.

[20] Epstein, B. (1960). Tests for the validity of the assumption that the underlying distribution of life is exponential, Part II, *Technometrics*, **2,** 167–183.

[21] Epstein, B. (1960). Estimation of the parameters of two parameter exponential distributions from censored samples, *Technometrics*, **2,** 403–406.

[22] Epstein, B. (1960). Statistical life test acceptance procedures, *Technometrics*, **2,** 435–446.

[23] Epstein, B. (1960). Estimation from life test data, *Technometrics*, **2,** 447–454.

[24] Epstein, B. (1961). Estimates of bounded relative error for the mean life of an exponential distribution, *Technometrics*, **3,** 107–109.

[25] Epstein, B. (1962). Simple estimates of the parameters of exponential distributions, (pp. 361–371 of *Contributions to Order Statistics* [Ed. A. E. Sarhan and B. G. Greenberg], New York: John Wiley & Sons, Inc.).

[26] Epstein, B. and Sobel, M. (1953). Life testing, *Journal of the American Statistical Association*, **48,** 486–502.

[27] Epstein, B. and Sobel, M. (1954). Some theorems relevant to life testing from an exponential distribution, *Annals of Mathematical Statistics*, **25,** 373–381.

[28] Epstein, B. and Sobel, M. (1955). Sequential life tests in the exponential case, *Annals of Mathematical Statistics*, **26,** 82–93.

[29] Ferguson, T. S. (1964). A characterization of the exponential distribution, *Annals of Mathematical Statistics*, **35,** 1199–1207.

[30] Fukuta, J. (1961). The use of sample quasi-ranges from the exponential population, *Research Reports of the Faculty of Engineering, Gifu University*, No. 11, 40–45.

[31] Gercbah (Gertsbach), I. B. (1967). On estimation of parameters of a distribution with an exponential factor, *Teoriya Veroyatnostei i ee Primeneniya*, **7,** 121–123. (English translation: *Theory of Probability and Applications*, **7,** 110–111.)

[32] Gnedenko, B. V., Belyaev Yu,K. and Solovev, A. D. (1965). *Mathematical Methods in Reliability Theory*, Moscow. (English translation: Academic Press, 1968.)

[33] Govindarajulu, Z. (1966). Characterization of the exponential and power distributions, *Skandinavisk Aktuarietidskrift*, **49,** 132–136.

[34] Gumbel, E. J. (1939). La dissection d'une répartition, *Annales de l'Université de Lyon, 3-me Séries, Section A*, **2,** 39–51.

[35] Guerrieri, G. (1965). Some characteristic properties of the exponential distribution, *Giornale degli Economisti e Annali di Economia*, **24,** 427–437.

[36] Harter, H. L. (1961). Estimating the parameters of negative exponential populations from one or two order statistics, *Annals of Mathematical Statistics*, **32,** 1078–1090.

[37] Harter, H. L. (1964). Exact confidence bounds, based on one order statistic, for the parameter of an exponential population, *Technometrics*, **6,** 301–317.

[38] Huzurbazar, V. S. (1955). Confidence intervals for the parameter of a distribution admitting a sufficient statistic when the range depends on the parameter, *Journal of the Royal Statistical Society, Series B*, **17,** 86–90.

[39] Joffe, A. D. (1964). Mixed exponential estimation by the method of half moments, *Applied Statistics*, **13,** 91–98.

[40] Knight, W. (1965). A method of sequential estimation applicable to the hypergeometric, binomial, Poisson, and exponential distributions, *Annals of Mathematical Statistics*, **36,** 1494–1503.

[41] Kulldorff, G. (1962). *On the asymptotic optimum spacings for the estimation of the scale parameter of an exponential distribution based on sample quantiles*, Mimeographed Report, University of Lund.

[42] Kulldorff, G. (1963). Estimation of one or two parameters of the exponential distribution on the basis of suitably chosen order statistics, *Annals of Mathematical Statistics*, **34,** 1419–1431.

[43] Kulldorff, G. (1961). *Contributions to the Theory of Estimation from Grouped and Partially Grouped Samples*, Stockholm: Almqvist and Wiksell, New York: John Wiley & Sons, Inc.

[44] Lieberman, A. (1959). Sequential life testing plans for the exponential distribution, *Industrial Quality Control*, **16,** 14–18.

[45] Likeš, J. (1967). Distributions of some statistics in samples from exponential and power-function populations, *Journal of the American Statistical Association*, **62,** 259–271.

[46] Likeš, J. (1968). Differences of two ordered observations divided by standard deviation in the case of negatively exponential distribution, *Metrika*, **12,** 161–172.

[47] Lloyd, E. H. (1952). Least-squares estimation of location and scale parameters using order statistics, *Biometrika*, **39,** 88–95.

[48] Marsaglia, G. (1961). Generating exponential random variables, *Annals of Mathematical Statistics*, **32,** 899–900.

[49] McGill, W. J. and Gibbon, J. (1965). The general-gamma distribution and reaction time, *Journal of Mathematical Psychology*, **2,** 1–18.

[50] Mendenhall, W. and Hader, R. J. (1958). Estimation of parameters of mixed exponentially distributed failure time distributions from censored life test data, *Biometrika*, **45,** 504–520.

[51] Nadler, J. (1960). Inverse binomial sampling plans when an exponential distribution is sampled with censoring, *Annals of Mathematical Statistics*, **31,** 1201–1204.

[52] Neumann, J. von (1951). *Various techniques in connection with random digits, Monte Carlo methods*, National Bureau of Standards, *Applied Mathematics Series*, **12,** 36–38, Washington, D. C.: U.S. Government Printing Office.

[53] Ogawa, J. (1951). Contributions to the theory of systematic statistics, I. *Osaka Mathematical Journal*, **3,** 175–213.

[54] Ogawa, J. (1960). Determination of optimum spacings for the estimation of the scale parameters of an exponential distribution based on sample quantiles, *Annals of the Institute of Statistical Mathematics, Tokyo*, **12,** 135–141.

[55] Ogawa, J. (1962). Optimum spacing and grouping for the exponential distribution (pp. 371–380 of *Contributions to Order Statistics*, (Ed. A. E. Sarhan and B. G. Greenberg), New York: John Wiley & Sons, Inc.).

[56] Patil, G. P. (1963). A characterization of the exponential-type distribution, *Biometrika*, **50,** 205–207.

[57] Patil, G. P. and Seshadri, V. (1964). Characterization theorems for some univariate probability distributions, *Journal of the Royal Statistical Society, Series B*, **26**, 286–292.

[58] Petigny, B. (1966). Extension de la distribution de Schuhl, *Annales des Ponts et Chaussées*, **136**, 77–84.

[59] Pugh, E. L. (1963). The best estimate of reliability in the exponential case, *Operations Research*, **11**, 57–61.

[60] Raghavachari, M. (1965). Operating characteristic and expected sample size of a sequential probability ratio test for the simple exponential distribution, *Bulletin of the Calcutta Statistical Association*, **14**, 65–73.

[61] Ray, W. D. (1957). Sequential confidence intervals for the mean of a normal population with unknown variance, *Journal of the Royal Statistical Society, Series B*, **19**, 133–143.

[62] Reinhardt, H. E. (1968). Characterizing the exponential distribution, *Biometrics*, **24**, 437–438.

[63] Rényi, A. (1953). On the theory of order statistics, *Acta Mathematica Academiae Scientarium Hungaricae*, **4**, 191–232.

[64] Rider, P. R. (1959). Quasi-ranges of samples from an exponential population, *Annals of Mathematical Statistics*, **30**, 252–254.

[65] Rider, P. R. (1961). The method of moments applied to a mixture of two exponential distributions, *Annals of Mathematical Statistics*, **32**, 143–147.

[66] Rossberg, H.-J. (1966). Charakterisierungsprobleme, die sich aus der von A. Rényi in die Theorie der Ranggrössen eingeführten Methode ergeben, *Monatsberichte der Deutschen Akademie der Wissenschaften zu Berlin*, **8**, 561–572.

[67] Saleh, A. K. M. E. (1966). Estimation of the parameters of the exponential distribution based on optimum order statistics in censored samples, *Annals of Mathematical Statistics*, **37**, 1717–1735.

[68] Saleh, A. K. M. E. and Ali, M. M. (1966). Asymptotic optimum quantiles for the estimation of the parameters of the negative exponential distribution, *Annals of Mathematical Statistics*, **37**, 143–151.

[69] Sarhan, A. E. (1954). Estimation of the mean and standard deviation by order statistics, *Annals of Mathematical Statistics*, **25**, 317–328.

[70] Sarhan, A. E. (1955). Estimation of the mean and standard deviation by order statistics. Part III, *Annals of Mathematical Statistics*, **26**, 576–592.

[71] Sarhan, A. E. and Greenberg, B. G. (1958). Estimation problems in the exponential distribution using order statistics, *Proceedings in Statistics of the Technical Missile Evaluation Symposium*, Blacksburg, Virginia, pp. 123–173. (See also [72], where the results quoted in the text are reported.)

[72] Sarhan, A. E. and Greenberg, B. G. (1962). *Contributions to Order Statistics*, New York: John Wiley & Sons, Inc.

[73] Sarhan, A. E., Greenberg, B. G. and Ogawa, J. (1963). Simplified estimates for the exponential distribution, *Annals of Mathematical Statistics*, **34**, 102–116.

[74] Sibuya, M. (1962). On exponential and other random variable generators, *Annals of the Institute of Statistical Mathematics, Tokyo*, **13**, 231–237.

[75] Sibuya, M. (1969). Maximization with respect to partition of interval and its application to the estimation of the exponential distribution, *Annals of the Institute of Statistical Mathematics, Tokyo*, (To appear).

[76] Siddiqui, M. M. (1963). Optimum estimators of the parameters of negative exponential distributions from one or two order statistics, *Annals of Mathematical Statistics*, **34,** 117–121.

[77] Srivastava, M. S. (1967). A characterization of the exponential distribution, *American Mathematical Monthly*, **74,** 414–416.

[78] Sukhatme, P. V. (1937). Tests of significance for samples of the χ^2-population with two degrees of freedom, *Annals of Eugenics, London*, **8,** 52–56.

[79] Susara, V. and Pathala, K. S. (1965). A probability distribution for the time of first birth, *Journal of Scientific Research, Banaras Hindu University*, **16,** 59–62.

[80] Tallis, G. M. and Light, R. (1968). The use of fractional moments for estimating the parameters of a mixed exponential distribution, *Technometrics*, **10,** 161–175.

[81] Varde, S. D. (1969). Life testing and reliability estimation for the two parameter exponential distribution, *Journal of the American Statistical Association*, **64,** 621–631.

[82] Zabransky, F., Sibuya, M. and Saleh, A. K. M. E. (1966). Tables for the estimation of the exponential distribution (Review of an unpublished Mathematical Table, File No. 92), *Mathematics of Computation*, **20,** 621.

19

Pareto Distribution

1. Genesis

The Pareto distribution is named after an Italian-born Swiss professor of economics, Vilfredo Pareto (1848–1923). Pareto's Law, as formulated by him [30], dealt with the distribution of income over a population and can be stated as follows:

$$N = Ax^{-a}$$

where N is the number of persons having income $\geq x$, and A, a are parameters (a is known both as *Pareto's constant* and as a shape parameter). It was felt by Pareto that this law was universal and inevitable — regardless of taxation and social and political conditions. "Refutations" of the law have been made by several well known economists over the past 50 years (Pigou [31]). More recently attempts have been made to explain many empirical phenomena using the Pareto distribution or some related form (e.g., Steindl [36], Mandelbrot [24], [25], [26], Hagstroem [14]).

Harris [15] has pointed out that a mixture of exponential distributions, with parameter θ^{-1} having a gamma distribution, and with origin at zero, gives rise to a Pareto distribution.

In fact if

$$\Pr[X \leq x \mid \theta] = 1 - e^{-x/\theta}$$

and $\mu = \theta^{-1}$ has a gamma distribution (equation (2) of Chapter 17), then

$$\begin{aligned}\Pr[X \leq x] &= \frac{1}{\beta^{\alpha}\Gamma(\alpha)} \int_0^{\infty} t^{\alpha-1} e^{-t/\beta}(1 - e^{-tx})\, dt \\ &= 1 - \frac{1}{\beta^{\alpha}\Gamma(\alpha)} \int_0^{\infty} t^{\alpha-1} e^{-t(x+\beta^{-1})}\, dt\end{aligned}$$

$$= 1 - (\beta x + 1)^{-\alpha}$$

which is of the form (3) below.

2. Definition

The Pareto distribution has survived in its original form as

$$(1) \qquad P(x) = \Pr[X \geq x] = \left(\frac{k}{x}\right)^a \qquad k > 0, a > 0; x \geq k$$

where $P(x)$ is the probability that the income is equal to or greater than x and k represents some minimum income. As a consequence of (1), the cumulative distribution function of X, representing 'income,' may be written

$$(2) \qquad F_X(x) = 1 - \left(\frac{k}{x}\right)^a \qquad k > 0, a > 0; x \geq k.$$

This is, in fact, a special form of Pearson Type VI distribution.

The relation given by (2) is now more properly known as the "Pareto distribution of the first kind". Two other forms of this distribution were proposed by Pareto. One, now referred to as the "Pareto distribution of the second kind" (sometimes *Lomax distribution*), is given by

$$(3) \qquad F_X(x) = 1 - \frac{K_1}{(x + C)^a}.$$

This is also a Pearson Type VI distribution. The other — the "Pareto distribution of the third kind" — has the cumulative distribution function

$$(4) \qquad F_X(x) = 1 - \frac{k_2 e^{-bx}}{(x + C)^a}.$$

We will consider only distribution (2) here. The Pareto density function corresponding to (2), is

$$(5) \qquad p_X(x) = \frac{ak^a}{x^{a+1}} \qquad (a > 0, x \geq k > 0).$$

3. Moments

Provided r is less then a, the rth moment about zero is

$$(6) \qquad \mu'_r = \frac{ak^r}{a - r}.$$

The expected value is $ak/(a - 1)$ (if $a > 1$) and the variance is

$$ak^2(a - 1)^{-2}(a - 2)^{-1} \qquad (\text{if } a > 2).$$

The mean deviation is $2k(a-1)^{-1}(1-a^{-1})^{a-1}$, and

(7) $$\frac{\text{mean deviation}}{\text{standard deviation}} = 2(1-2a^{-1})^{\frac{1}{2}}(1-a^{-1})^{a-1} \qquad (a > 2).$$

The value of this ratio is 0.513 when $a = 3$; 0.597 when $a = 4$. As a tends to infinity, the ratio tends to $2e^{-1} = 0.736$.

4. Estimation of Parameters

In this section we will suppose that $X_1, X_2, \ldots, X_n$ are independent random variables each distributed as in (2).

4.1 *Least Squares Estimators*

Rearranging (2) and taking logarithms of both sides we obtain

(8) $$\log [1 - F_X(x)] = a \log k - a \log x.$$

The parameters a and k may be estimated by least squares from sample estimates of $F_X(x)$, using, as dependent variable, the logarithm of 1 minus the cumulative distribution of the sample. The least squares estimator of a is then

(9)
$$\tilde{a} = \frac{-n \sum_{i=1}^{n} \log X_i \log [1 - F_X(X_i)] + \left(\sum_{i=1}^{n} \log X_i\right)\left(\sum_{i=1}^{n} \log [1 - F_X(X_i)]\right)}{n \sum_{i=1}^{n} (\log X_i)^2 - \left(\sum_{i=1}^{n} \log X_i\right)^2}.$$

The corresponding least squares estimator of k may be obtained by substituting into (8) the arithmetic mean value of the dependent and independent variable along with the estimator $\tilde{a}$ and solving for k. Estimators of the parameters obtained by least squares methods have been shown to be consistent (Quandt [33]).

4.2 *Estimators from Moments*

Provided that $a > 1$ the mean of the Pareto distribution exists and may be obtained from (6). By equating this to the sample mean $\bar{X}$ and rearranging, we obtain the relation

(10) $$a^* = \frac{\bar{X}}{\bar{X} - k^*}$$

between estimators a^*, k^* of a and k. A formula for k^* may be found by the following argument (Quandt [33]): The probability that all n of the X_i's are greater than a particular value x is $(k/x)^{an}$. Let $F_{X_1'}(x)$ be the cumulative distribution function of the smallest sample value then,

(11) $$F_{X_1'}(x) = 1 - \left(\frac{k}{x}\right)^{an} \qquad (x \geq k).$$

The corresponding density function is

$$p_{X_1'}(x) = \frac{ank^{an}}{x^{an+1}}. \tag{12}$$

From (12) the expected value of X_1' (the smallest sample value) is

$$E(X_1') = \frac{ank}{an - 1}. \tag{13}$$

Equating X_1' to $E(X_1')$ in (13) the estimator of k is found to be

$$k^* = \frac{(a^*n - 1)X_1'}{a^*n}, \tag{14}$$

and substituting this in (10) we obtain

$$a^* = \frac{n\overline{X} - X_1'}{n(\overline{X} - X_1')}. \tag{15}$$

The estimators a^* and k^* are consistent (Quandt [33]).

4.3 *Estimation from Quantiles*

Select two numbers P_1 and P_2 between 0 and 1 and obtain estimators of the respective quantiles $\hat{X}_{P_1}$ and $\hat{X}_{P_2}$. (In large samples it is possible to take $\hat{X}_{P_j} = X'_{[(n+1)P_j]}$.) Estimators of a and k are then obtained by solving the two simultaneous equations

$$P_j = 1 - (k/\hat{X}_{P_j})^a \qquad (j = 1,2). \tag{16}$$

The estimator of a is

$$\frac{\log\left(\dfrac{1 - P_1}{1 - P_2}\right)}{\log\left(\dfrac{\hat{X}_{P_1}}{\hat{X}_{P_2}}\right)}. \tag{17}$$

The corresponding estimator of k can be obtained from (16).

It has been shown that the estimators of a and k obtained by this method are consistent (Quandt [33]).

4.4 *Maximum Likelihood Estimators*

The likelihood function for a sample $(X_1, \ldots, X_n)$ from a Pareto distribution is

$$L = \prod_{j=1}^{n} \frac{ak^a}{X_j^{a+1}}. \tag{18}$$

Taking logarithms of both sides, differentiating partially with respect to the parameter a and setting the result to zero we find the relation

$$\hat{a} = n\left[\sum_{j=1}^{n} \log(X_j/\hat{k})\right]^{-1} \tag{19}$$

between the maximum likelihood estimators $\hat{a}$, $\hat{k}$ of a, k respectively.

A second equation (corresponding to $\partial \log L/\partial k = 0$) cannot be obtained in the usual way since $\log L$ is unbounded with respect to k. Since k is a lower bound on the random variable X, $\log L$ must be maximized subject to the constraint:

$$\hat{k} \leq \min_i X_i. \tag{20}$$

By inspection, the value of $\hat{k}$ which maximizes (18) subject to (20) is

$$\hat{k} = \min_i X_i. \tag{21}$$

It has been shown (Quandt [33]), that both $\hat{a}$ and $\hat{k}$ are consistent estimators.

$\hat{a}$ may be expressed as a function of $\hat{k}$ and the geometric mean by substituting $G = \left(\prod_{i=1}^{n} X_i\right)^{1/n}$ into (19) to yield

$$\hat{a} = [\log (G/\hat{k})]^{-1} \tag{22}$$

where $\hat{k} = \min (X_1, \ldots, X_n) = X_1'$.

It is straightforward to show that the sample geometric mean G is a sufficient statistic for a when k is known, and $\hat{k}$ (in the sense of (21)) is a sufficient statistic for k when a is known. When both a and k are unknown $(\hat{a},\hat{k})$ is a joint set of sufficient statistics for (a,k): moreover, $\hat{a}$ is stochastically independent of $\hat{k}$ (Malik [22]).

The probability density function of $\hat{a}$ is

$$p_{\hat{a}}(x) = \frac{a^{n-1}n^{n-1}}{\Gamma(n-1)x^n} \exp(-na/x) \qquad (x \geq 0). \tag{23}$$

(Note that $2na/\hat{a}$ is distributed as χ^2 with $2n$ degrees of freedom.) The expected value of $\hat{a}$ is $na/(n-2)$; the variance is $n^2(n-2)^{-2}(n-3)^{-1}a^2$ and the distribution of $\hat{a}$ tends to normality as n tends to infinity. A $100(1-\alpha)\%$ confidence limit for a is given by the limits $\hat{a}\chi^2_{2n,\alpha/2}(2n)^{-1}$, $\hat{a}\chi^2_{2n,1-\alpha/2}(2n)^{-1}$.

5. Some Measures of Location and Dispersion and Their Estimators

Although the expected value does not always exist the following alternative measures of location may be considered.

5.1 *Mode*

The mode of the Pareto distribution is located at k. An estimator of the mode is the minimum among the sample values of $X_1, \ldots, X_n$.

5.2 *Geometric Mean*

The population geometric mean is

(24) $$\gamma = k \exp(1/a).$$

The maximum likelihood estimator of γ is G, the sample geometric mean (Muniruzzaman [29]).

For a random sample of size n from the Pareto distribution (2),

(25) $$E(G) = \frac{k}{\left(1 - \frac{1}{na}\right)^n}$$

and from (25)

$$\lim_{n\to\infty} E(G) = \gamma.$$

The variance of G is (to order n^{-1})

(26) $$n^{-1}(k/a)^2 e^{2/a}.$$

5.3 *Harmonic Mean*

The harmonic mean is located at

(27) $$h = k(1 + 1/a).$$

The maximum likelihood estimator of the harmonic mean, provided k is known, is

(28) $$\hat{h} = k\left(1 + \log\frac{G}{k}\right)$$ (Muniruzzaman [29]).

$\hat{h}$ is an unbiased estimator of h. Its variance is $n^{-1}(k/a)^2$.

5.4 *Median*

The median is

(29) $$m = k2^{1/a}.$$

The maximum likelihood estimate of m, if k is known, is

(30) $$\hat{m} = k2^{\log(G/k)}$$ (Muniruzzaman [29]).

(31.1) $$E(\hat{m}) = k\{1 - (na)^{-1}\log 2\}^{-n}$$

and $\lim_{n\to\infty} E(\hat{m}) = m$. The variance of $\hat{m}$ (to order n^{-1}) is

(31.2) $$n^{-1}(k/a)^2\, 2^{2/a}.$$

Note that

(32) $$\text{var}(G) \geq \text{var}(\hat{m}) \geq \text{var}(\hat{h}).$$

The equalities hold if and only if $\gamma = k$.

5.5 *Miscellaneous*

The distribution of $Y = \sum_{j=1}^{n} \log X_i$ has density function

$$p_Y(y) = \frac{a^n}{\Gamma(n)}(y - n \log k)^{n-1} e^{-a(y-n \log k)} \qquad (y > n \log k) \tag{33}$$

(Malik [19]).

The probability density function of $Z = e^Y = X_1 X_2 \cdots X_n$ is

$$p_Z(z) = \frac{a^n k^{na}}{\Gamma(n)} \left[\log\left(\frac{z}{k^n}\right)\right]^{n-1} z^{-(a+1)} \qquad (z > k^n). \tag{34}$$

The distribution of the sample geometric mean $G = Z^{1/n}$ has probability density function

$$p_G(g) = \frac{n^n a^n}{k\Gamma(n)} e^{-(1+an)\log(g/k)} \left(\log\left(\frac{g}{k}\right)\right)^{n-1} \qquad (k < g < \infty). \tag{35}$$

$2\,an\log(G/k)$ is distributed as χ^2 with $2n$ degrees of freedom (Malik [22]). The distributions of $\hat{h}$ and $\hat{m}$ can be directly derived from that of G.

A measure of dispersion has been suggested (Muniruzzaman [29]) which is termed "geometric standard deviation." This is

$$\lambda = \exp\{\sqrt{E[(\log X - \log \gamma)^2]}\}. \tag{36}$$

The moment generating function of $\log X$ is

$$k^t(1 - ta^{-1})^{-1}, \tag{37}$$

from which

$$\operatorname{var}(\log X) = a^{-2} = [\log(\gamma/k)]^2. \tag{38}$$

Thus the geometric standard deviation of X is γ/k $(= \lambda)$ and the Pareto distribution can be completely specified by the location parameter γ and scale parameter λ. The maximum likelihood estimator of λ is G/X_1'.

The probability density function of the sample median M in samples of size $n = 2p + 1$ is

(39)

$$p_M(m) = \frac{(2p+1)!\,ak^{a(p+1)}}{(p!)^2} m^{-(ap+a+1)} \left[1 - \frac{m^{-a}}{k^{-a}}\right]^p \qquad (k \le m \le \infty).$$

The rth moment of M is

$$\mu_r'(M) = \frac{\Gamma(n+1)\Gamma\left(\frac{n+1}{2} - \frac{r}{a}\right)}{\Gamma\left(\frac{n+1}{2}\right)\Gamma\left(n + 1 - \frac{r}{a}\right)} \cdot k^r. \tag{40}$$

Note that

$$E(M) = \frac{\Gamma(n+1)\Gamma(\frac{1}{2}(n+1) - a^{-1})}{\Gamma(\frac{1}{2}(n+1))\Gamma(n+1-a^{-1})} \cdot k. \tag{41}$$

(Note also from (29) that the population median value is $2^{1/a}k$.)

Applying Stirling's formula to (41), using logarithmic expansion and retaining terms to the order of $1/n$, we obtain

$$E(M) \doteqdot k2^{1/a}\left(1 + \frac{a+1}{2na^2}\right); \quad \text{var}(M) \doteqdot k^2 2^{2/a} a^{-2} n^{-1}. \tag{42}$$

It is apparent from (29) that

$$\lim_{n \to \infty} E(M) = m.$$

From (31.2) and (42) we see that

$$\frac{V(\hat{m})}{V(M)} \doteqdot 1 \qquad \text{(Muniruzzaman [29]).}$$

On setting the parameter k in (2) equal to 1, we obtain

$$F_X(x) = 1 - x^{-a} \qquad (x \geq 1). \tag{43}$$

This may be regarded as a *standard form* of the Pareto distribution. A location parameter ϵ can be introduced to yield

$$F_X(x) = 1 - (x - \epsilon)^{-a} \qquad (x \geq 1 + \epsilon). \tag{44}$$

It can be shown (Gumbel [13]) that if T has an exponential distribution with scale parameter θ, then $X = \exp(T) + \epsilon$ has the Pareto distribution (44) with

$$a = 1/\theta, \quad a \geq 1$$
$$x > 1 + \epsilon, \quad \epsilon \geq 0.$$

A single-order statistic estimator using only X'_m of a, given ϵ, is

$$\tilde{a} \mid \epsilon = \frac{1}{\tilde{\theta}} = \frac{1}{C_{mn} \log (X'_m - \epsilon)} \tag{45}$$

where

$$C_{mn} = \frac{1}{\sum_{i=1}^{m} (n - i + 1)^{-1}}. \qquad \text{(Moore and Harter [27])}$$

$\tilde{a} \mid \epsilon$ and $\tilde{\theta}$ are consistent estimators of the shape parameter of the Pareto distribution and the scale parameter of the exponential distribution respectively (Moore and Harter [27]).

It can be shown (Moore and Harter [27]) that an exact confidence interval for a based on X'_m alone, if ϵ be known, is

$$\{D_{l,m,n} \log (X'_m - \epsilon)\}^{-1} > a > \{D_{u,m,n} \log (X'_m - \epsilon)\}^{-1} \tag{46}$$

where the coefficients $D_{l,m,n}$ and $D_{u,m,n}$ have been tabulated (Harter [16]) for $n = 1(1)20(2)40$ for m optimal. (In the sense that the value of m is selected which maximizes the efficiency of the confidence interval.)

6. Order Statistics

We have already noted the distribution of the median of $n(= 2p + 1)$ independent random variables each having the Pareto distribution. The jth smallest order statistic X'_j has density function

(47)
$$p_{X'_j}(x) = \frac{n!}{(j-1)!(n-j)!}\left(\frac{a}{x}\right)\left(\frac{k}{x}\right)^{a(n-j+1)}\left\{1 - \left(\frac{k}{x}\right)^a\right\}^{j-1} \qquad (k \leq x)$$

and its rth moment about zero is

$$\mu'_r(X'_j) = \frac{\Gamma(n+1)\Gamma(n-j+1-ra^{-1})}{\Gamma(n+1-ra^{-1})\Gamma(n-j+1)} k^r. \tag{48}$$

(cf. (40) which is obtained by putting $j = \frac{1}{2}(n + 1)$ in (48).) The joint probability density function of two order statistics X'_r, X'_s $(r < s)$ is

$$\begin{aligned} p_{X'_r,X'_s}(x_r,x_s) &= \frac{n!a^2k^{2a}}{(r-1)!(s-r-1)!(n-s)!} \\ &\quad \times \left[1 - \left(\frac{k}{x_r}\right)^a\right]\left[\left(\frac{k}{x_s}\right)^a - \left(\frac{k}{x_r}\right)^a\right]^{r-s-1} \\ &\quad \times \left(\frac{k}{x_s}\right)^a (x_r x_s)^{-(a+1)} \qquad (k \leq x_r \leq x_s), \end{aligned} \tag{49}$$

and their covariance can be calculated from

$$E[X'_r X'_s] = \frac{\Gamma(n+1)\Gamma(n-r-2a^{-1}+1)\Gamma(n-s-a^{-1}+1)}{\Gamma(n-s+1)\Gamma(n-r-a^{-1}+1)\Gamma(n-2a^{-1}+1)} k^2.$$

Malik [19] has published tables of

(i) $E[X'_j]$ for $n \leq 12$
(ii) $E[X'_r X'_s]$ for $n \leq 12$
(iii) $\text{var}(X'_j)$ and $\text{cov}(X'_r X'_s)$ for $n \leq 8$

to four decimal places (for all possible values of j, r, s in each case).

7. Characterization

Let $X'_1 \leq X'_2 \leq \cdots \leq X'_n$ be the order statistics from an absolutely continuous distribution function $F_X(x)$ where the lower bound on the random variable X is given by k, that is, $F_X(k) = 0$. It can then be shown for a sample of size n from this distribution that the statistics X'_1 and $(X'_1 + \cdots + X'_n)/X'_1$, or equivalently $(X'_2 + \cdots + X'_n)/X'_1$, are independent if and only if the random variable X has the Pareto distribution (Srivastava [35]).

8. Applications and Related Distributions

The discrete form of the Pareto distribution (Zipf-Estoup's law) has been discussed in Chapter 10, Section 3.1. In many cases the Pareto distribution may be used as an approximation to the Zipf distribution. Many socio-economic and other naturally occurring quantities are distributed according to certain statistical distributions with very long right tails. Examples of some of these empirical phenomena are distributions of city population sizes, occurrence of natural resources, stock price fluctuations, size of firms, personal incomes, and error clustering in communication circuits. Many distributions have been developed in an attempt to explain these empirical data.

The Pareto and lognormal distributions have played a major part in these investigations. It has been observed that while the fit of the Pareto curve may be rather good at the extremities of the income range, the fit over the whole range is often rather poor. On the other hand, the lognormal (Gibrat) distribution (Chapter 14) fits well over a large part of the income range but diverges markedly at the extremities (Fisk [10]). (Direct comparison of lognormal and Pareto distributions in another context will be found in reference [11].)

Truncated Pareto distributions, with an upper limit (k') as well as a lower limit (k), have been found by Goldberg [12] to fit the distribution of oil fields, in a specified area, by size. A truncated Pareto distribution has density function

$$p_X(x) = \frac{a}{k}[1 - (k/k')^a]^{-1}(k/x)^{a+1} \qquad (k \le x \le k')$$

and cumulative distribution function

$$F_X(x) = \begin{cases} 0 & (x \le k) \\ [1 - k/k')^a]^{-1}[(k/x)^a - (k/k')^a] & (k < x < k') \\ 1 & (x \ge k'). \end{cases}$$

A mixture of two Pareto distributions is sometimes called a *double Pareto* distribution.

Modifications of the Pareto distribution have been developed to give better representation of income distributions.

One of the best known income distributions is the *Champernowne distribution* (Champernowne [7]). It is supposed that the random variable X, termed "income-power," and defined by $X = \log T$, where T is the actual income, has a density function of form

$$p_X(x) = \frac{n}{\cosh\{\alpha(x - x_0)\} + \lambda} \tag{50}$$

where n, α, x_0 and λ are parameters (n is, of course, a function of α and λ).* This is included in Perks' family of distributions, described in Section 11 of Chapter 22.

The curve given by (50) is symmetrical and x_0 is the median value of the income-power. If we let $\log t_0 = x_0$ then the density function of income ($T = e^X$) is

$$p_T(t) = \frac{n}{t\left[\frac{1}{2}\left(\frac{t}{t_0}\right)^{-\alpha} + \lambda + \frac{1}{2}\left(\frac{t}{t_0}\right)^{\alpha}\right]} \tag{51}$$

and t_0 is the median value of the income.

In order to find the proportion of persons with income greater than t we integrate (51). The form which the integral takes will depend on the value of λ. For $-1 < \lambda < 1$ the cumulative distribution function is

$$F_T(t) = 1 - \frac{1}{\theta}\tan^{-1}\left\{\frac{\sin\theta}{\cos\theta + \left(\frac{t}{t_0}\right)^{\alpha}}\right\} \qquad (t > 0), \tag{52}$$

where

$$0 < \theta < \pi \qquad \text{and} \qquad \cos\theta = \lambda.$$

For $\lambda = 1$

$$F_T(t) = 1 - \frac{t_0^{\alpha}}{t_0^{\alpha} + t^{\alpha}} \qquad (t > 0). \tag{53}$$

For $\lambda > 1$

$$F_T(t) = 1 - \frac{1}{2\eta}\log\left\{\frac{t^{\alpha} + e^{\eta}t_0^{\alpha}}{t^{\alpha} + e^{-\eta}t_0^{\alpha}}\right\} \quad (t > 0), \tag{54}$$

where

$$\cosh\eta = \lambda.$$

In (52), (53) and (54), α is equal to Pareto's constant (Section 1), and t_0 is the median value of the income. No simple interpretation exists for the parameters θ and η. The latter may be regarded as a parameter used for adjusting the kurtosis of the fitted density function (50) along the income power scale (Champernowne [7]).

The three forms of the density function corresponding to (52), (53), and (54) may be obtained by differentiation and are respectively

$$\frac{\alpha\sin\theta}{\theta t\left\{\left(\frac{t}{t_0}\right)^{\alpha} + 2\cos\theta + \left(\frac{t_0}{t}\right)^{\alpha}\right\}}; \tag{55}$$

*We have retained the unusual notation (n) used by Champernowne. This symbol does not have anything to do with "sample size" in the present context.

(56)
$$\frac{\alpha}{t\left\{\left(\frac{t}{t_0}\right)^{\alpha/2}+\left(\frac{t_0}{t}\right)^{\alpha/2}\right\}^2};$$

(57)
$$\frac{\alpha \sinh \eta}{\eta t\left\{\left(\frac{t}{t_0}\right)^{\alpha}+2\cosh \eta+\left(\frac{t_0}{t}\right)^{\alpha}\right\}}.$$

(These are all of form (51), of course.) Methods of fitting the Champernowne distribution may be found in [7].

In general, it is possible to improve the fit of a distribution by the incorporation of extra parameters. To this end Champernowne [7] proposed a five parameter model of the form:

$$F_T(t) = 1 - \frac{1}{(1+\sigma)\theta}\left[(\sigma-1)\theta + 2\tan^{-1}\left\{\frac{\sin\theta}{\cos\theta + (t/t_0)^{\sigma\alpha}}\right\}\right] \quad \text{for } 0 \le t \le t_0.$$

and

$$F_T(t) = 1 - \frac{2\sigma}{(1+\sigma)\theta}\tan^{-1}\left\{\frac{\sin\theta}{\cos\theta + (t/t_0)^{\alpha}}\right\} \quad \text{for } t \ge t_0$$

The fifth parameter σ may be considered as a measure of skewness, in that when $\sigma > 1$ the curve exhibits positive skewness, and when $\sigma < 1$ the curve is negatively skewed. When $\sigma = 1$, the distribution (52) is obtained.

The limiting form of the density function of the Champernowne distribution as $\theta \to 0$ is

(58)
$$\frac{(t/t_0)^{\alpha-1}}{t_0[1+(t/t_0)^\alpha]^2}.$$

On making the transformation $(T/t_0)^\alpha = e^\phi$ the *logistic* or *sech square density function*

(59)
$$p_\phi(t) = \frac{e^t}{[1+e^t]^2} = \frac{e^{-t}}{[1+e^{-t}]^2}$$

is obtained (see Chapter 22, Equation (7)) (Fisk [10]). The shape of this distribution is quite similar to that of the normal density function. However, in the tails the sech square density is greater than the normal. For economists concerned with the upper tails of distributions, the Pareto, Champernowne and sech square distributions are probably more useful than the lognormal which generally gives a poor fit in the tails (Aitchison and Brown [2]).

The cumulative distribution function corresponding to (58) is

(60)
$$F_T(t) = \frac{(t/t_0)^\alpha}{1+(t/t_0)^\alpha},$$

that is,

(61)
$$F_T(t) = 1 - \frac{(t/t_0)^{-\alpha}}{1+(t/t_0)^{-\alpha}}.$$

For small t, (61) is close to the Pareto distribution (2).

A number of papers [24], [25], [26] have recently been published by Mandelbrot concerning a class of distributions which has been termed "stable Paretian". In the development of the theory he distinguishes between two forms of the Pareto law (Mandelbrot [24]):

Strong Law of Pareto — The distribution is of the form given by (2) that is,

$$1 - F_X(x) = \left(\frac{x}{k}\right)^{-\alpha} \qquad x \geq k$$
$$= 1 \qquad x < k.$$

Weak or Asymptotic form of the law of Pareto — The form of this distribution is

$$1 - F_X(x) \sim \left(\frac{x}{k}\right)^{-\alpha} \quad \text{as} \quad x \to \infty. \tag{62}$$

This implies that if $\log [1 - F_X(x)]$ is plotted against $\log x$ the resulting curve should be asymptotic to a straight line with slope equal to $-a$ as x approaches infinity. It has been shown (Lévy [17]) that there is a class of distributions which follow the asymptotic form of the law of Pareto and are characterized by the fact that $0 < a < 2$. These are known as the *stable Paretian* or *stable non-Gaussian* distributions. The normal distribution ($a = 2$) is also a member of the family of stable laws and has the property that it is only one with a finite variance. A further property of the non-Gaussian stable laws when $1 < \alpha < 2$ is that the expected value is finite.

It has been shown by Lévy [17] that the logarithm of the characteristic function of the stable Paretian distribution is

$$\log \phi(t) = i\,\delta t - \gamma|t|^{\alpha}[1 - i\beta(t/|t|)\tan(\alpha\pi/2)]. \tag{63}$$

The parameters of (63) are α, β, γ, and δ. The location parameter is δ, and if $\alpha > 1$ then δ is the mean of the distribution. β is an index of skewness such that when equal to zero the curve is symmetric; when $\beta > 0$ the curve is skewed to the right and when $\beta < 0$ it is skewed to the left. The scale parameter is γ; α is intimately related to the Pareto exponent (in the sense that the limit of $x^{\alpha-1}\Pr[X > x]$ is finite and non-zero) and controls the amount of probability found in the tails of the distribution. When $0 < \alpha < 2$, the extreme tails of the stable distributions are higher than those of the normal with the total probability in the tails increasing as α moves from 2 to 0. Explicit expressions for the density functions of stable Paretian distributions are known for only three cases; the Cauchy ($\alpha = 1, \beta = 0$), the normal ($\alpha = 2$), and the "coin tossing" (Bernoulli) case ($\alpha = \frac{1}{2}, \beta = 1, \delta = 0, \gamma = 1$) (Fama [8]).

Putting $\beta = 0$ in (63) we have

$$\log \phi(t) = i\delta t - \gamma|t|^{\alpha}.$$

For the variable $Y = (X - \delta)\gamma^{-1/\alpha}$ we have the "standard form" of symmetric stable distribution with

$$\log \phi_Y(t) = -|t|^\alpha.$$

Bergström [4] has shown that, for $\alpha > 1$,

$$p_Y(y) = \frac{1}{\pi\alpha} \sum_{j=0}^{\infty} (-1)^j \frac{\Gamma((2j+1)\alpha^{-1})}{(2j)!} y^{2j}$$

so that

$$F_Y(y) = \frac{1}{2} + \sum_{j=1}^{\infty} (-1)^{j-1} \frac{\Gamma((2j-1)\alpha^{-1})}{(2j-1)!} y^{2j-1}.$$

For large y, the asymptotic series

$$1 + \frac{1}{\pi} \sum_{j=1}^{\infty} (-1)^j \frac{\Gamma(j\alpha)}{j!\, y^{\alpha j}} \sin\left(\tfrac{1}{2} j\alpha\pi\right)$$

is useful for calculating $F_Y(y)$, when $\alpha > 1$.

Fama and Roll [9] give tables of values of $F_Y(y)$, to 4 decimal places, for $\alpha = 1.0(0.1)1.9(0.05)2.00$ and $y = 0.05(0.05)1.00(0.1)2.0(0.2)4.0(0.4)6.0(1)8$, 10, 15, 20. They also give values of y_P satisfying $F_Y(y_P) = P$, to 3 decimal places, for the same values of α and $P = 0.52(0.02)0.94(0.01)0.97(0.005)0.995$, 0.9995. It is notable that $y_{0.72}$ varies but little with α. Its value increases from 0.827 when $\alpha = 1$ (Cauchy) to about 0.830 ($\alpha = 1.3$–1.6) and then decreases to 0.824 for $\alpha = 2$ (normal).

Fama and Roll [9] suggest that this stability may be used to construct an estimator of $\gamma^{1/\alpha}$:

$$(0.827)^{-1} \times [\text{upper } 28\% \text{ order statistic} - \text{lower } 28\% \text{ order statistic}]$$

which should be robust with respect to changes in α.

Maximum likelihood estimation of α and estimation by moments will be described in Section 4 of Chapter 24. Estimation from order statistics has been studied by Malik [22], using formulas for moments and product moments of order statistics derived in Malik [19]. He showed, inter alia, that the characteristic function of the kth smallest order statistic in a random sample of size n is

$$\frac{\Gamma(n+1)}{\Gamma(k)} \sum_{j=0}^{\infty} \frac{\Gamma(k + j\alpha^{-1})}{\Gamma(n + j\alpha^{-1} + 1)} \frac{(it)^j}{j!}, \tag{65}$$

and obtained the interesting recurrence relationships

$$(k-1)E[(Y'_k)^r] = (k + r\alpha^{-1} - 1)E[(Y'_{k-1})^r] \qquad (k > 1) \tag{66.1}$$

$$(2\alpha^{-1} + l - 1)E[Y'_k Y'_l] = (\alpha^{-1} + l - 1)E[Y'_k Y'_{l-1}] \qquad (k < l). \tag{66.2}$$

Aigner [1] has investigated the use of the Pareto distribution itself as an approximation to the upper tails (or portions thereof) of distributions.

Mandelbrot [24] has suggested that if α is only 'slightly less' than 2, a useful approximation (when $\delta = \beta = 0$) to the density function may be of the form

$$p_X(x) = \begin{cases} \alpha - 1.5 & \text{for } |x| \leq 1 \\ \alpha(2 - \alpha)|x|^{-(\alpha+1)} & \text{for } |x| > 1. \end{cases}$$

A class of '*generalized Pareto*' distributions has been described by Ljubo [18]. The cumulative distribution function is

$$F_X(x) = \begin{cases} 1 - \left(\dfrac{k + \alpha}{x + \alpha}\right)^a e^{-\beta(x-k)} & (x \geq k) \\ 0 & (x < k) \end{cases}$$

with $\beta > 0$.

If X has the probability density function (5), then $Y = X^{-1}$ has the density function

$$p_Y(y) = ak^a y^{a-1} \qquad (0 < y < k^{-1}). \tag{64}$$

This distribution, which is a special Pearson Type I distribution, is called the *power-function distribution.* Its moments are, of course, simply the negative moments of the corresponding Pareto distribution, so that

$$\mu_r'(Y) = \frac{ak^{-r}}{a + r}.$$

Moments of order statistics for this distribution were calculated by Malik [20]. Rider [34] has obtained the distribution of products and quotients of maximum values of sets of independent random variables having distribution (64). He suggests such distributions might be used in approximate representation of the lower tail of the distribution of a random variable having a fixed lower bound (as in a gamma distribution, for example).

REFERENCES

[1] Aigner, D. J. (1967). *A technique for calculating the moment contributions of an open interval in a J-shaped frequency distribution*, Unpublished manuscript, University of Illinois and University of Wisconsin.

[2] Aitchison, J. and Brown, J. A. C. (1957). *The Lognormal Distribution*, London: Cambridge University Press.

[3] Berger, J. M. and Mandelbrot, B. (1963). A new model for error clustering in telephone circuits, *IBM Journal of Research and Development*, **7,** 224–236.

[4] Bergström, H. (1952). On some expansions of stable distributions, *Arkiv för Matematik*, **2,** 375–378.

[5] Bhattacharya, N. (1963). A property of the Pareto distribution, *Sankhyā, Series B.*, **25,** 195–196.

[6] Brillinger, D. (1963). Necessary and sufficient conditions for a statistical problem to be invariant under a Lie group, *Annals of Mathematical Statistics*, **34,** 492–500.

[7] Champernowne, D. G. (1952). The graduation of income distributions, *Econometrica*, **20,** 591–615.

[8] Fama, E. F. (1963). Mandelbrot and the stable Paretian hypothesis, *Journal of Business, University of Chicago*, **36,** 420–429.

[9] Fama, E. F. and Roll, R. (1968). Some properties of symmetric stable distributions, *Journal of the American Statistical Association*, **63,** 817–836.

[10] Fisk, P. R. (1961). The graduation of income distributions, *Econometrica*, **29,** 171–185.

[11] Freiling, E. C. (1966). *A comparison of the fallout mass-size distributions calculated by lognormal and power-law models*, San Francisco: U. S. Naval Radiological Defense Laboratory.

[12] Goldberg, G. (1967). The Pareto law and the pyramid distribution, *Publication 505*, Houston, Texas: Shell Development.

[13] Gumbel, E. J. (1958). *Statistics of Extremes*, New York: Columbia University Press.

[14] Hagstroem, K.-G. (1960). Remarks on Pareto distributions, *Skandinavisk Aktuarietidskrift*, 59–71.

[15] Harris, C. M. (1968). The Pareto distribution as a queue service discipline, *Operations Research*, **16,** 307–313.

[16] Harter, H. L. (1964). Exact confidence bounds based on one order statistic, for the parameter of an exponential population, *Technometrics*, **6,** 301–317.

[17] Lévy, P. (1925). *Calcul des Probabilités*, (Chapter 6), Paris: Gauthier-Villars.

[18] Ljubo, M. (1965). Curves and concentration indices for certain generalized Pareto distributions, *Statistical Review*, **15,** 257–260. (In Serbo-Croatian, English summary.)

[19] Malik, H. J. (1966). Exact moments of order statistics from the Pareto distribution, *Skandinavisk Aktuarietidskrift*, 144–157.

[20] Malik, H. J. (1967). Exact moments of order statistics for a power-function population, *Skandinavisk Aktuarietidskrift*, 64–69.

[21] Malik, H. J. (1968). Estimation of the parameters of the power-function population, *Metron*, (To appear).

[22] Malik, H. J. (1970). Estimation of the parameters of the Pareto distribution, *Metrika*, **16** (To appear).

[23] Malik, H. J. (1970). Distribution of product statistics from a Pareto population, *Metrika*, **16** (To appear).

[24] Mandelbrot, B. (1960). The Pareto-Lévy law and the distribution of income, *International Economic Review*, **1,** 79–106.

[25] Mandelbrot, B. (1963). The variation of certain speculative prices, *Journal of Business, University of Chicago*, **36,** 394–419.

[26] Mandelbrot, B. (1967). The variation of some other speculative prices, *Journal of Business, University of Chicago*, **40,** 393–413.

[27] Moore, A. H. and Harter, H. L. (1967). One-order statistics conditional estimators of the shape parameters of the limited and Pareto distributions, and the scale parameters of the Type II asymptotic distributions of smallest and largest values, *IEEE Transactions on Reliability*, **16,** 100–103.

[28] Muniruzzaman, A. N. M. (1950). On some distributions in connection with Pareto's law, *Proceedings of the 1st Pakistan Statistical Conference*, 90–93.

[29] Muniruzzaman, A. N. M. (1957). On measures of location and dispersion and tests of hypotheses on a Pareto population, *Bulletin of the Calcutta Statistical Association*, **7,** 115–123.

[30] Pareto, V. (1897). *Cours d'Economie Politique*, Lausanne and Paris: Rouge and Cie.

[31] Pigou, A. C. (1932). *The Economics of Welfare*, London: the Macmillan Company.

[32] Quandt, R. E. (1964). Statistical discrimination among alternative hypotheses and some economic regularities, *Journal of Regional Science*, **5,** 1–23.

[33] Quandt, R. E. (1966). Old and new methods of estimation and the Pareto distribution, *Metrika*, **10,** 55–82.

[34] Rider, P. R. (1964). Distribution of product and of quotient of maximum values in samples from a power-function population, *Journal of the American Statistical Association*, **59,** 877–880.

[35] Srivastava, M. S. (1965). A characterization of Pareto's distribution and $(k + 1)x^k/\theta^{(k+1)}$, (Abstract), *Annals of Mathematical Statistics*, **36,** 361–362.

[36] Steindl, J. (1965). *Random Processes and the Growth of Firms*, New York: Hafner.

20

Weibull Distribution

1. Definition and Historical Remarks

A random variable X has a *Weibull distribution* if there are values of the parameters $c(> 0)$, $\alpha(> 0)$ and ξ_0 such that

$$Y = \{(X - \xi_0)/\alpha\}^c$$

has the exponential distribution, with probability density function

$$p_Y(y) = e^{-y} \qquad (0 < y). \tag{1}$$

The probability density function of X is

$$p_X(x) = c\alpha^{-1}\{(x - \xi_0)/\alpha\}^{c-1} \exp\left[- \{(x - \xi_0)/\alpha\}^c\right] \qquad (\xi_0 < x). \tag{2}$$

(The standard Weibull distribution is obtained by putting $\alpha = 1$, $\xi_0 = 0$ — see (4) on page 252.)

The distribution is named after Waloddi Weibull, a Swedish physicist, who used it in 1939 to represent the distribution of the breaking strength of materials [76], [77]. A more widely available account by the same author was published in 1951 [78]. In Russian literature it is sometimes called the Weibull-Gnedenko distribution [28]. Some remarks on the history and also on the relation between extreme-value and Weibull distributions are in Mann [56].

The use of the distribution in reliability and quality control work has been advocated by Kao [45], [46]. In the late fifties and early sixties articles on this distribution began to appear in the journal *Industrial Quality Control* (Berrettoni [3]). Some indication of the recent popularity of the Weibull distribution is seen in Splitstone's thesis [72]. The distribution is often suitable where the conditions of 'strict randomness' of the exponential distribution are not satisfied. Discussion of applications will be found in Section 8.

2. Genesis

When a theoretical model, incorporating certain assumptions about independence, leads to an exponential distribution to represent the variation of an observable statistic, this distribution may be replaced by a Weibull distribution (at any rate in the initial stages of the analysis) to allow for possible inaccuracy in the model. Usually, there is no explicit theoretical reasoning indicating that a Weibull distribution should be used — it is just that a power transformation is a practical, convenient way of introducing some flexibility in the model.

There are situations in which the Weibull distribution can be derived by probabilistic arguments. For example, if $c = 2$ and $\xi_0 = 0$, then

$$p_X(x) = 2\alpha^{-1}(x/\alpha)e^{-(x/\alpha)^2} \qquad (x > 0) \tag{3}$$

and so X is distributed as $(\alpha/\sqrt{2}) \times$ (χ with two degrees of freedom). Gnedenko *et al.* [28] have shown that a distribution of this form arises as a limiting distribution in the following case: Consider a device consisting of n doubled elements. For each *pair*, failure occurs when *both* elements have failed. The combined '*device*' fails if at least one pair fails. It is supposed that the distribution of lifetime (t) for each element in the jth pair is exponential with

$$\Pr[t > t_0] = e^{-\lambda_j t_0}$$

and all lifetimes are mutually independent. If n is increased indefinitely and the λ's are varied in such a way that

$$\sum_{j=1}^{n} \lambda_j^2 = \lambda^2 \text{ (fixed), while } \sum_{j=1}^{n} \lambda_j^3 \to 0,$$

then the lifetime of the device has a limiting distribution of form (3) with $\alpha = \lambda^{-1}$.

A further example is the connection between the Weibull and extreme value distributions (see Section 10). Such examples of probabilistic bases for the Weibull distribution are, however, not commonly encountered in situations in which the distribution is, in fact, employed. Gittus [27], however, has described situations in which the Weibull distribution is likely to arise.

3. Properties

Some graphs of the probability density function (2), for various values of c, are shown in Figure 1. In each case $\alpha = 1$ and $\xi_0 = 0$. (The value zero for ξ_0 is by far the most frequently used, especially in representing distributions of lifetimes.)

For $c > 1$, the probability density function tends to zero as x tends to zero, and there is a single mode at

$$x = \alpha[(c - 1)/c]^{1/c} + \xi_0.$$

This value tends to $(\alpha + \xi_0)$ very rapidly as c tends to infinity. For $0 < c < 1$,

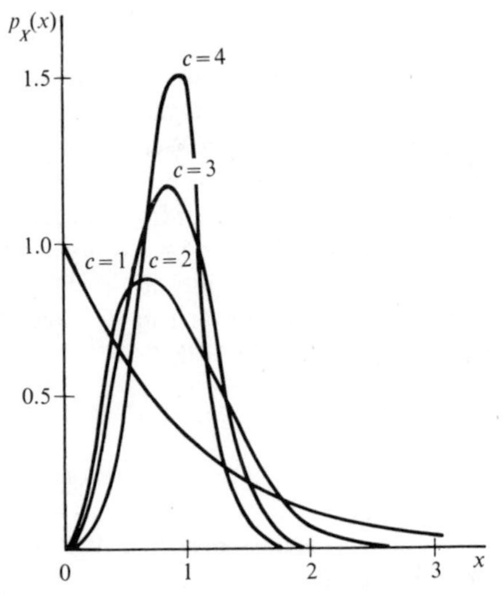

FIGURE 1

Weibull Density Functions

the mode is at zero, and $p_X(x)$ is a decreasing function of x for all $x > \xi_0$. The median is $\alpha[\log_e 2]^{1/c} + \xi_0$.

The cumulative distribution function is

$$F_X(x) = 1 - \exp [- \{(x - \xi_0)/\alpha\}^c].$$

Note that, whatever the value of c,

$$F_X(\xi_0 + \alpha) = 1 - e^{-1} \doteqdot 0.63.$$

In the remainder of this section all the distributions to be discussed will have $\xi_0 = 0$, $\alpha = 1$, so that we have the standard density function

$$p_X(x) = cx^{c-1} \exp (-x^c) \qquad (x > 0). \tag{4}$$

The distribution of X now depends only on the single parameter c. The moment-ratios, coefficient of variation, and standardized cumulants $\kappa_k/\kappa_2^{k/2}$ of this distribution are, of course, the same as those of the distribution (2), which latter are thus seen to depend only on c, and not on ξ_0 or α. Moments corresponding to (2) are easily obtained from those corresponding to (4) by using the transformation $X' = \xi_0 + \alpha X$.

Since X^c has the exponential distribution (1), the rth moment about zero of X is also the (r/c)-th moment about zero of a variable with distribution (1). Hence

$$\mu_r' = E(X^r) = \Gamma(rc^{-1} + 1). \tag{5}$$

The expected value of X is $\Gamma(c^{-1} + 1)$ and the variance of X is equal to $\Gamma(2c^{-1} + 1) - [\Gamma(c^{-1} + 1)]^2$. Expressions for the moment ratios can easily be obtained but are not given here explicitly. When c is large

(6)*

$$E(X) \doteqdot 1 - \gamma c^{-1} + \frac{1}{2}\left(\frac{\pi^2}{6} + \gamma^2\right) c^{-2} \doteqdot 1 - 0.57722c^{-1} + 0.98905c^{-2}$$

$$\text{var}(X) \doteqdot (\pi^2/6)c^{-2} \doteqdot 1.64493c^{-2}.$$

Table 1 gives some numerical values of the mean, the standard deviation, $\sqrt{\beta_1}$ and β_2 for various values of c. It will be noted that for $c = 3.6$ (approx.) $\sqrt{\beta_1}$ is zero. For values of c in the neighborhood of 3.6, the Weibull distribution is similar in shape to a normal distribution. Dubey [21] has made a detailed study of the closeness of agreement between the two distributions. Note also that β_2 has a minimum value of about 2.71 when $c = 3.35$ (approx.).

TABLE 1

Moments of Weibull Distributions

c	Mean	Standard Deviation	$\sqrt{\beta_1}$	β_2
1.2	0.9407	0.7872	1.52	6.24
1.4	0.9114	0.6596	1.20	4.84
1.6	0.8966	0.5737	0.96	4.04
1.8	0.8893	0.5112	0.78	3.56
2.0	0.8862	0.4633	0.63	3.25
2.2	0.8856	0.4249	0.51	3.04
2.4	0.8865	0.3935	0.40	2.91
2.6	0.8882	0.3670	0.32	2.82
2.8	0.8905	0.3443	0.24	2.76
3.0	0.8930	0.3245	0.17	2.73
3.2	0.8957	0.3072	0.11	2.71
3.4	0.8984	0.2918	0.05	2.71
3.6	0.9011	0.2780	0.00	2.72
3.8	0.9038	0.2656	−0.04	2.73
4.0	0.9064	0.2543	−0.09	2.75

From (5), the moment generating function of $\log X$ is

(7) $$E(e^{t \log X}) = E(X^t) = \Gamma(tc^{-1} + 1).$$

Hence the rth cumulant of $\log X$ is $c^{-r}\psi^{(r)}(1)$. In particular

(8) $$E(\log X) = -\gamma/c; \quad \text{var}(\log X) = (\pi^2/6)c^{-2}.$$

(See also Chapter 21—$\log X$ has an 'extreme value' distribution.)

The information generating function is

(9) $$T(u) = E[(p(X))^u] = c^u \int_0^\infty x^{u(c-1)} \exp(-ux^c)\, dx.$$

*γ is Euler's constant (Chapter 1, Section 3).

Putting $ux^c = v$ (taking $u > 0$ and remembering $c > 0$), we obtain

$$(10) \qquad T(u) = c^{u-1}\Gamma\left(\frac{1 + u(c-1)}{c}\right) \cdot u^{-[1+u(c-1)]/c},$$

whence the entropy is

$$(11) \qquad -T'(1) = \frac{(c-1)\gamma}{c} - \log_e c + 1.$$

A detailed account of these and other properties is given in Lehman [51].

4. Characterization

Dubey [14] has obtained the following result: Let $X_1, X_2, \ldots, X_n$ be independent and identically distributed random variables; then $\min(X_1, X_2, \ldots, X_n)$ has a Weibull distribution if and only if the common distribution of the X_i's is a Weibull distribution.

This result can be easily established using the formula for the distribution of $\min(X_1, X_2, \ldots, X_n)$. (See next Section.)

5. Order Statistics

If $X_1, X_2, \ldots, X_n$ are independent random variables, each with the same Weibull probability density function

$$(12) \qquad p_{X_i}(x) = c\alpha^{-1}\{(x - \xi_0)/\alpha\}^{c-1} \exp[-\{(x - \xi_0)/\alpha\}^c] \qquad (x > \xi_0)$$

and $Y = \min(X_1, X_2, \ldots, X_n)$, then

$$\begin{aligned} \Pr[Y > y] = \{\Pr[X_i > y]\}^n &= \exp[-n\{(y - \xi_0)/\alpha\}^c] \\ &= \exp[-\{n^{1/c}(y - \xi_0)/\alpha\}^c]. \end{aligned}$$

Hence Y has the same Weibull distribution as each X_i, except that α is replaced by $\alpha n^{-1/c}$.

Since the cumulative distribution function of each X_i is

$$(13) \qquad F_{X_i}(x) = 1 - \exp[-\{(x - \xi_0)/\alpha\}^c] \qquad (x > \xi_0),$$

the probability density functions of the order statistics $X_1' \le X_2' \le \cdots \le X_n'$ can be expressed explicitly in terms of elementary functions. In fact

$$\begin{aligned} (14) \quad p_{X_r'}(x) &= \frac{n!}{(r-1)!(n-r)!}\{1 - \exp[-\{(x - \xi_0)/\alpha\}^c]\}^{r-1} \\ &\quad \times \exp[-(n - r + 1)\{(x - \xi_0)/\alpha\}^c] \cdot \frac{c\{(x - \xi_0)/\alpha\}^{c-1}}{\alpha} \\ &= \frac{n!c\{(x - \xi_0)/\alpha\}^{c-1}}{(r-1)!(n-r)!\alpha} \\ &\quad \times \sum_{j=0}^{r-1} \binom{r-1}{j} (-1)^j \exp[-(n - r + 1 - j)\{(x - \xi_0)/\alpha\}^c] \\ &\qquad\qquad (x > \xi_0). \end{aligned}$$

Since

$$\int_0^\infty z^{c-1+l} \exp(-\beta z^c)\, dz = c^{-1}\beta^{-(1+l/c)}\Gamma(1 + l/c),$$

it follows from (14) that

$$E[(X_r' - \xi_0)^l] = \frac{n!\alpha^l}{(r-1)!(n-r)!}\Gamma(1 + l/c) \times \sum_{j=0}^{r-1} \binom{r-1}{j} (-1)^j (n - r + 1 - j)^{-(1+l/c)}. \tag{15}$$

Lieblein [53] has given formulas for product moments, obtained in a similar way.

Weibull [79] gives expected values, variances and covariances, to 5 decimal places, for all order statistics in random samples of sizes 5(5)20, from distribution (4) with $c^{-1} = 0.1(0.1)1.0$. Govindarajulu and Joshi [30] constructed similar tables for sample sizes 2(1)10 and $c = 1, 2, 2.5, 3(1)10$.

6. Estimation

If c and ξ_0 are known then $Z = (X - \xi_0)^c$ can be calculated. The remaining parameter α can then be estimated using the fact that Z has the exponential distribution with probability density function

$$p_Z(z) = \alpha^{-1}e^{-z/\alpha} \qquad (0 < z). \tag{16}$$

Thus the methods developed for the exponential distribution (Chapter 18, Sections 5 and 6) may be used.

6.1 *Maximum Likelihood*

The most usual situation is ξ_0 known to be zero, but c unknown, so that both c and α must be estimated. Given sample values from a random sample of size n, represented by n mutually independent random variables $X_1, X_2, \ldots, X_n$, each having the same probability density function

$$p_{X_i}(x) = c\alpha^{-1}(x/\alpha)^{c-1} \exp[-(x/\alpha)^c] \qquad (x > 0), \tag{17}$$

the maximum likelihood estimators $\hat{c}$, $\hat{\alpha}$ of c and α respectively, satisfy the equations

$$\hat{\alpha} = \left[n^{-1} \sum_{i=1}^n X_i^{\hat{c}}\right]^{1/\hat{c}} \tag{18.1}$$

$$\hat{c} = \left[\left(\sum_{i=1}^n X_i^{\hat{c}} \log X_i\right)\left(\sum_{i=1}^n X_i^{\hat{c}}\right)^{-1} - n^{-1}\sum_{i=1}^n \log X_i\right]^{-1}. \tag{18.2}$$

(If ξ_0 is not equal to zero, each X_i is replaced by $(X_i - \xi_0)$.) The value of c has to be obtained from (18.2) and then used in (18.1) to obtain $\hat{\alpha}$. It will be noted that if c were in fact *known* to be equal to $\hat{c}$ then $\hat{\alpha}$ would be the maximum likelihood estimator of α.

If ξ_0 also has to be estimated, X_i is replaced by $X_i - \hat{\xi}_0$ in (18.1) and (18.2) and there is a third equation

$$(18.3) \qquad (\hat{c} - 1) \sum_{i=1}^{n} (X_i - \hat{\xi}_0)^{-1} = \hat{c}\hat{\alpha}^{-\hat{c}} \sum_{i=1}^{n} (X_i - \hat{\xi}_0)^{\hat{c}-1}.$$

If the value $\hat{\xi}_0$ satisfying the equations (18) with X_i replaced by $X_i - \hat{\xi}_0$ is greater than $\min(X_1, X_2, \ldots, X_n)$ then it is the maximum likelihood estimator of ξ_0. Otherwise the maximum likelihood estimator is $\min(X_1, X_2, \ldots, X_n)$, and (18.1) and (18.2) must then be solved for $\hat{c}$ and $\hat{\alpha}$ with X_i replaced by $(X_i - \min(X_1, X_2, \ldots, X_n))$.

It should be noted that the maximum likelihood estimators are 'regular' (in the sense of having the usual asymptotic distribution) only for $c > 2$. If it is *known* that $0 < c < 1$, then $\min(X_1, \ldots, X_n)$ is a "super-efficient" estimator for ξ_0.

If the maximum likelihood estimators are regular, then (Dubey [13]) for large n (when ξ_0 is known)

$$n \operatorname{var}(\hat{\alpha}) \doteqdot \left[1 + \frac{\{\psi(2)\}^2}{\psi'(1)}\right](\alpha/c)^2 = 1.087(\alpha/c)^2,$$

$$n \operatorname{var}(\hat{c}) \doteqdot c^2[\psi'(1)]^{-1} = 6c^2/\pi^2 = 0.608c^2,$$

$$\operatorname{corr}(\hat{\alpha}, \hat{c}) \doteqdot \frac{\psi(2)}{[\psi'(1) + \{\psi(2)\}^2]^{\frac{1}{2}}} = 0.313.$$

Haan and Beer [32] have suggested that Equation (18.2) (with X_i replaced by $X_i - \hat{\xi}_0$) be solved for $\hat{c}$, for each of a series of trial values of $\hat{\xi}_0$. The value of $\hat{\alpha}$ is then easily obtained from (18.1) and the value of the likelihood can be calculated. An alternative method would be to compare the values of $\hat{\alpha}$ obtained from (18.1) and (18.3).

Monte Carlo sampling experiments (Aroian [1], Miller [61]) have established that $\hat{c}$ is a biased estimator of c, the relative bias depending on sample size but *not* on c. For sample size 170, the bias is about 0.8%.

In life-testing work, it is quite often convenient to discontinue observations as soon as a prespecified number of items, k say, out of n under test, fail. The observed lifetimes, $X_1' \leq X_2' \leq \cdots \leq X_k'$ so obtained may then be regarded as a censored sample, excluding the $(n - k)$ highest values. From the joint probability density function

(19)

$$\frac{n!}{(n-k)!} \prod_{j=1}^{k} [c\alpha^{-1}\{(x_j' - \xi_0)/\alpha\}^{c-1}$$
$$\times \exp\left[-\sum_{j=1}^{k} \{(x_j' - \xi_0)/\alpha\}^c - (n-k)\{(x_k' - \xi_0)/\alpha\}^c\right]$$
$$(\xi_0 \leq x_1' \leq x_2' \leq \cdots \leq x_k')$$

maximum likelihood equations for estimators of ξ_0, α and c can be obtained (Cohen [10]). If the value of ξ_0 is known to be zero, then the equations obtained are of similar form to (18.1) and (18.2) but with

$$n^{-1}, \quad \sum_{i=1}^{n} X_i^{\hat{c}}, \quad \sum_{i=1}^{n} X_i^{\hat{c}} \log X_1, \quad \sum_{i=1}^{n} \log X_i$$

replaced by

$$k^{-1}, \quad \sum_{i=1}^{k} X_i'^{\hat{c}} + (n-k)X_k'^{\hat{c}},$$

$$\sum_{i=1}^{k} X_i'^{\hat{c}} \log X_i' + (n-k)X_k' \log X_k' \quad \text{and} \quad \sum_{i=1}^{k} \log X_i'$$

respectively.

For samples of size 10, censored by omission of the 5 highest values, McCool [57], has given estimated percentile points of the distribution of $\hat{c}/c$, based on sampling experiments. The estimates are based on five sets of 1000 samples each, for $c = 0.9, 1.1, 1.3, 1.5, 1.7$ respectively. It is noteworthy that the estimated value of $E[\hat{c}/c]$ is 1.56, indicating that $\hat{c}$ should be divided by 1.56 to obtain an approximately unbiased estimator of c.

McCool also showed that the distribution of $\hat{c} \log (\hat{X}_p/X_p)$, where $\hat{X}_p$ is the maximum likelihood estimator of

$$X_p = \alpha[-\log(1-p)]^{1/c}$$

does not depend on any of the parameters.

6.2 *Methods of Moments*

From the first three moments, it is possible to determine the values of ξ_0, α and c. The value of the first moment-ratio $(\sqrt{\beta_1})$ depends only on c, and given $\sqrt{\beta_1}$, c can be determined numerically. Using this value of c, α is calculated from the standard deviation and finally $\xi_0 = (\text{expected value}) - \alpha\Gamma(c^{-1}+1)$. Replacing population values of moments by estimates from a sample, estimators of c, α and ξ_0 are obtained. If ξ_0 is known, then c can be estimated from the ratio (standard deviation)/[(mean) $-\ \xi_0$]. (A nomogram for this purpose, provided by Kotel'nikov [50] will be shown in Section 7.) Dubey [18] [20] has provided tables of asymptotic relative efficiency of moment estimators.

6.3 *Estimators Based on Distribution of log X*

Using formulas (5) it is possible to construct simple estimators of c^{-1}, applicable when ξ_0 is known. Menon [60] suggested the estimator

$$\widetilde{(c^{-1})} = \frac{\sqrt{6}}{\pi} \cdot (\text{sample standard deviation of } (\log X_1, \log X_2, \ldots, \log X_n)), \tag{20}$$

which is an asymptotically normal and unbiased estimator of c^{-1}. Its variance is $(1.1 + O(n^{-1}))(c^{-2}/n)$, and its asymptotic efficiency, relative to the Cramér-Rao lower bound, is 55%. (A generalization of Menon's estimator with a high asymptotic efficiency was constructed by Kagan [44].)

Weibull [80] has proposed that estimation of the parameters might be based on the first and second moments *about the smallest sample value*

$$R_l = n^{-1} \sum_{j=1}^{n} (X_j - X_1')^l \qquad (l = 1,2).$$

As n tends to infinity the expected value of R_2/R_1^2 tends to a function $f(c)$ of the shape parameter c. By solving the equation

$$f(c^*) = R_2/R_1^2$$

an asymptotically unbiased estimator of c is obtained. The table below (based on [80]) gives some values of $f(c)$:

c	$f(c)$
0.0	1.0000
0.1	1.0145
0.2	1.0524
0.3	1.1093
0.4	1.1831
0.5	1.2732
0.6	1.3801
0.7	1.5045
0.8	1.6480
0.9	1.8124
1.0	2.0000

For $c \geq 0.5$, this estimator appears to be markedly more accurate than the estimator based on central moments. The asymptotic efficiency (compared with maximum likelihood estimation) decreases from 97.6% when $c = 0.5$ to 60.8% when $c = 1.0$, while that of a central moment estimator (obtained by equating the sample value of $\sqrt{b}$, to the population value of $\sqrt{\beta_1}$) decreases from 17.1% to 7.6% over the same range of values of c. Of course the moment estimator uses the third moment, while R_2/R_1^2 uses only the first two moments, so one might expect the former to be relatively inaccurate.

(21) γ^{-1} [$\log \alpha$ − (arithmetic mean of $\log X_1, \ldots, \log X_n$)]

is an unbiased estimator of c^{-1}, with variance

$$\gamma^{-2}(\pi^2/6)(c^2 n)^{-1} = 4.93c^{-2}/n.$$

This estimator does not seem to have much to recommend it, apart from its unbiasedness (Menon [60]). Its efficiency is 12%. On the other hand, the estimator of α derived from (8) and (20)

$$\tilde{\alpha} = \exp\left[n^{-1}\sum_{i=1}^{n}\log X_i + \gamma\widetilde{(c^{-1})}\right]$$

is asymptotically unbiased and has an asymptotic efficiency of 95%. ($\text{Var}(\tilde{\alpha}) = 1.2c^{-2}/n + c^{-2}O(n^{-\frac{3}{2}})$.)

Methods of this kind are also discussed in Section 7 of Chapter 21, in connection with an 'extreme value' distribution, which is, in fact, the distribution of log X, as we noted in Section 3 (see Bain and Antle [2], Miller and Freund [62], White [82] [83], Wilson [85]).

6.4 *Estimators Based on Order Statistics*

For the case when ξ_0 can be assumed to be zero, Vogt [75] has constructed a median-unbiased estimator of c. Since $Y'_j = (X'_j/\alpha)^c$ ($j = 1,2,\ldots,n$) are order statistics corresponding to independent standard exponential variables, it can be shown that

$$\Pr[Y'_n | Y'_1 \leq w] = n\sum_{j=0}^{n-1}(-1)^j\binom{n-1}{j}\{j(w-1)+n\}^{-1}.$$

There is a unique value of $w(>0)$, say $w_{\frac{1}{2}}$, such that the probability is equal to $\frac{1}{2}$. Then since the events $Y'_n/Y'_1 \leq w_{\frac{1}{2}}$ and $c \leq (\log X'_n - \log X'_1)^{-1}(\log w_{\frac{1}{2}})$ are identical, it follows that c is the median of the distribution of

$$(\log X'_n - \log X'_1)^{-1}(\log w_{\frac{1}{2}}).$$

Some values of $\log_{10} w_{\frac{1}{2}}$, based on [75], are shown in Table 2.

TABLE 2

Values of $g_n = \log_{10} w_{\frac{1}{2}}$ *such that* $g_n(\log_{10} X'_n - \log_{10} X'_1)^{-1}$ *is a median-unbiased estimator of* c.

n	2	3	4	5	6	7	8	9	10
g_n	0.477	0.802	1.009	1.159	1.278	1.375	1.458	1.530	1.594

For large samples, when ξ_0 is known to be zero, Dubey [22] suggests using the estimated 16.7% and 97.4% quantiles, y_1, y_2 respectively, and estimating c by the formula:

$$c^* = 2.989\,[\log(\hat{y}_2/\hat{y}_1)]^{-1}. \tag{22}$$

This estimator is asymptotically unbiased and normal, with approximate variance $0.916c^2/n$.

The construction of best linear unbiased estimators for c^{-1} and $\log\alpha$ using the first k order statistics has been studied by White [82].

For the special case when $k = 2$, the following estimators have been suggested by Leone *et al.* [52].

(i) If ξ_0 is known to be zero, c is estimated by solving the equation

$$\frac{X_2'}{X_1'} = n\left(\frac{n}{n-1}\right)^{1/c} - n + 1. \tag{23}$$

(ii) If the value of ξ_0 is not known, but that of c is known, then ξ_0 is estimated as $X_1' - n^{-1}(X_2' - X_1')\left[\left(\frac{n}{n-1}\right)^{1/c} - 1\right]^{-1}$.

(iii) If the values of both α and c are known, then ξ_0 is estimated as

$$X_1' - \alpha(n+1)^{-1/c}.$$

This is an unbiased estimator of ξ_0.

(iv) If the value of c is known, α is estimated as

$$\frac{1}{n}(X_2' - X_1')\frac{(n-1)^{1/c}}{1 - n^{-1/c}(n-1)^{1/c}\Gamma(c^{-1}+1)}. \tag{24}$$

(v) Finally, if the values of both ξ_0 and c are known α may be estimated as

$$(X_{\mu+1}' - \xi_0)[\log\{(n+1)/(n-\mu)\}]^{-1/c} \tag{25}$$

using the single order statistic $X_{\mu+1}'$ where μ is the largest integer not exceeding $n(1 - e^{-1})$.

Five decimal place coefficients of best linear unbiased estimators of ξ_0 and α, c being known, have been given by Govindarajulu and Joshi [30]. Up to sample size (n) five, all possible tail censorings (leaving at least 2 observations) are included; for $n = 6(1)17$, only censoring from above is included, and fewer decimal places are given (four for $6 \le n \le 10$, three for $11 \le n \le 17$). Values of c are 1.5(0.5)3(1)8.

Musson [65] gave similar tables, for the case when a restricted number of order statistics is to be used, when $n \le 10$. Stump [73], using an approximate method, extended Musson's tables to $n \le 40$.

Note that a linear function of the order statistics $\{X_j'\}$ is also a linear function of c^{-1}-th powers $\{Z_j'^{1/c}\}$ of order statistics corresponding to n independent exponentially distributed variables. If all the sample were available then it would be best to use a linear function of the Z_j''s (i.e. of $X_j'^c$'s). While this may not be so if incomplete sets of order statistics are used, it appears quite likely that, if c be known, functions of form $\sum \lambda_j X_j'^c$ will be more useful than those of form $\sum \lambda_j X_j'$.

For distribution (2), the cumulative distribution function is

$$F_X(x) = 1 - \exp[-\{(x - \xi_0)/\alpha\}^c]. \tag{26}$$

This equation can be written

$$x = \xi_0 + \alpha[-\log\{1 - F_X(x)\}]^{1/c}.$$

Suppose that $\hat{X}_{p_i}$, $\hat{X}_{p_j}$, $\hat{X}_{p_k}$ are estimators of the values of X corresponding to $F_X(x) = p_i, p_j, p_k$ respectively. (In a large sample, X_{p_i} could be the $[(n+1)p_i]$-th order statistic.) Then the three equations

$$\hat{X}_{p_s} = \xi_0^* + \alpha^*[-\log(1-p_s)]^{1/c^*} \qquad (s = i,j,k) \tag{27}$$

could be solved, yielding the estimators ξ^*, α^*, c^*.

The equation for c^* is

$$\frac{\hat{X}_{p_i} - \hat{X}_{p_j}}{\hat{X}_{p_j} - \hat{X}_{p_k}} + \frac{[-\log(1-p_i)]^{1/c^*} - [-\log(1-p_j)]^{1/c^*}}{[-\log(1-p_j)]^{1/c^*} - [-\log(1-p_k)]^{1/c^*}}. \tag{28}$$

If it can be arranged that

$$-\log(1-p_j) = \sqrt{(-\log(1-p_i))(-\log(1-p_k))},$$

then (28) simplifies to

$$c^* = \frac{1}{2}\left\{\log\left[\frac{-\log(1-p_k)}{-\log(1-p_j)}\right]\right\} \Big/ \left\{\log\left(\frac{\hat{X}_{p_i} - \hat{X}_{p_j}}{\hat{X}_{p_j} - \hat{X}_{p_k}}\right)\right\}. \tag{28$'$}$$

Dubey [22] has discussed the optimal choice of p_i, p_j and p_k.

If the value of ξ_0 is known (e.g. zero) then a variety of methods of estimating c and α is available.

The function $Y_j = 1 - \exp[-(X_j/\alpha)^c]$ is the probability integral transform of X_j if X_j has distribution (2) with $\xi_0 = 0$. If $Y_1' \le Y_2' \le \cdots \le Y_n'$ are the corresponding order statistics, then

$$Y_j' = 1 - \exp[-(X_j'/\alpha)^c] \qquad (j = 1,2,\ldots,n)$$

and the Y_j''s are distributed as order statistics in a random sample from the standard uniform distribution over the interval 0 to 1 (Equation (1) of Chapter 25 with $a = h = \frac{1}{2}$), and so $E[Y_j'] = j/(n+1)$.

Estimators α^*, c^* of α, c, obtained by minimizing

$$\sum_{j=1}^{n} [\log(-\log(1-Y_j')) - \log(-\log(1-E[Y_j']))]^2$$
$$= \sum_{j=1}^{n}\left[c\log(X_j'/\alpha) - \log\left(-\log\left(\frac{n-j+1}{n+1}\right)\right)\right]^2$$

with respect to c and α, are

$$c^* = \sum_{j=1}^{n}(K_j - \bar{K})(\log X_j' - \overline{\log X}) \Big/ \sum_{j=1}^{n}(\log X_j' - \overline{\log X})^2 \tag{29.1}$$

where

$$K_j = \log\left[-\log\left(\frac{n-j+1}{n+1}\right)\right],$$

$$\bar{K} = n^{-1}\sum_{j=1}^{n} K_j; \quad \overline{\log X} = n^{-1}\sum_{j=1}^{n}\log X_j'$$

and

$$(29.2) \qquad \alpha^* = \left(\prod_{j=1}^{n} X_j\right)^{1/n} \left[\prod_{j=1}^{n} \left\{-\log\left(\frac{n-j+1}{n+1}\right)\right\}\right]^{-1/(nc^*)}.$$

These estimators were proposed by Gumbel [31].

Estimators of similar form, with K_j replaced by $\log\left[\sum_{i=1}^{j}(n-i+1)^{-1}\right]$, were proposed by Bain and Antle [2], and with K_j replaced by

$$\log\left[-\log\frac{n-j+\frac{3}{2}}{n-j}\right]$$

by Miller and Freund [62]. (A further estimator was obtained by Bain and Antle [2], replacing K_j by *minus* expected value of the $(n-j+1)$-th order statistic of random samples of size n from the standard Type 1 extreme value distribution (16) of Chapter 21.) (See also Kimball [49] and Mann [56].) Bain and Antle [2] show that for all estimators of this form the distribution of c^*/c does not depend on any parameter (and so confidence intervals for c can be constructed from an observed value of c^*). They further found, from a Monte Carlo investigation, that all the estimators (with various values of K_j) just described are comparable with (and rather better than) Menon's estimator (20).

The c^*'s are biased estimators of c, but the bias could be removed by an appropriate corrective multiplier (equal to $\{E[c^*/c]\}^{-1}$). Bain and Antle give values for this multiplier.

Sometimes it is desired to estimate the 'reliability', that is,

$$R(x) = \Pr[X > x] = \exp[-\{(x-\xi_0)/\alpha\}^c]$$

or, if ξ_0 be known to be zero, $\exp[-(x/\alpha)^c]$. Of course, if $\hat{\alpha}$, $\hat{c}$ are maximum likelihood estimators of α, c respectively, then $\exp[-(x/\hat{\alpha})^{\hat{c}}]$ is a maximum likelihood estimator of $R(x)$.

The reader is referred to papers by Clark [8], Clark and Williams [9], Dubey [12]–[23], Harter and Moore [36]–[38], Jaech [39], [40], Mann [54]–[56], and White [81]–[84] for additional details on estimation procedures.

7. Tables and Graphs

Although the Weibull cumulative distribution function is easily calculable, it is often convenient to have tables, or some other source, from which values can be obtained quickly, without direct calculation.

For the case $\xi_0 = 0$, $\alpha = 1$ (as in (4)), Plait [67] has given values of the probability density function $(cx^{c-1}\exp(-x^c))$ to 8 decimal places for $c = 0.1(0.1)3.0(1)10$ and of the cumulative distribution function $(1 - \exp(-x^c))$ to 7 decimal places for $c = 0.1(0.1)4.0$. This latter function is also tabulated, less extensively, in Dourgnon and Reyrolle [11]. Harter and Dubey [35] give values of the mean, variance and first six standardized cumulant ratios $\kappa_k/(\kappa_2)^{k/2}$, $k = 3, \ldots, 8$ to eight decimal places for $c = 1.1(0.1)10.0$.

Expressing the cumulative distribution function (with $\xi_0 = 0$) in terms of the expected value, μ_1', we have

(26)′ $$F_X(x) = 1 - \exp\{[-x\Gamma(c^{-1}+1)/\mu_1'^c]\} \qquad (x > 0).$$

Kotel'nikov [50] used this formula to give a nomogram (Figure 2) for calculating $F_X(x)$, given μ_1' and the standard deviation σ. One part of the nomogram determines c from μ_1' and σ (see also Section 6.2). This is done by finding the intersection of the straight line joining the appropriate points on the μ_1' and σ scales with the c scale. Then it is necessary to place the scale at the foot of the Figure along the appropriate line μ_1' and move it till the dotted line on the nomogram passes through the appropriate σ value on the scale. (Note that the edge of the scale to be used depends on c.) Then for any given value of x, the value of $F_X(x)$ is the "$F_X(x)$ curve" passing through the value x on the appropriate edge of the scale. (In practice it is convenient for the movable scale to be transparent.)

Weibull probability paper can be constructed in several different ways (Kao [46], Nelson [66], Plait [67]). Starting from the formula

$$F_X(x) = 1 - \exp[-\{(x-\xi_0)/\alpha\}^c] \qquad (x \geq \xi_0)$$

for the cumulative distribution function, we have

$$\log\log(1 - F_X(x)) = -c\log\{(x-\xi_0)/\alpha\}$$

or

$$\log\log\left[\frac{1}{1-F_X(x)}\right] = c\log(x-\xi_0) - c\log\alpha.$$

Putting $w = \log\log\left[\frac{1}{1-F_X(x)}\right]$, $v = \log(x-\xi_0)$ we have the linear relation

$$w = cv - c\log\alpha \qquad (\text{with } c > 0).$$

So if $(1 - F_X(x))$ or $(1 - F_X(x))^{-1}$ be plotted against $(x - \xi_0)$ on (log log *vs.* log) paper a straight line is obtained. The slope of the graph is $-c$ or c, respectively, and the intercept with the Y axis is $-c\log\alpha$. Figure 3 is reproduced from [66].

Harter and Dubey [35] have constructed extensive tables, based on Monte Carlo (sampling) experiments, of the distribution of sample variances calculated from random samples from Weibull populations, and also of the "Weibull-t", that is

$$\frac{\sqrt{n}\,[(\text{arithmetic mean of sample}) - (\text{population mean})]}{(\text{sample variance})^{\frac{1}{2}}},$$

as a function of independent random variables each having the same Weibull distribution. They also calculated tables of the distribution of the sample arithmetic mean, using a Cornish-Fisher expansion (Chapter 12, Section 5).

Tables of moments of order statistics have been described in Section 5.

FIGURE 2

Kotel'nikov's Nomogram for the Weibull Distribution

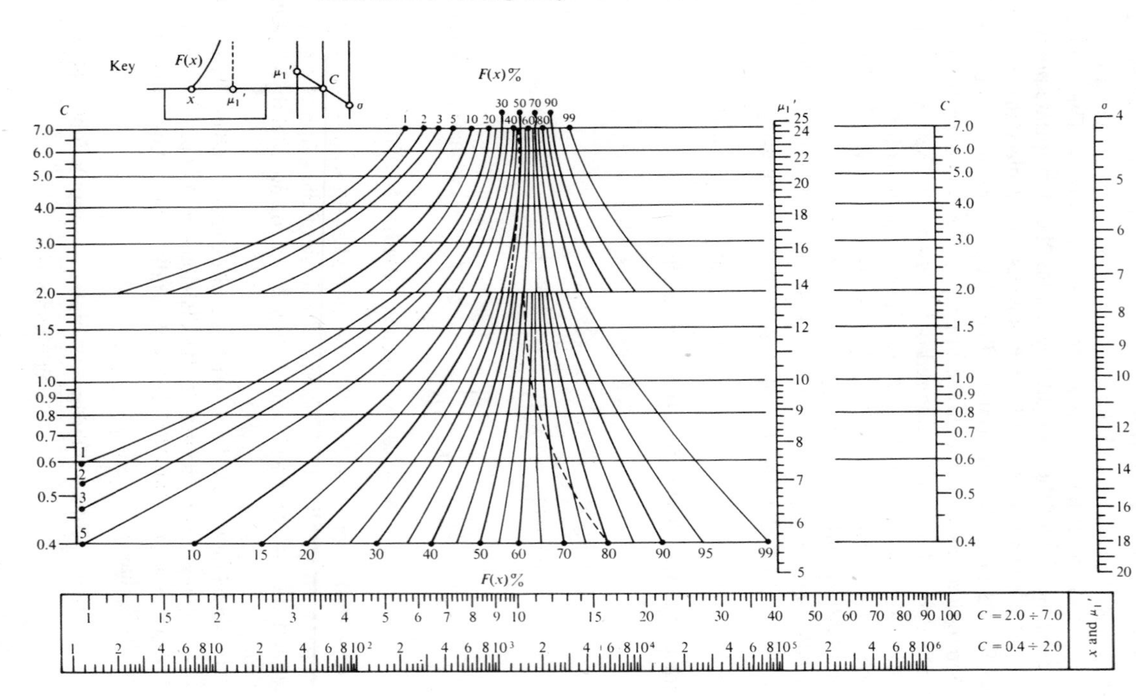

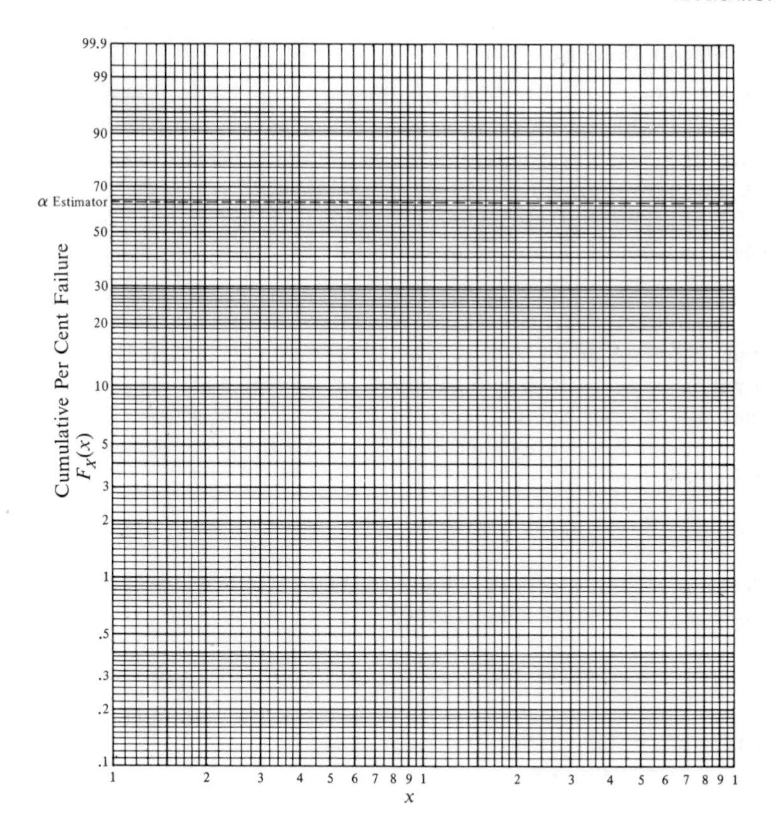

FIGURE 3

Weibull Probability Paper

8. Applications

Berrettoni [3] has described many applications of the Weibull distribution, using graphical methods in most cases. As stated, this method adds flexibility when an exponential distribution might possibly be adequate. Sometimes the exponential distribution will be found to suffice, and the Weibull distribution will be dispensed with (see, e.g. Kao [45]). However, the Weibull distribution may provide just the extra flexibility needed to make a model sufficiently accurate for use in an analysis.

The Weibull distribution is sometimes used as a tolerance distribution in the analysis of quantal response data. The explicit form of its cumulative distribution function (26) makes it especially suitable for this purpose, though care must be taken in judging whether it is appropriate.

Other examples of applications will be found in papers by Plait [67], Johnson [42], Freudenthal and Gumbel [25], and Jaech [40]. (See, however, Gorski [29] for a negative reaction.)

9. Related Distributions

The fact that there is a power transformation of any variable with a Weibull distribution which produces an exponentially distributed variable has been mentioned earlier in this chapter. Here we will be concerned with relationships between the Weibull and two other distributions.

If X has a Weibull distribution with $\xi_0 = 0$, as in (17), then the probability density function of $Y = -c \log (X/\alpha)$ is

$$p_Y(y) = e^{-y}e^{-e^{-y}}$$

which is a form of the extreme value distribution (Chapter 21). This transformation is the basis of some methods of estimating the parameters c and α which will be described in Chapter 21.

The probability density function (17) of X can be conveniently written in the form

$$p_X(x) = c\theta x^{c-1} \exp(-\theta x^c) \qquad (c > 0) \tag{17$'$}$$

where $\theta = \alpha^{-c}$. If c is fixed, but α varies so that the newly defined parameter θ be assumed to have the gamma probability density function

$$p_\theta(t) = \frac{\delta^p t^{p-1} e^{-\delta t}}{\Gamma(p)} \qquad (\theta > 0;\ p,\delta > 0)$$

then the probability density function of X is

$$\begin{aligned} p_X(x) &= \frac{\delta^p c x^{c-1}}{\Gamma(p)} \int_0^\infty t^p \exp(-t(x^c + \delta))\, dt \\ &= p\, \delta^p c x^{c-1}(x^c + \delta)^{-(p+1)} \qquad (x > 0). \end{aligned} \tag{30}$$

It follows that X^c has a Pareto distribution (Dubey [12]). See also reference [12] of Chapter 19. Note also that (30) is of the form discussed by Burr (see Chapter 12, Section 4.5) with a scale parameter δ.

Mixtures of Weibull distributions have been discussed by Kao [46] and Rider [69]; truncated Weibull distributions have been discussed by Aroian [1].

Harris and Singpurwalla [34]* construct compound Weibull (including, of course, exponential) distributions, by assigning (i) uniform, (ii) two-point, (iii) two-parameter gamma distributions to α^{-c}. They also discuss two-component mixtures with respect to the shape parameter c.

*See also "On estimation in Weibull distributions with random scale parameters", *Naval Research Logistics Quarterly*, **16**, 405–410, (1969), by the same authors.

REFERENCES

[1] Aroian, L. A. (1965). Some properties of the conditional Weibull distribution, *Transactions of the 19th Technical Conference of the American Society for Quality Control*, 361–368.

[2] Bain, L. J. and Antle, C. E. (1967). Estimation of parameters in the Weibull distribution, *Technometrics*, **9**, 621–627.

[3] Berrettoni, J. N. (1964). Practical applications of the Weibull distribution, *Industrial Quality Control*, **21**, 71–79.

[4] Bhattacharya, S. K. (1962). On a probit analogue used in a life-test based on the Weibull distribution, *Australian Journal of Statistics*, **4**, 101–105.

[5] Blischke, W. R., Johns, M. V., Truelove, A. J. and Murdle, P. B. (1965). *Estimation of the Location Parameters of the Pearson Type III and Weibull Distributions in the Non-Regular Case and Other Results in Non-Regular Estimation*, ARL 66–0233, Aerospace Research Laboratories, Wright-Patterson Air Force Base, Ohio.

[6] Blom, G. (1958). *Statistical Estimates and Transformed Beta Variables*, New York: John Wiley & Sons, Inc.

[7] Brownlee, J. (1923). *Tracts for Computers*, **9**, London: Cambridge University Press.

[8] Clark, L. J. (1964). *Estimation of the Scale Parameter of the Weibull Probability Density Function by the Use of One Order and of M Order Statistics*, Unpublished thesis, Air Force Institute of Technology, Dayton Air Force Base, Ohio.

[9] Clark, E. C. and Williams, G. T. (1958). Distribution of the members of an ordered sample, *Annals of Mathematical Statistics*, **29**, 862–870.

[10] Cohen, A. C. (1965). Maximum likelihood estimation in the Weibull distribution based on complete and on censored samples, *Technometrics*, **7**, 579–588.

[11] Dourgnon, F. and Reyrolle, J. (1966). Tables de la fonction de répartition de la loi de Weibull, *Revue de Statistique Appliquée*, **14**, No. 4, 83–116.

[12] Dubey, S. D. (1968). A compound Weibull distribution, *Naval Research Logistics Quarterly*, **15**, 179–188.

[13] Dubey, S. D. (1965). Asymptotic properties of several estimators of Weibull parameters, *Technometrics*, **7**, 423–434.

[14] Dubey, S. D. (1966). Characterization theorems for several distributions and their applications, *Journal of Industrial Mathematics*, **16**, 1–22

[15] Dubey, S. D. (1966). Comparative performance of several estimators of the Weibull parameters, *Proceedings of the 20th Technical Conference of the American Society for Quality Control*, 723–735.

[16] Dubey, S. D. (1966). Some test functions for the parameters of the Weibull distributions, *Naval Research Logistics Quarterly*, **13**, 113–128.

[17] Dubey, S. D. (1966). On some statistical inferences for Weibull laws, *Naval Research Logistics Quarterly*, **13**, 227–251.

[18] Dubey, S. D. (1966). Hyper-efficient estimator of the location parameter of the Weibull laws, *Naval Research Logistics Quarterly*, **13**, 253–264.

[19] Dubey, S. D. (1966). Asymptotic efficiencies of the moment estimators for the parameters of the Weibull laws, *Naval Research Logistics Quarterly*, **13,** 265–288.

[20] Dubey, S. D. (1966). Transformations for estimation of parameters, *Journal of the Indian Statistical Association*, **4,** 109–124.

[21] Dubey, S. D. (1967). Revised tables for asymptotic efficiencies of the moment estimators for the parameters of the Weibull laws, *Naval Research Logistics Quarterly*, **14,** 261–267.

[22] Dubey, S. D. (1967). Normal and Weibull distributions, *Naval Research Logistics Quarterly*, **14,** 69–79.

[23] Dubey, S. D. (1967). Some percentile estimators for Weibull parameters, *Technometrics*, **9,** 119–129.

[24] Dubey, S. D. (1967). Monte Carlo study of the moment and maximum likelihood estimators of Weibull parameters, *Trabajos de Estadistica*, **18,** (II & III), 131–141.

[25] Freudenthal, A. M. and Gumbel, E. J. (1954). Minimum life in fatigue, *Journal of the American Statistical Association*, **49,** 575–597.

[26] Fukuta, J. (1963). *Estimation of Parameters in the Weibull Distribution and its Efficiency*, Research Report No. 13, Faculty of Engineering, Gifu University, Japan.

[27] Gittus, J. H. (1967). On a class of distribution functions, *Applied Statistics*, **16,** 45–50.

[28] Gnedenko, B. V., Beljaev, Yu, K. and Solov'ev, A. D. (1965). *Mathematical Methods in Reliability Theory*, Moscow. (In Russian. English translation: Academic Press, 1968.)

[29] Gorski, A. C. (1968). Beware of the Weibull euphoria, *Transactions of IEEE — Reliability*, **17,** 202–203.

[30] Govindarajulu, Z. and Joshi, M. (1968). Best linear unbiased estimation of location and scale parameters of Weibull distribution using ordered observations, *Statistical Applications Research, JUSE*, **15,** 1–14.

[31] Gumbel, E. J. (1958). *Statistics of Extremes*, New York: Columbia University Press.

[32] Haan, C. T. and Beer, C. E. (1967). Determination of maximum likelihood estimators for the three parameter Weibull distribution, *Iowa State Journal of Science*, **42,** 37–42.

[33] Hahn, G. J., Godfrey, J. T. and Renzi, N. A. (1960). *Weibull Density Computer Programs*, General Electric Company Report No. 60GL235.

[34] Harris, C. M. and Singpurwalla, N. D. (1968). Life distributions derived from stochastic hazard functions, *Transactions of IEEE — Reliability*, **17,** 70–79.

[35] Harter, H. L. and Dubey, S. D. (1967). *Theory and tables for tests of hypotheses concerning the mean and the variance of a Weibull population*, ARL 67–0059, Aerospace Research Laboratories, Wright-Patterson Air Force Base, Ohio.

[36] Harter, H. L. and Moore, A. H. (1965). Point and interval estimators, based on *m* order statistics, for the scale parameter of a Weibull population with known shape parameter, *Technometrics*, **7,** 405–422.

[37] Harter, H. L. and Moore, A. H. (1965). Maximum-likelihood estimation of the

parameters of gamma and Weibull populations from complete and from censored samples, *Technometrics*, **7**, 639–643, (Correction *Technometrics*, **9**, (1967), 195).

[38] Harter, H. L. and Moore, A. H. (1967). Asymptotic variances and covariances of maximum likelihood estimators, from censored samples, of the parameters of Weibull and gamma populations, *Annals of Mathematical Statistics*, **38**, 557–570.

[39] Jaech, J. L. (1964). Estimation of Weibull distribution shape parameter when no more than two failures occur per lot, *Technometrics*, **6**, 415–422.

[40] Jaech, J. L. (1968). *Estimation of Weibull parameters from grouped failure data*, presented at the Pittsburgh meeting of the American Statistical Association, August, 1968.

[41] Johns, M. V. and Lieberman, G. J. (1966). An exact asymptotically efficient confidence bound for reliability in the case of the Weibull distribution, *Technometrics*, **8**, 135–175.

[42] Johnson, L. G. (1968). The probabilistic basis of cumulative damage, *Transactions of the 22nd Technical Conference of the American Society of Quality Control*, 133–140.

[43] Johnson, N. L. (1966). Cumulative sum control charts and the Weibull distribution, *Technometrics*, **8**, 481–491.

[44] Kagan, A. M. (1965). Zamechaniya o razdyelyayushchikh razbyeniakh, (Remarks on separating subdivisions) *Trudy Matamaticheskogo Instituta imeni Steklova*, **79**, 26–31. (In Russian. English translation published by the American Mathematical Society, 1968.)

[45] Kao, J. H. K. (1958). Computer methods for estimating Weibull parameters in reliability studies, *Transactions of IRE-Reliability and Quality Control*, **13**, 15–22.

[46] Kao, J. H. K. (1959). A graphical estimation of mixed Weibull parameters in life-testing electron tubes, *Technometrics*, **1**, 389–407.

[47] Khirosi, S. and Mieko, N. (1963). On the graphical estimation of the parameter of the Weibull distribution from small samples, *Bulletin of the Electrotechnical Laboratory*, **27**, 655–663.

[48] Khirosi, S., Sideru, T. and Minoru, K. (1966). On the accuracy of estimation of the parameters of the Weibull distribution from small samples, *Bulletin of the Electrotechnical Laboratory*, **30**, 753–765.

[49] Kimball, B. F. (1960). On the choice of plotting positions on probability paper, *Journal of the American Statistical Association*, **55**, 546–560.

[50] Kotel'nikov, V. P. (1964). A nomogram connecting the parameters of Weibull's distribution with probabilities, *Teoriya Veroyatnostei i ee Primeneniya*, **9**, 743–746. (English translation, **9**, 670–673)

[51] Lehman, E. H. (1962). Shapes, moments and estimators of the Weibull distribution, *Transactions of IEEE-Reliability*, **11**, 32–38.

[52] Leone, F. C., Rutenberg, Y. H. and Topp, C. W. (1960). *Order Statistics and Estimators for the Weibull Distribution*, Case Statistical Laboratory, Publication No. 1026.

[53] Lieblein, J. (1955). On moments of order statistics from the Weibull distribution, *Annals of Mathematical Statistics*, **26**, 330–333.

[54] Mann, Nancy R. (1966). *Exact three-order-statistics confidence bounds on reliability parameters under Weibull assumptions*, (Appendix B to Aerospace Research Laboratories Report ARL67–0023, Wright-Patterson Air Force Base, Ohio).

[55] Mann, Nancy R. (1967). Tables for obtaining the best linear invariant estimates of parameters of the Weibull distribution, *Technometrics*, **9,** 629–645.

[56] Mann, Nancy R. (1968). Point and interval estimation procedures for the two-parameter Weibull and extreme-value distributions, *Technometrics*, **10,** 231–256.

[57] McCool, J. I. (1966). *Inference on Weibull percentiles and shape parameter from maximum likelihood estimates*, Report No. AL68PO23, SKF Industries Research Laboratory, King of Prussia, Pennsylvania.

[58] Mendenhall, W. (1958). A bibliography on life testing and related topics, *Biometrika*, **45,** 521–543.

[59] Mendenhall, W. and Lehman, E. H. (1960). An approximation to the negative moments of the positive binomial useful in life testing, *Technometrics*, **2,** 227–242.

[60] Menon, M. V. (1963). Estimation of the shape and scale parameters of the Weibull distribution, *Technometrics*, **5,** 175–182.

[61] Miller, D. W. (1966). *Degree of normality of maximum likelihood estimates of the shape parameter of the Weibull failure distribution*, Unpublished thesis, Washington University, St. Louis, Missouri.

[62] Miller, I. and Freund, J. E. (1965). *Probability and Statistics for Engineers*, Englewood Cliffs, New Jersey: Prentice-Hall.

[63] Moore, A. H. and Harter, H. L. (1965). One-order-statistic estimation of the scale parameters of Weibull populations, *Transactions of IEEE-Reliability*, **14,** 100–106.

[64] Moore, A. H. and Harter, H. L. (1966). Point and interval estimation, from one-order statistic, of the location parameter of an extreme-value distribution with known scale parameter, and of the scale parameter of a Weibull distribution with known shape parameter, *Transactions of IEEE-Reliability*, **15,** 120–126.

[65] Musson, T. A. (1965). *Linear Estimation of the Location and Scale Parameters of the Weibull and Gamma Distributions by the use of Order Statistics*, Unpublished thesis, Air Force Institute of Technology, Dayton Air Force Base, Ohio.

[66] Nelson, L. S. (1967). Weibull probability paper, *Industrial Quality Control*, **23,** 452–453.

[67] Plait, A. (1962). The Weibull distribution — with tables, *Industrial Quality Control*, **19,** 17–26.

[68] Quayle, R. J. (1963). *Estimation of the scale parameter of the Weibull probability density function by use of one-order statistic*, Unpublished thesis, Air Force Institute of Technology, Wright-Patterson Air Force Base, Dayton, Ohio.

[69] Rider, P. R. (1961). Estimating the parameters of mixed Poisson, binomial, and Weibull distributions by the method of moments, *Bulletin de l'Institut International de Statistique*, **39,** 225–232.

[70] Soland, R. M. (1967). Bayesian analysis of the Weibull process with unknown scale parameter, *Transactions of IEEE-Reliability*, **17,** 84–90.

[71] Soland, R. M. (1966). *Use of the Weibull distribution in Bayesian decision theory*, Technical Paper RAC-TP-225, Research Analysis Corporation, McLean, Virginia.

[72] Splitstone, D. (1967). *Estimation of the Weibull Shape and Scale Parameters*, M.Sc. thesis, Iowa State University.

[73] Stump, F. B. (1968). *Nearly best unbiased estimation of the location and scale parameters of the Weibull distribution by the use of order statistics*, Master's thesis, Air Force Institute of Technology, Wright-Patterson Air Force Base, Ohio.

[74] Tate, R. F. (1959). Unbiased estimation: Functions of location and scale parameters, *Annals of Mathematical Statistics*, **30,** 341–366.

[75] Vogt, H. (1968). Zur Parameter- und Prozentpunktschätzung von Lebensdauerverteilungen bei kleinem Stichprobenumfang, *Metrika*, **14,** 117–131.

[76] Weibull, W. (1939). A statistical theory of the strength of material, *Ingeniörs Vetenskaps Akademiens Handligar, Stockholm*, **No. 151.**

[77] Weibull, W. (1939). The phenomenon of rupture in solids, *Ingeniörs Vetenskaps Akademiens Handligar, Stockholm*, **No. 153.**

[78] Weibull, W. (1951). A statistical distribution function of wide applicability, *Journal of Applied Mechanics*, **18,** 293–297.

[79] Weibull, W. (1967). *Estimation of distribution parameters by a combination of the best linear order statistic method and maximum likelihood*, Technical Report AFML-TR-67-105, Air Force Materials Laboratory, Wright-Patterson Air Force Base, Ohio.

[80] Weibull, W. (1967). *Moments about Smallest Sample Value*, Technical Report AFML-TR-67-375, Air Force Materials Laboratory, Wright-Patterson Air Force Base, Ohio.

[81] White, J. S. (1964). Least-squares unbiased censored linear estimation for the log Weibull (extreme value) distribution, *Journal of Industrial Mathematics*, **14,** 21–60.

[82] White, J. S. (1965). *Linear Estimation for the Log Weibull Distribution*, General Motors Research Publication, GMR-481.

[83] White, J. S. (1966). *A Technique for Estimating Weibull Percentage Points*, General Motors Research Publication, GMR-572.

[84] White, J. S. (1967). *Estimating Reliability from the First Two Failures*, General Motors Research Publication, GMR-669.

[85] Wilson, R. B. (1965). *Two Notes on Estimating Shape Parameters*, RAND Corporation Memorandum, RM-4459-PR.

21

Extreme Value Distributions

1. Definition

Extreme value distributions are generally considered to comprise the three following families:

(1) *Type 1:* $\Pr[X \le x] = \exp\{-e^{-(x-\xi)/\theta}\}$

(2) *Type 2:*

$$\Pr[X \le x] = \begin{cases} 0 & (x < \xi) \\ \exp\left\{-\left(\dfrac{x-\xi}{\theta}\right)^{-k}\right\} & (x \ge \xi) \end{cases}$$

(3) *Type 3:*

$$\Pr[X \le x] = \begin{cases} \exp\left\{-\left(\dfrac{\xi - x}{\theta}\right)^{k}\right\} & (x \le \xi) \\ 1 & (x > \xi) \end{cases}$$

where ξ, $\theta(> 0)$ and $k(> 0)$ are parameters.

The corresponding distributions of $(-X)$ are also called extreme value distributions.

Of these three families of distributions, Type 1 is by far the one most commonly referred to in discussions of 'extreme value' distributions. Indeed, some authors call (1) "the" extreme value distribution. In view of this, and the fact that distributions (2) and (3) can be transformed to Type 1 distributions by the simple transformations

$$Z = \log(X - \xi); \qquad Z = -\log(\xi - X)$$

respectively, we will, for the greater part of this chapter, confine ourselves to discussion of Type 1 distributions. We may also note that the Type 3 distribution of $(-X)$ is a *Weibull* distribution. Such distributions have been discussed in Chapter 20, and so there is no need to discuss them in detail here.

Of course, Types 1 and 2 are also closely related to the Weibull distribution, by the simple formulas relating Z and X, just quoted. Type 1 is sometimes called the log-Weibull distribution (e.g. White [100]).

Type 1 distributions are sometimes called 'double exponential' distributions, on account of the functional form of (1). We do not use this to avoid confusion with Laplace distributions (Chapter 23) which are also called double exponential.

The name 'extreme value' is attached to these distributions because they can be obtained as limiting distributions (as $n \to \infty$) of the greatest value among n independent random variables each having the same continuous distribution (see Section 3). By replacing X by $-X$, limiting distributions of *least* values are obtained. As already mentioned, these are also extreme value distributions, but do not need separate treatment.

Although the distributions are labelled 'extreme value,' it is to be borne in mind (i) that they do not represent distributions of *all* kinds of 'extreme values' (for example, in samples of finite size), and (ii) they can be used empirically (i.e. without an 'extreme value' model) in the same way as other distributions.

In this last connection, we note that the Type 1 distribution may be regarded as an approximation to a Weibull distribution with large value of c. (Compare Equation (6) of Chapter 20 and Equations (19) and (20) of this chapter.) Also note that if X has a Type 1 distribution, $Z = \exp[-(X - \xi)/\theta]$ has an exponential distribution with probability density function:

$$p_Z(z) = e^{-z} \qquad (0 \leq z).$$

2. Historical Remarks and Applications

Extreme value theory seems to have first arisen mainly in connection with the needs of astronomers in utilizing (or possibly rejecting) outlying observations. Earlier published papers (e.g. Fuller [21], Griffith [26]) on the subject were highly specialized both in fields of application and in methods of mathematical analysis. Systematic development of general theory may be regarded as starting in 1922, with a paper by Bortkiewicz [5] on the distribution of range in random samples from a normal distribution, (already referred to in Chapter 13, Section 6.2). As can be seen from that section, subsequent development of this particular topic was quite rapid. From our present point of view, the importance of [5] resides in the fact that the concept of 'distribution of largest value' was clearly introduced. In the next year, von Mises [78] evaluated the expected value of this distribution, and Dodd [9] calculated its median, and also considered some non-normal parent distributions. Of more direct relevance to the present chapter is a paper by Fréchet [19] published in 1927, in which asymptotic distributions of largest values are considered. In the following year, Fisher and Tippett [18] published results of an independent inquiry into the same kind of problems.

The theoretical developments of the 1920's were followed, in the 1930's, by practical applications using extreme value statistics of distributions of human life-times* radioactive emissions (Gumbel [28], [29]), and strength of materials (Weibull [96]). In this last field Weibull effectively advocated the use of reversed Type 3 distributions which have now become well-known as *Weibull* distributions (and have been discussed, as such, in Chapter 20).

From the late 1930's onwards, many further fields of application have been found, including rainfall (Potter [85]), flood flows (Gumbel [30], [31], [32], [34], Rantz and Riggs [87]), earthquakes (Nordquist [82]), general meteorological data (Jenkinson [54], Thom [91]), aircraft load (Press [86]), corrosion (Aziz [1], Eldredge [13]), and microorganism survival times (Velz [95]). In this process a leading part was played by Gumbel who contributed a major part of the theoretical developments, and in 1958 published ([38]) a consolidated account of theory and practice which contains a good bibliography. It may be usefully studied with later works [42], [43] of Gumbel to gain a deeper knowledge of extreme value distributions. Other fields of application are included in the References (Clough and Kotz [6], Epstein [14], [16], Epstein and Brooks [17], Frenkel and Kontorova [20], Gumbel and Mustafi [45], King [62], Lieblein and Zelen [68], Posner [84], Weibull [97], Winer [101]).

3. Genesis

Extreme value distributions were obtained as limiting distributions of greatest (or least) values in random samples of increasing size. To obtain a nondegenerate limiting distribution, it is necessary to 'reduce' the actual greatest value by applying a linear transformation with coefficients which depend on the sample size. This process is analogous to standardization (as in central limit theorems — Chapter 13, Section 2) though not restricted to this particular sequence of linear transformations.

If $X_1, X_2, \ldots, X_n$ are independent random variables with common probability density function

$$p_{X_j}(x) = f(x) \qquad (j = 1,2,\ldots,n)$$

then the cumulative distribution function of $X_n' = \max(X_1, X_2, \ldots, X_n)$ is

$$F_{X_n'}(x) = [F(x)]^n \tag{4}$$

where

$$F(x) = \int_{-\infty}^{x} f(t)\,dt.$$

As n tends to infinity, it is clear that for any fixed value of x

$$\lim_{n\to\infty} F_{X_n'}(x) = \begin{cases} 1 & \text{if } F(x) = 1 \\ 0 & \text{if } F(x) < 1. \end{cases}$$

Even if it is proper, this limiting distribution would be "trivial" and of no

*In fact the *Gompertz* distribution of life time, already in use for about a century, is a Type 1 distribution, though not generally regarded as such.

special interest. If there is a limiting distribution of interest, we must find it as the limiting distribution of some sequence of transformed 'reduced' values, such as $(a_n X'_n + b_n)$, where a_n, b_n may depend on n, but not on x.

To distinguish the limiting cumulative distribution of the "reduced" greatest value from $F(x)$, we will denote it by $G(x)$. Then since the greatest of Nn values $X_1, X_2, \ldots, X_{Nn}$ is also the greatest of the N values

$$\max (X_{(j-1)n+1}, X_{(j-1)n+2}, \ldots, X_{jn}) \qquad (j = 1,2,\ldots,N)$$

it follows that $G(x)$ must satisfy the equation

$$[G(x)]^N = G(a_N x + b_N). \tag{5}$$

This equation was obtained by Fréchet [19] and also used by Fisher and Tippett [18]. It is sometimes called the *stability postulate*.

Type 1 distributions are obtained by taking $a_N = 1$; Types 2 and 3 by taking $a_N \neq 1$. In this latter case

$$x = a_N x + b_N \qquad \text{if} \qquad x = b_N(1 - a_N)^{-1}$$

and from (5) it follows that $G(b_N(1 - a_N)^{-1})$ must equal either 1 or 0. Type 2 corresponds to 1, Type 3 to 0.

We now consider the case $a_N = 1$ (Type 1) in a little detail. Equation (5) is now:

$$[G(x)]^N = G(x + b_N). \tag{6}$$

Since $G(x + b_N)$ must also satisfy (5),

$$[G(x)]^{NM} = [G(x + b_N)]^M = G(x + b_N + b_M). \tag{7}$$

But, also from (5),

$$[G(x)]^{NM} = G(x + b_{NM}) \tag{8}$$

and from (7) and (8) we can have

$$b_N + b_M = b_{NM}$$

whence

$$b_N = \theta \log N \qquad (\theta \text{ a constant}). \tag{9}$$

Taking logarithms of (6) *twice* and inserting the value of b_N from (9), we have (noting that $G \leq 1$)

$$\log N + \log(-\log G(x)) = \log(-\log G(x + \theta \log N)). \tag{10}$$

In other words, when the argument of

$$h(x) = \log(-\log G(x))$$

increases by $\theta \log N$, $h(x)$ decreases by $\log N$. Hence

(11) $$h(x) = h(0) - x/\theta.$$

Since $h(x)$ decreases as x increases, $\theta > 0$. From (11)

$$\begin{aligned} -\log G(x) &= \exp[-(x - \theta h(0))/\theta] \\ &= \exp[-(x - \xi)/\theta] \end{aligned}$$

where $\xi = \theta \log(-\log G(0))$. Hence

$$G(x) = \exp[-e^{-(x-\xi)/\theta}]$$

in agreement with (1).

We will not enter into details of derivation for Types 2 and 3.

Gnedenko [24], in a paper of fundamental importance in this field, established certain correspondences between the parent distribution ($F(x)$ in the above analysis) and the Type to which the limiting distribution belongs. It should be noted that the conditions relate essentially to the behavior of $F(x)$ for high (low) values of x if the limiting distribution of greatest (least) values is to be considered. It is quite possible for greatest and least values, corresponding to the same parent distribution, to have different limiting distributions.

We now summarize Gnedenko's results:

For Type 1 distribution: Defining X_α by the equation

$$F(X_\alpha) = \alpha,$$

the condition is

(12) $$\lim_{n\to\infty} n[(1 - F(X_{1-n^{-1}} + y(X_{1-(ne)^{-1}} - X_{1-n^{-1}}))] = e^{-y}.$$

For Type 2 distribution:

(13) $$\lim_{x\to\infty} \frac{1 - F(x)}{1 - F(cx)} = c^k \qquad (c > 0; k > 0).$$

For Type 3 distribution:

(14) $$\lim_{x\to 0} \frac{1 - F(cx + \omega)}{1 - F(x + \omega)} = c^k \qquad (c > 0; k > 0)$$

where $F(\omega) = 0$, $F(x) < 1$ for $x < \omega$.

Gnedenko has also shown that these conditions are necessary, as well as sufficient, and, further, that there are *no other* distributions satisfying the stability postulate.

An alternative interpretation of these conditions has been given by Clough and Kotz [6].

Among distributions satisfying the Type 1 condition (11) are normal, logistic and exponential; Type 2 condition (12) is satisfied by Cauchy distributions; Type 3 conditions are satisfied by nondegenerate distributions with range of variation bounded above. Alternative names for Types 1, 2 and 3 are *exponential* (also *Gumbel*), *Cauchy* (also *Fréchet*), and *bounded* (also *Weibull*) *extreme value distributions* respectively. The names in parentheses refer to discoverers and users, rather than to shape of parent distribution.

Adaption of the above results to limiting distributions of *least* observations is straightforward.

It is easily proved that if $Y_1, Y_2, \ldots,$ are independent variables, each having the exponential distribution:

$$\Pr(Y \leq y) = 1 - e^{-y} \qquad (y > 0)$$

and if L is the zero-truncated Poisson variable

$$\Pr(L = l) = (e^{\lambda} - 1)^{-1}\lambda^l/l! \qquad (l = 1,2,\ldots),$$

the random variable defined by

$$X = \max(Y_1,\ldots,Y_L)$$

has the extreme value distribution:

$$\Pr(X \leq x) = (e^{\lambda} - 1)^{-1}[\exp\{\lambda(1 - e^{-x})\}] = c\exp[-\lambda e^{-x}].$$

(In a similar way, the Fréchet distribution can be generated from the Pareto distribution (Chapter 19) and the truncated Weibull from the power function distribution.)

Sibuya [90] has suggested a method of generating random numbers of extreme value distributions, based on these results.

Tables of the random numbers of these distributions (i.e., (1)–(3)) have been prepared by Goldstein [25] (Section 5).

4. Moments and Order Statistics

In this section we will consider Type 1 distributions (1) exclusively. Corresponding to (1) is the probability density function

$$p_X(x) = \theta^{-1}e^{-(x-\xi)/\theta}\exp[-e^{-(x-\xi)/\theta}]. \tag{15}$$

If $\xi = 0$ and $\theta = 1$ or if equivalently, considering the distribution of $Y = (X - \xi)/\theta$, we have the *standard form*

$$p_Y(y) = \exp(-y - e^{-y}). \tag{16}$$

Since (as pointed out in Section 1) the variable $Z = \exp[-(X - \xi)/\theta] = e^{-Y}$ has the exponential distribution

$$p_Z(z) = e^{-z} \qquad (z \geq 0),$$

it follows that

$$E[e^{t(X-\xi)/\theta}] = E[Z^{-t}] = \Gamma(1 - t)$$

for $t < 1$. Replacing t by θt, we see that the moment generating function of

X is

$$E[e^{tX}] = e^{t\xi}\Gamma(1 - \theta t) \qquad (\theta|t| < 1) \tag{17}$$

and the cumulant generating function is

$$\Psi(t) = \xi t + \log \Gamma(1 - \theta t). \tag{18}$$

The cumulants of X are

$$\kappa_1(X) = \xi - \theta\psi(1) = \xi + \gamma\theta = \xi + 0.57722\theta \tag{19.1}$$

$$\kappa_r(X) = (-\theta)^r\psi^{(r-1)}(1) \qquad (r \geq 2). \tag{19.2}$$

In particular

$$\text{var}(X) = \tfrac{1}{6}\pi^2\theta^2 = 1.64493\theta^2 \tag{20}$$

and the moment ratios are

$$\alpha_3^2(X) = \beta_1(X) = 1.29857; \quad \alpha_4(X) = \beta_2(X) = 5.4. \tag{21}$$

Note that ξ and θ are purely location and scale parameters, respectively. All distributions (15) have the same shape.

The distribution is unimodal. Its mode is at $X = \xi$, and there are points of inflection at

$$X = \xi \pm \theta \log [\tfrac{1}{2}(3 + \sqrt{5})] = \xi \pm 0.96242\theta.$$

The median $X_{0.5}$ satisfies the equation

$$\exp [-\exp \{-(X_{0.5} - \xi)/\theta\}] = 0.5,$$

whence

$$X_{0.5} = \xi - \theta \log \log 2 \doteqdot \xi + 0.36611\theta.$$

(For additional details see Lehman [63].)

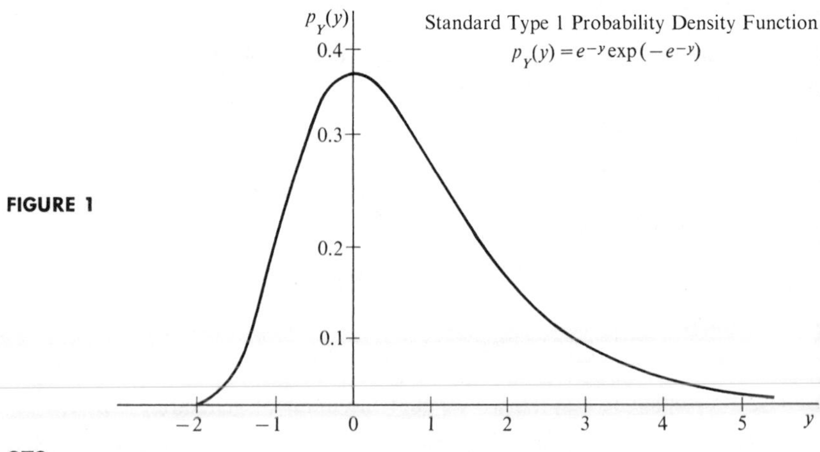

FIGURE 1

The standard probability density function (16) is shown in Figure 1. Its shape is very closely mimicked by a log-normal distribution with $e^{\sigma^2} = 1.1325$ (using the notation of Chapter 14). (The β_1, β_2 values of this lognormal distribution are 1.300, 5.398 respectively.) In Table 1 the *standardized* cumulative distribution functions are compared.

If $Y_1' \leq Y_2' \leq \cdots \leq Y_n'$ are the order statistics corresponding to n independent random variables each having the distribution (16), then the probability density function of Y_r' is

$$p_{Y_r'}(y) = \frac{n!}{(r-1)!(n-r)!} e^{-y} \times [\exp(-re^{-y})] \sum_{j=0}^{n-r} (-1)^j \binom{n-r}{j} \exp(-je^{-y}). \tag{22}$$

The expected value of $Y_r'^\beta$ is (Lieblein [64])

$$E(Y_r'^\beta) = \frac{n!}{(r-1)!(n-r)!} \sum_{j=0}^{n-r} (-1)^j \binom{n-r}{j} g_\beta(r+j) \tag{23}$$

where

$$g_\beta(c) = \int_{-\infty}^{\infty} y^\beta \exp(-y - ce^{-y})\, dy = (-1)^\beta \frac{d^\beta}{dt^\beta} (c^{-t}\Gamma(t))\bigg|_{t=1}.$$

In particular $g_1(c) = c^{-1}(\gamma + \log c)$; $g_2(c) = c^{-1}[\frac{1}{6}\pi^2 + (\gamma + \log c)^2]$.

TABLE 1

Standardized Cumulative Distribution Functions

x	$F(x)$ (a)	$F(x)$ (b)
−2.0	0.00068	0.00022
−1.5	0.02140	0.01959
−1.0	0.1321	0.1342
−0.5	0.3443	0.3471
0.0	0.5704	0.5700
0.5	0.7440	0.7423
1.0	0.8558	0.8546
1.5	0.92237	0.92096
2.0	0.95774	0.95792
2.5	0.97752	0.97730
3.0	0.98810	0.98837
3.5	0.99371	0.99389
4.0	0.99668	0.99677

(*a*) *Type 1 Extreme Value Distribution:* $F(x) = \exp[-\exp\{-1.28254x - 0.57722\}]$;
(*b*) *Lognormal Distribution:* $F(x) = (\sqrt{2\pi})^{-1} \int_{-\infty}^{u(x)} \exp(-\frac{1}{2}u^2)\, du$ *with* $u(x) = 6.52771 \log_{10}(x + 2.74721) - 2.68853$.

Using these values, $E[Y_r']$ and $\text{var}(Y_r')$ can be calculated. It is interesting to note that the variance of the largest value is $\pi^2/6$, whatever be the value of n (White [100]).

Kimball [56], [58] expressed the expected value of Y'_{n-r+1} as

$$\gamma + \sum_{j=1}^{r-1} (-1)^j \binom{n}{j} \Delta^j \log n, \tag{24}$$

where Δ^i represents forward difference of ith order (cf. Chapter 1, Section 2).

5. Tables and Probability Paper

The following tables are included in Gumbel [36]:

(i) Values of the standard cumulative distribution function, $\exp(-e^{-y})$, and probability density function, $\exp(-y - e^{-y})$, to 7 decimal places for $y = -3(0.1) - 2.4(0.05)0.00(0.1)4.0(0.2)8.0(0.5)17.0$

(ii) The inverse of the cumulative distribution function (i.e. percentiles), $y = -\log(-\log F)$ to 5 decimal places for $F = 0.0001(0.0001)0.0050(0.001)$ $0.988(0.0001)0.9994(0.00001)0.99999$.

In Owen's tables [83] there is a similar table, to 4 decimal places, for

$$\begin{aligned} F = {} & 0.0001(0.0001)0.0010(0.0010)0.0100(0.005)0.100(0.010)0.90(0.005) \\ & 0.990(0.001)0.999(0.0001)0.9999, 0.99995, 0.99999, 0.999995, \\ & 0.999999, 0.9999995, 0.9999999, 0.99999995. \end{aligned}$$

(The especial interest in very high values of F, in both Gumbel [36] and Owen [83] may be associated with the genesis of the distribution, though it seems rather risky to rely on practical applicability so far out in the tails of a distribution.)

Reference [36] contains other tables. In particular there are two relating to asymptotic distribution of range (see Section 8), and a table giving the probability density function in terms of the cumulative distribution function ($p = -F \log F$) to 5 decimal places for $F = 0.0001(0.0001)0.0100(0.001)0.999$.

Table 2 (below) gives *standardized* percentile points (i.e. for a Type 1 extreme value distribution with expected value zero and standard deviation 1, corresponding to $\theta = \sqrt{6}/\pi = 0.77970$; $\zeta = -\gamma\theta = -0.45006$). The positive skewness of the distribution is clearly indicated by these values.

Lieblein and Salzer [67] have published a table of the expected values (to 7 decimal places) of the mth largest among n independent random variables having the standard Type 1 extreme value distribution (16), for

$$m = 1(1)\min(26, n); \qquad n = 1(1)10(5)60(10)100.$$

Lieblein and Zelen [68] gave the variances and covariances (also to 7 decimal places) for sets of 2, 3, 4, 5 and 6 independent Type 1 variables. (These values are also given in [66].) Mann [72] gave similar tables for the Type 1 smallest value distribution for up to 25 variables.

These tables have been extended by White [100], who gives (to 7 decimal

TABLE 2

Standardized Percentiles for Type 1 Extreme-Value Distribution

α	Percentile
0.0005	−2.0325
0.0001	−1.9569
0.0025	−1.8460
0.005	−1.7501
0.01	−1.6408
0.025	−1.4678
0.05	−1.3055
0.1	−1.1004
0.25	−0.7047
0.5	−0.1643
0.75	0.5214
0.9	1.3046
0.95	1.8658
0.975	2.4163
0.99	3.1367
0.995	3.6791
0.9975	4.2205
0.999	4.9355
0.9995	5.4761

places) expected values and variances of all order statistics for sample sizes 1(1)50(5)100.

From (1) it follows that

$$-\log(-\log \Pr[X < x]) = (x - \xi)/\theta. \tag{25}$$

Hence if the cumulative observed *relative frequency* F_x, equal to (number of observations less than or equal to x)/(total number of observations), is calculated, and $-\log(-\log F_x)$ is plotted against x, an approximately straight line relation should be obtained, with slope θ^{-1} and intersecting the horizontal (x) axis at $x = \xi$. Using graph paper with vertical scale giving $-\log(-\log F_x)$ directly, it is not necessary to use tables of logarithms. Such graph paper is sometimes called *extreme value probability paper.* It is also quite common to use such paper with the x-axis vertical, and for practical purposes it is sometimes convenient to have the $-\log(-\log F_x)$ marked, not with F_x, but with the 'return period' $(1 - F_x)^{-1}$ — (see Gumbel [34], and Kimball [61]). Such paper is called *extremal probability paper.*

Tables of 500 random numbers (to 3 decimal places) representing values chosen at random from the standard Type 1 distribution, and 500 each from three standard distributions of each of Types 2 and 3 ($k^{-1} = 0.2,0.5,0.8$ in Equations (13) and (14)) have been given by Goldstein [25].

6. Characterization

If X has a Type 1 extreme value distribution, then e^X has a Weibull distribution, and $e^{X/\theta}$ has an exponential distribution, and (exp $(X - \xi)/\theta$) has a *standard* exponential distribution. It is clear that some characterization theorems for exponential distributions may also be used for Type 1 extreme value distributions, simply by applying them to $e^{X/\theta}$, or exp $((X - \xi)/\theta)$. Dubey [11] characterizes this distribution by the property that $Y_n = \min(X_1, X_2, \ldots, X_n)$ is a Type 1 random variable if and only if $X_1, X_2, \ldots, X_n$ are independent identically distributed Type 1 random variables.

Sethuraman [89] has obtained characterizations of all three types of extreme value distribution, in terms of 'complete confounding' of random variables. If X and Y are independent and the distributions of Z, Z given $Z = X$, and Z given $Z = Y$ are the same (for example, Z might be equal to $\min(X,Y)$ as in the cases described in [89]), they are said to *completely confound* each other with respect to the third. Sethuraman showed that if all pairs from the variables X, Y and Z completely confound each other with respect to the third, and Y, Z have the same distributions as $a_1X + b_1$, $a_2X + b_2$ respectively (with $(a_1,b_1) \neq (a_2,b_2)$) then the distribution of X is one of the three extreme-value (minimum) distributions (provided we limit ourselves to the cases when $\Pr[X > Y] > 0$; $\Pr[Y > X] > 0$ etc.). The type of distribution depends on the values of a_1, a_2, b_1, b_2.

7. Estimation

Given n independent random variables $X_1, X_2, \ldots, X_n$, with probability density function (15), the Cramér-Rao lower bounds for variances of unbiased estimators of ξ and θ are

$$(26) \qquad \begin{aligned} (1 + 6(1 - \gamma)^2\pi^{-2})\theta^2 n^{-1} &= 1.10867\theta^2 n^{-1} \qquad (\text{for } \xi) \\ 6\pi^{-2}\theta^2 n^{-1} &= 0.60793\theta^2 n^{-1} \qquad (\text{for } \theta) \end{aligned}$$

(e.g. Downton [10]).

These formulas also give the asymptotic variances (as n increases) of the maximum likelihood estimators $\hat{\xi}$, $\hat{\theta}$, which satisfy the equations

$$(27.1) \qquad \sum_{j=1}^{n} \exp(-(X_j - \hat{\xi})/\hat{\theta}) = n$$

and

$$(27.2) \qquad \sum_{j=1}^{n} (X_j - \hat{\xi})[1 - \exp(-(X_j - \hat{\xi})/\hat{\theta})] = n\hat{\theta}.$$

($\hat{\xi}$ and $\hat{\theta}$ are biased estimators of ξ and θ respectively — see also below.)

The asymptotic correlation between $\hat{\xi}$ and $\hat{\theta}$ is

$$[1 - \{1 + 6\pi^{-2}(1 - \gamma)^2\}^{-1}]^{\frac{1}{2}} = 0.313.$$

Equation (27.1) can be written

$$\hat{\xi} = -\hat{\theta}\log\left[n^{-1}\sum_{j=1}^{n} e^{-X_j/\hat{\theta}}\right] \tag{28.1}$$

and, inserting this expression for $\hat{\xi}$ in (27.2), the following equation in $\hat{\theta}$ is obtained:

$$\hat{\theta} = n^{-1}\sum_{j=1}^{n} X_j - \left[\sum_{j=1}^{n} X_j e^{-X_j/\hat{\theta}}\right]\left[\sum_{j=1}^{n} e^{-X_j/\hat{\theta}}\right]^{-1}. \tag{28.2}$$

It is necessary to solve (28.2) for $\hat{\theta}$ numerically. An iterative method is usually the most convenient. (We may note that if $\hat{\theta}$ is large compared with X_j then the right hand side of (28.2) is approximately

$$\overline{X}\{1 - \hat{\theta}^{-1}S^2/\overline{X}\} \tag{28.3}$$

where $\overline{X} = n^{-1}\sum_{j=1}^{n} X_j$; $S^2 = n^{-1}\sum_{j=1}^{n}(X_j - \overline{X})^2$. Occasionally, this may be useful in obtaining an approximate solution of (28.2).)

Kimball [60] has suggested a modification of (28.2) leading to an equation which is simpler to solve. The equation may be written (using (28.1))

$$\begin{aligned}\hat{\theta} &= \overline{X} - n^{-1}\sum_{j=1}^{n} X_j e^{-(X_j-\hat{\xi})/\hat{\theta}} \\ &= \overline{X} + n^{-1}\sum_{j=1}^{n} X_j \log \hat{F}(X_j)\end{aligned}$$

where $\hat{F}(X_j)$ is the *estimated* value of the cumulative distribution function. Now $\hat{F}(X_j)$ may also be estimated from the sample values, leading to the equation

$$\hat{\theta}^* = \overline{X} + n^{-1}\sum_{j=1}^{n} X'_j\left\{\sum_{i=j}^{n} i^{-1}\right\} \tag{29}$$

where X'_j is the jth order statistic corresponding to $X_1, X_2, \ldots, X_n$. (This equation is obtained by replacing $\log \hat{F}(X'_j)$ in (28.2) by the expected value of $\log F_X(X_j)$.)

A good approximation is

$$\hat{\theta}^* \doteqdot \overline{X} + \sum_{j=1}^{n} X'_j \log\{(j - \tfrac{1}{2})/(n + \tfrac{1}{2})\}. \tag{29$'$}$$

This estimator (29) is a linear function of the order statistics. It is natural to compare it with the best linear unbiased estimator of θ and with approximations thereto obtained by the methods of Blom [3] and Weiss [98].

Downton [10] has carried out a number of comparisons of this kind. He studied the Type 1 distribution appropriate to least sample values (with cumulative distribution function of form $(1 - \exp[-\exp\{(x - \xi)/\theta\}])$) but his results also apply (with some simple modification) to distribution (1). Downton's results are all in terms of efficiencies, i.e. ratios of the values of variances

TABLE 3

Efficiencies of Linear Unbiased Estimators of ξ for the Extreme Value Distribution*

n	2	3	4	5	6	∞
Best linear	84.05	91.73	94.45	95.82	96.65	100.00
Blom's approximation	84.05	91.72	94.37	95.68	96.45	100.00
Weiss' approximation	84.05	91.73	94.41	95.74	96.53	—
Kimball's approximation	84.05	91.71	94.45	95.82	96.63	—

TABLE 4

Efficiencies of Linear Unbiased Estimators of θ for the Extreme Value Distribution*

n	2	3	4	5	6	∞
Best linear	42.70	58.79	67.46	72.96	76.78	100.00
Blom's approximation	42.70	57.47	65.39	70.47	74.07	100.00
Weiss' approximation	42.70	58.00	66.09	71.04	74.47	—
Kimball's approximation	42.70	57.32	65.04	69.88	73.25	—

*Efficiencies are expressed in percentages.

given by (26) to corresponding values for the estimator in question. Tables 3 and 4 (taken from [10]) give efficiencies for estimators of ξ as well as θ, where in each case

$$\text{estimator of } \xi = \overline{X} - \gamma \cdot (\text{estimator of } \theta). \tag{30}$$

For the small values of n shown, the asymptotic formulas used may not be very accurate. However, the tables probably do give a good idea of relative accuracy.

It can be seen that it is likely that ξ can be estimated with quite good accuracy using a linear function of order statistics. The situation is rather less satisfactory for estimation of θ.

White [99] has given tables of coefficients of best linear unbiased estimators using order statistics from samples of sizes up to and including 20 (both for complete and censored samples). His tables are an extension of tables of Lieblein [66], and have, in turn, been extended to samples of size 25 by Mann

[70]. Approximate best linear unbiased estimators have been devised by McCool [75] and Hassanein [51].

Harter and Moore [50] have presented the results of a sampling experiment comparing maximum likelihood estimators based on singly and doubly censored samples with best linear unbiased estimators using the same order statistics. They find that maximum likelihood estimators have considerable advantages over best linear unbiased estimators for censored extreme-value distribution, even though they have a bias (which increases with the amount of censoring).

Kimball [60] gave a table of corrective multipliers to render $\hat{\theta}^*$ (see (29)) unbiased, from which it appears that, for $n \geq 10$, the estimator

$$\left(1 + \frac{2.3}{n}\right)^{-1} \hat{\theta}^*$$

is very nearly unbiased.

If θ is *known*, the maximum likelihood estimator of ξ is

$$\hat{\xi}_\theta = -\theta \log\left[n^{-1} \sum_{j=1}^{n} e^{-X_j/\theta}\right]. \tag{31}$$

This is not an unbiased estimator of ξ. In fact (Kimball [60])

$$E[\hat{\xi}_\theta \mid \theta] = \xi + \theta\{\gamma + \log n - 1 - \tfrac{1}{2} - \cdots - (n-1)^{-1}\} \tag{32}$$

and

$$\operatorname{var}(\hat{\xi}_\theta \mid \theta) = \theta^2(\tfrac{1}{6}\pi^2 - 1^{-2} - 2^{-2} - \cdots - (n-1)^{-2}). \tag{33}$$

It may be noted that while $\hat{\xi}_\theta$ is a biased estimator of ξ, $e^{-\hat{\xi}_\theta/\theta}$ is an unbiased estimator of $e^{-\xi/\theta}$.

We have made no mention yet of estimators based on sample moments. These are very simply calculated from the formulas

$$\tilde{\theta} = (\sqrt{6}/\pi) \times (\text{sample standard deviation}) \tag{34.1}$$

and

$$\tilde{\xi} = (\text{sample mean}) - \gamma\tilde{\theta} \qquad (\text{see also (30)}). \tag{34.2}$$

The variance of $\tilde{\xi}$ is approximately

$$\left[\tfrac{1}{6}\pi^2 + \tfrac{1}{4}\gamma^2(\beta_2 - 1) - \frac{1}{\sqrt{6}}\gamma\pi\sqrt{\beta_1}\right]\theta^2 n^{-1} = 1.1678\theta^2 n^{-1} \tag{35.1}$$

(β_2 and β_1 given by (21)) and the variance of $\tilde{\theta}$ is approximately

$$\tfrac{1}{4}(\beta_2 - 1)\theta^2 n^{-1} = 1.1\theta^2 n^{-1} \qquad (\text{Tiago de Oliveira [92]†}). \tag{35.2}$$

†Oliveira also gave formulas for calculating the joint distribution of the sample mean and standard deviation.

On comparison with (26) we see that the efficiency of $\tilde{\xi}$ (relative to the maximum likelihood estimator $\hat{\xi}$) is about 95%, while that of $\tilde{\theta}$ relative to $\hat{\theta}$ is only about 55%.

Oliveira [92] showed that the joint asymptotic standardized distribution of $\tilde{\xi}$ and $\tilde{\theta}$ is bivariate normal with correlation coefficient

$$\text{(36)}\qquad \tfrac{1}{12}\pi^2[\sqrt{\beta_1} - \tfrac{3}{2}\gamma(\beta_2 - 1)\pi^{-1}] \times \left[\tfrac{1}{6}\pi^2 + \tfrac{1}{4}\gamma^2(\beta_2 - 1) - \frac{1}{\sqrt{6}}\gamma\pi\sqrt{\beta_1}\right]^{-\frac{1}{2}} [\tfrac{1}{4}(\beta_2 - 1)]^{-\frac{1}{2}} = 0.123.$$

Using these results, asymptotic confidence regions for (ξ,θ) can be constructed.

If the value of θ be known, then $e^{-X/\theta}$ has an exponential distribution with expected value $e^{-\xi/\theta}$. Confidence intervals for this quantity, and so for ξ, can be constructed using methods described in Chapter 18, Section 6.

In view of the genesis of extreme value distributions, it is not surprising that considerable attention has been devoted to estimating percentile points of the distribution. The upper $100\alpha\%$ point, $X_{1-\alpha}$, for example, can be regarded as a value which will be exceeded with an average return period of α^{-1} units of time. For a Type 1 distribution,

$$\text{(37)}\qquad X_{1-\alpha} = \xi - \theta \log\{-\log(1 - \alpha)\}$$

which is a linear function of ξ and θ. (The coefficients $-\log\{-\log(1 - \alpha)\}$ are tabulated in Gumbel [38].) Lieblein and Zelen [68] gave values of variances of best linear unbiased estimators of the median and the lower 10% point (to 7 decimal places) for samples of sizes 2–6.

As noted in Section 1, if X has a Weibull distribution with probability density function

$$p_X(x) = c\beta^{-1}\{(x - \xi_0)/\beta\}^{c-1} \exp[-\{(x - \xi_0)/\beta\}^c] \qquad (x > \xi_0)$$

(see (2) of Chapter 20), then $\log(X - \xi_0)$ has a Type 1 extreme value distribution. If ξ_0 is known, therefore, the methods of estimation developed for Type 1 distributions can also be used for Weibull distributions. Conversely, methods developed primarily for Weibull distributions (with ξ_0 known) can also be used for Type 1 distributions. Bain and Antle [2], and White [99], [100] refer especially to methods of the latter kind.

Using the standard large-sample approximations for variances and covariances of order statistics, Hassanein [52] has obtained approximately optimal sets of $k(=1(1)15)$ order statistics to use in estimating ξ when θ is known. These are $X'_{[n\lambda_i]+1}(i = 1,\ldots,k)$ where

$$2(\log\lambda_i + 1) = \frac{\lambda_{i+1}\log\lambda_{i+1} - \lambda_i\log\lambda_i}{\lambda_{i+1} - \lambda_i} - \frac{\lambda_i\log\lambda_i - \lambda_{i-1}\log\lambda_{i-1}}{\lambda_i - \lambda_{i-1}}$$

(with $\lambda_0 = 0, \lambda_{k+1} = 1$).

Harter and Moore [50] have presented results of an empirical enquiry into the variability and bias of maximum likelihood estimators of ξ and θ based on doubly censored random samples. Generally, they found that both $\hat{\xi}$ and $\hat{\theta}$ are negatively biased, except that when the left censoring is much heavier than the right $\hat{\xi}$ has a positive bias. Harter and Moore [50] also give an extensive and useful list of references.

8. Related Distributions

There is clearly a close connection between Type 1 (and Type 2) extreme value distributions and Weibull distributions (which are Type 3 extreme value distributions). The standard Type 1 distribution can also be regarded as a transitional limiting form between Weibull (Type 3) distributions and Type 2 distributions. This can be seen by writing the (standard) cumulative distribution function in a form suggested by Jenkinson [54]:

$$F_Y(y) = \exp[-(1 - ky)^{k^{-1}}]. \tag{38}$$

Then $k > 0$ corresponds to Type 3
$k < 0$ corresponds to Type 2
and $k = 0$ corresponds to Type 1.

A rather unexpected relation holds between the logistic and Type 1 distributions. If two independent random variables each have the same Type 1 distribution, their difference has a logistic distribution (Gumbel [39]). Gumbel [40] [41] has also studied the distribution of products and ratios of independent variables having extreme value distributions. Tables of the distribution of the 'extremal quotient' ((greatest)/(—least), $X'_n/(-X'_1)$) have been published by Gumbel and Pickands [46].

Limiting distributions of second, third, etc. greatest (or least) values may also be regarded as being related to extreme value distributions. Gumbel [38] has shown that under the same conditions as those leading to the Type 1 extreme value distribution, the limiting distribution of the rth greatest value, Y'_{n-r+1}, has the standard form of probability density function

$$p_{Y'_{n-r+1}}(y) = r^r[(r - 1)!]^{-1} \exp[-ry - re^{-y}]. \tag{39}$$

$100\alpha\%$ points of this distribution are given in [38] to 5 decimal places for

$r = 1(1)15(5)50$
$\alpha = 0.005, 0.01, 0.025, 0.05, 0.1, 0.25, 0.5, 0.75, 0.9, 0.95, 0.975, 0.99, 0.995$

The moment generating function of distribution (39) is

$$r^t\Gamma(r - t)/\Gamma(r).$$

The cumulant generating function is

$$t \log r + \log \Gamma(r - t) - \log \Gamma(r)$$

so the cumulants are

$$\kappa_1 = \log r - \psi(r) \tag{40}$$
$$\kappa_s = (-1)^s \psi^{(s-1)}(r) \qquad (s \geq 2).$$

The limiting distribution (39), which corresponds to a fixed value of r, should be distinguished from distributions obtained by allowing r to vary with n (usually in such a way that r/n is nearly constant), or by keeping r constant but varying the argument value. Borgman [4], for example, has shown that if x_n be defined by $F_X(x_n) = 1 - w/n$, for given fixed w (where $F_X(x)$ is the cumulative distribution function of the population distribution), then

$$\lim_{n\to\infty} \Pr[X'_{n-r+1} \leq x_n] = 1 - [(r-1)!]^{-1} \int_0^w t^{r-1} e^{-t}\, dt. \tag{41}$$

The right hand side of (41) can also be written in terms of a χ^2 distribution, as $\Pr[\chi^2_{2r} > 2w]$.

The asymptotic distribution of *range* is naturally closely connected with extreme value distributions. If both the greatest and least values have limiting distributions of Type 1, then (Gumbel [33]) the limiting distribution of the range, R, is of form

$$\Pr[R \leq r] = 2e^{-r/2} K_1(2e^{-r/2}) \tag{42}$$

with probability density function

$$p_R(r) = 2e^{-r} K_0(2e^{-r/2})$$

where K_0, K_1 are modified Bessel functions of the second kind of orders zero, one respectively. Gumbel gives the values

$$E[R] = 2\gamma = 1.15443;$$
$$\text{median } R = 0.92860;$$
$$\text{modal } R = 0.50637.$$

Also

$$\operatorname{var}(R) = \pi^2/3 = 3.2899.$$

In Gumbel [35] there are tables of $\Pr[R \leq r]$ and $p_R(r)$ to 7 decimal places for

$$r = -4.6(0.1)-3.3(0.05)11.00(0.5)20.0,$$

and of percentile points R_α, to 4 decimal places, for

$$\alpha = 0.0002(0.0001)0.0010(0.001)0.010(0.01)0.95(0.001)0.998$$

and to 3 decimal places for

$$\alpha = 0.0001, 0.999(0.0001)0.9999.$$

Some forms of "*generalized*" and *compound* Type 1 extreme value distributions have been constructed by Dubey [12]. He generalizes the distribution by introducing an extra parameter τ, defining the cumulative distribution function by the equation

$$\Pr[X \leq x] = \exp[-\tau\theta \exp\{-(x - \xi)/\theta\}]. \tag{43}$$

However, since

$$\tau\theta \exp\{-(x - \xi)/\theta\} = \exp\{-(x - \xi')/\theta\}$$

with $\xi' = \xi + \theta \log \tau\theta$, it can be seen that X still has an ordinary Type 1 distribution. This "generalized" distribution is, however, introduced only as an intermediate step in the construction of a *compound* Type 1 extreme value distribution, which can be denoted formally:

$$\text{"Generalized" Type 1 Extreme Value } (\xi,\theta,\tau) \underset{\tau}{\wedge} \text{Gamma } (p,\beta).$$

Here τ is supposed to have probability density function:

$$p_\tau(t) = \frac{\beta^p}{\Gamma(p)} t^{p-1} e^{-\beta t} \qquad (t > 0; p > 0,\ \beta > 0).$$

The resulting compound distribution has cumulative distribution function:

$$\begin{aligned} \Pr[X \leq x] &= [\beta^p/\Gamma(p)] \int_0^\infty t^{p-1} \exp[-\tau\{\beta + \theta \exp(-(x - \xi)/\theta)\}]\, dt \\ &= [1 + \theta\beta^{-1} \exp\{-(x - \xi)/\theta\}]^{-p}. \end{aligned} \tag{44}$$

This can also be regarded as a *generalized logistic distribution*. (See Hald [47] and Chapter 22, Section 11.)

REFERENCES

[1] Aziz, P. M. (1955). Application of the statistical theory of extreme values to the analysis of maximum pit depth data for aluminum, *Corrosion*, **12,** 495–506.

[2] Bain, L. J. and Antle, C. E. (1967). Estimation of parameters in the Weibull distribution, *Technometrics*, **9,** 621–627.

[3] Blom, G. (1958). *Statistical Estimates and Transformed Beta-Variables*, Stockholm: Almquist and Wiksell.

[4] Borgman, L. E. (1961). The frequency distribution of near extreme, *Journal of Geophysical Research*, **66,** 3295–3307.

[5] Bortkiewicz, L. von (1922). Variationsbreite und mittlerer Fehler, *Sitzungsberichte der Berliner Mathematischen Gesellschaft*, **21,** 3–11.

[6] Clough, D. J. and Kotz, S. (1965). Extreme value distributions with a special queueing model application, *CORS Journal*, **3,** 96–109.

[7] Coelho, D. P. and Gil, T. P. (1963). Studies on extreme double exponential distribution. I. The location parameter, *Revista da Faculdade de Ciências de Lisboa*, **10,** 37–46.

[8] Daniels, H. E. (1942). A property of the distribution of extremes, *Biometrika*, **32,** 194–195.

[9] Dodd, E. L. (1923). The greatest and least variate under general laws of error, *Transactions of the American Mathematical Society*, **25,** 525–539.

[10] Downton, F. (1966). Linear estimates of parameters in the extreme value distribution, *Technometrics*, **8,** 3–17.

[11] Dubey, S. D. (1966). Characterization theorems for several distributions and their applications, *Journal of Industrial Mathematics*, **16,** 1–22.

[12] Dubey, S. D. (1969). A new derivation of the logistic distribution, *Naval Research Logistics Quarterly*, **16,** 37–40.

[13] Eldredge, G. G. (1957). Analysis of corrosion pitting by extreme value statistics and its application to oil well tubing caliper surveys, *Corrosion*, **13,** 51–76.

[14] Epstein, B. (1948). Application to the theory of extreme values in fracture problems, *Journal of the American Statistical Association*, **43,** 403–412.

[15] Epstein, B. (1958). The exponential distribution and its role in life testing, *Industrial Quality Control*, **15,** 5–9.

[16] Epstein, B. (1960). Elements of the theory of extreme values, *Technometrics*, **2,** 27–41.

[17] Epstein, B. and Brooks, H. (1948). The theory of extreme values and its implications in the study of the dielectric strength of paper capacitors, *Journal of Applied Physics*, **19,** 544–550.

[18] Fisher, R. A. and Tippett, L. H. C. (1928). Limiting forms of the frequency distribution of the largest or smallest member of a sample, *Proceedings of the Cambridge Philosophical Society*, **24,** 180–190.

[19] Frenkel, J. I. and Kontorova, T. A. (1943). A statistical theory of the brittle strength of crystals, *Journal of Physics, USSR*, **7,** 108–114.

[20] Fréchet, M. (1927). Sur la loi de probabilité de l'écart maximum, *Annales de la Société Polonaise de Mathématique, Cracovie*, **6,** 93–116.

[21] Fuller, W. E. (1914). Flood flows, *Transactions of the American Society of Civil Engineering*, **77**, 564.

[22] Geffroy, J. (1958). Contribution a la théorie des valeurs extrêmes, *Publications de l'Institut de Statistique de l'Université de Paris*, **7**, No. 3/4, 37–121.

[23] Geffroy, J. (1959). Contribution a la théorie des valeurs extrêmes, II, *Publications de l'Institut de Statistique de l'Université de Paris*, **8**, No. 1, 3–65.

[24] Gnedenko, B. (1943). Sur la distribution limite du terme maximum d'une série aléatoire, *Annals of Mathematics*, **44**, 423–453.

[25] Goldstein, N. (1963). Random numbers from the extreme value distributions, *Publications de l'Institut de Statistique de l'Université de Paris*, **12**, 137–158.

[26] Griffith, A. A. (1920). The phenomena of rupture and flow in solids, *Philosophical Transactions of the Royal Society of London, Series A*, **221**, 163–198.

[27] Gumbel, E. J. (1935). Les valeurs extrêmes des distributions statistique, *Annales de l'Institut Henri Poincaré*, **4**, 115–158.

[28] Gumbel, E. J. (1937). La durée extrême de la vie humaine, *Actualités Scientifiques et Industrielles*, Paris: Hermann et Cie.

[29] Gumbel, E. J. (1937). Les intervalles extrêmes entre les émissions radioactives, *Journal de Physique et le Radium*, **8**, 446–452.

[30] Gumbel, E. J. (1941). The return period of flood flows, *Annals of Mathematical Statistics*, **12**, 163–190.

[31] Gumbel, E. J. (1944). On the plotting of flood discharges, *Transactions of the American Geophysical Union*, **25**, 699–719.

[32] Gumbel, E. J. (1945). Floods estimated by probability methods, *Engineering News-Record*, **134**, 97–101.

[33] Gumbel, E. J. (1947). The distribution of the range, *Annals of Mathematical Statistics*, **18**, 384–412.

[34] Gumbel, E. J. (1949). *The Statistical Forecast of Floods*, Bulletin No. **15**, pp. 1–21, Ohio Water Resources Board, Columbus, Ohio.

[35] Gumbel, E. J. (1949). Probability tables for the range, *Biometrika*, **36**, 142–148.

[36] Gumbel, E. J. (1953). Introduction to *Probability Tables for the Analysis of Extreme-Value Data*, National Bureau of Standards, Applied Mathematics Series, **22**; Washington, D.C.: U.S. Government Printing Office.

[37] Gumbel, E. J. (1954). *Statistical Theory of Extreme Values and Some Practical Applications*, National Bureau of Standards, Applied Mathematics, Series, **33**; Washington, D.C.: U.S. Government Printing Office.

[38] Gumbel, E. J. (1958). *Statistics of Extremes*, New York: Columbia University Press.

[39] Gumbel, E. J. (1961). Sommes et différences de valeurs extrêmes indépendentes, *Comptes Rendus de l'Académie des Sciences, Paris*, **253**, 2838–2839.

[40] Gumbel, E. J. (1962). Produits et quotients de deux plus grandes valeurs indépendantes, *Comptes Rendus de l'Académie des Sciences, Paris*, **254**, 2132–2134.

[41] Gumbel, E. J. (1962). Produits et quotients de deux plus petites valeurs indépendantes, *Publications de l'Institut de Statistique de l'Université de Paris*, **11**, 191–193.

[42] Gumbel, E. J. (1962). Statistical theory of extreme values (main results). (Chapter 6 of *Contributions to Order Statistics* (Ed. A. E. Sarhan and B. G. Greenberg), New York: John Wiley & Sons, Inc.).

[43] Gumbel, E. J. (1962). Statistical estimation of the endurance limit — An application of extreme-value theory. (In *Contributions to Order Statistics* (Ed. A. E. Sarhan and B. G. Greenberg), pp. 406–431; New York: John Wiley & Sons, Inc.)

[44] Gumbel, E. J. (1965). A quick estimation of the parameters in Fréchet's distribution, *Review of the International Statistical Institute*, **33,** 349–363.

[45] Gumbel, E. J. and Mustafi, C. K. (1966). Comments to: Edward C. Posner, "The application of extreme value theory to error free communication," *Technometrics*, **8,** 363–366.

[46] Gumbel, E. J. and Pickands, J. (1967). Probability tables for the extremal quotient, *Annals of Mathematical Statistics*, **38,** 1541–1551.

[47] Hald, A. (1952). *Statistical Theory with Engineering Applications*, New York: John Wiley & Sons, Inc.

[48] Harter, H. L. and Moore, A. H. (1967). A note on estimation from a Type 1 extreme-value distribution, *Technometrics*, **9,** 325–331.

[49] Harter, H. L. and Moore, A. H. (1968). Conditional maximum-likelihood estimators, from singly censored samples, of the scale parameters of Type II extreme-value distributions, *Technometrics*, **10,** 349–359.

[50] Harter, H. L. and Moore, A. H. (1968). Maximum likelihood estimation, from doubly censored samples, of the parameters of the first asymptotic distribution of extreme values, *Journal of the American Statistical Association*, **63,** 889–901.

[51] Hassanein, K. M. (1965). *Estimation of the parameters of the extreme value distribution by order statistics*, National Bureau of Standards, Institute of Applied Technology, Project No. 2776–M.

[52] Hassanein, K. M. (1968). Analysis of extreme-value data by sample quantiles for very large samples, *Journal of the American Statistical Association*, **63,** 877–888.

[53] Irwin, J. O. (1942). The distribution of the logarithm of survival times when the true law is exponential, *Journal of Hygiene, Cambridge*, **42,** 328–333.

[54] Jenkinson, A. F. (1955). The frequency distribution of the annual maximum (or minimum) values of meteorological elements, *Quarterly Journal of the Royal Meteorological Society*, **81,** 158–171.

[55] Kao, J. H. K. (1958). Computer methods for estimating Weibull parameters in reliability studies, *IRE Transactions on Reliability and Quality Control*, **13,** 15–22.

[56] Kimball, B. F. (1946). Sufficient statistical estimation functions for the parameters of the distribution of maximum values, *Annals of Mathematical Statistics*, **17,** 299–309.

[57] Kimball, B. F. (1946). Assignment of frequencies to a completely ordered set of sample data, *Transactions of the American Geophysical Union*, **27,** 843–846. (Discussion in **28,** (1947) 951–953.)

[58] Kimball, B. F. (1949). An approximation to the sampling variance of an estimated maximum value of given frequency based on fit of doubly exponential distribution of maximum values, *Annals of Mathematical Statistics*, **20,** 110–113.

[59] Kimball, B. F. (1955). Practical applications of the theory of extreme values, *Journal of the American Statistical Association*, **50,** 517–528. (Correction: **50,** 1332.)

[60] Kimball, B. F. (1956). The bias in certain estimates of the parameters of the extreme-value distribution, *Annals of Mathematical Statistics*, **27**, 758–767.

[61] Kimball, B. F. (1960). On the choice of plotting positions on probability paper, *Journal of the American Statistical Association*, **55,** 546–560.

[62] King, J. R. (1959). Summary of extreme-value theory and its relation to reliability analysis, *Proceedings of the 12th Annual Conference of the American Society for Quality Control*, 163–167.

[63] Lehman, E. H. (1963). Shapes, moments and estimators of the Weibull distribution, *Transactions of the IEEE-Reliability*, **12,** 32–38.

[64] Lieblein, J. (1953). On the exact evaluation of the variances and covariances of order statistics in samples from the extreme-value distribution, *Annals of Mathematical Statistics*, **24,** 282–287.

[65] Lieblein, J. (1954). *A new method of analyzing extreme-value data*, National Advisory Committee, Aeronautics, Technical Note No. 3053.

[66] Lieblein, J. (1962). Extreme-value distribution. (In *Contributions to Order Statistics* (Ed. A. E. Sarhan and B. G. Greenberg), pp. 397–406, New York: John Wiley & Sons, Inc.)

[67] Lieblein, J. and Salzer, H. E. (1957). Table of the first moment of ranked extremes, *Journal of Research of the National Bureau of Standards*, **59,** 203–206.

[68] Lieblein, J. and Zelen, M. (1956). Statistical investigation of the fatigue life of deep-groove ball bearings, *Journal of Research of the National Bureau of Standards*, **57,** 273–316.

[69] Mann, Nancy R. (1965). *Point and interval estimates for reliability parameters when failure times have the two-parameter Weibull distribution* (Ph.D. thesis, University of California, Los Angeles).

[70] Mann, Nancy R. (1967). *Results on Location and Scale Parameter Estimation with Application to the Extreme-value Distribution*, Report ARL 67–0023, Aerospace Research Laboratories, Wright-Patterson Air Force Base, Ohio.

[71] Mann, Nancy R. (1968). Point and interval estimation procedures for the two-parameter Weibull and extreme-value distributions, *Technometrics*, **10,** 231–256.

[72] Mann, Nancy R. (1968). *Results on Statistical Estimation and Hypothesis Testing with Application to the Weibull and Extreme Value Distributions*, Report ARL 68–0068, Aerospace Research Laboratories, Wright-Patterson Air Force Base, Ohio.

[73] Maritz, J. S. and Munro, A. H. (1967). On the use of the generalised extreme-value distribution in estimating extreme percentiles, *Biometrics*, **23,** 79–103.

[74] Massonie, J. P. (1966). Estimation de l'exposant d'une fonction de distribution tronquée, *Comptes Rendus de l'Académie des Sciences, Paris*, **262,** 350–352.

[75] McCool, J. I. (1965). The construction of good linear unbiased estimates from the best linear estimates for a smaller sample size, *Technometrics*, **7,** 543–552.

[76] McCord, J. R. (1964). On asymptotic moments of extreme statistics, *Annals of Mathematical Statistics*, **35,** 1738–1745.

[77] Miller, I. and Freund, J. (1965). *Probability and Statistics for Engineers*, Englewood Cliffs, New Jersey: Prentice-Hall.

[78] Mises, R. von (1923). Über die Variationsbreite einer Beobachtungsreihe, *Sitzungsberichte der Berliner Mathematischen Gesellschaft*, **22,** 3–8.

[79] Moore, A. H. and Harter, H. L. (1966). Point and interval estimation, from one order statistic, of the location parameter of an extreme value distribution with known scale parameter, and of the scale parameter of a Weibull distribution with the known shape parameters, *Transactions of the IEEE-Reliability*, **15,** 120–126.

[80] Moore, A. H. and Harter, H. L. (1967). One-order-statistic conditional estimators of shape parameters of limited and Pareto distributions and scale parameters of Type II asymptotic distributions of smallest and largest values, *Transactions of the IEEE-Reliability*, **16,** 100–103.

[81] Mustafi, C. K. (1963). Estimation of parameters of the extreme value distribution with limited type of primary probability distribution, *Bulletin of the Calcutta Statistical Association*, **12,** 47–54.

[82] Nordquist, J. M. (1945). Theory of largest values, applied to earthquake magnitudes, *Transactions of the American Geophysical Union*, **26,** 29–31.

[83] Owen, D. B. (1962). *Handbook of Statistical Tables*, Reading, Massachusetts: Addison-Wesley.

[84] Posner, E. C. (1965). The application of extreme-value theory to error-free communication, *Technometrics*, **7,** 517–529.

[85] Potter, W. D. (1949). *Normalcy tests of precipitation and frequency studies of runoff on small watersheds*, U.S. Department of Agriculture Technical Bulletin, No. 985, Washington, D.C.: U.S. Government Printing Office.

[86] Press, H. (1949). *The application of the statistical theory of extreme value to gust-load problems*, National Advisory Committee, Aeronautics, Technical Note No. 1926.

[87] Rantz, S. F. and Riggs, H. C. (1949). *Magnitude and frequency of floods in the Columbia River Basin*, U.S. Geological Survey, Water Supply Paper 1080, 317–476.

[88] Sen, P. K. (1961). A note on the large-sample behaviour of extreme sample values from distribution with finite end-points, *Bulletin of the Calcutta Statistical Association*, **10,** 106–115.

[89] Sethuraman, J. (1965). On a characterization of the three limiting types of the extreme, *Sankhyā, Series A*, **27,** 357–364.

[90] Sibuya, M. (1967). On exponential and other random variable generators, *Annals of the Institute of Statistical Mathematics, Tokyo*, **13,** 231–237.

[91] Thom, H. C. S. (1954). Frequency of maximum wind speeds, *Proceedings of the American Society of Civil Engineers*, **80,** 104–114.

[92] Tiago de Oliveira, J. (1963). Decision results for the parameters of the extreme value (Gumbel) distribution based on the mean and the standard deviation, *Trabajos de Estadística*, **14,** 61–81.

[93] Tippett, L. H. C. (1925). On the extreme individuals and the range of samples taken from a normal population, *Biometrika*, **17,** 364–387.

[94] Uzgören, N. T. (1954). The asymptotic development of the distribution of the extreme values of a sample, *Studies in Mathematics and Mechanics*, presented to R. von Mises, Academic Press, New York, 346–353.

[95] Velz, C. J. (1947). Factors influencing self-purification and their relation to pollution abatement, *Sewage Works Journal*, **19,** 629–644.

[96] Weibull, W. (1939). The phenomenon of rupture in solids, *Ingenior Vetenskaps Akademiens Handlingar*, **153,** 2.

[97] Weibull, W. (1949). A statistical representation of fatigue failures in solids, *Kunglig Tekniska Högskolans Handlingar*, **27.**

[98] Weiss, L. (1961). On the estimation of scale parameters, *Naval Research Logistics Quarterly*, **8,** 245–256.

[99] White, J. S. (1964). Least-square unbiased censored linear estimation for the log Weibull (extreme value) distribution, *Journal of Industrial Mathematics*, **14,** 21–60.

[100] White, J. S. (1969). The moments of log-Weibull order statistics, *Technometrics*, **11,** 373–386.

[101] Winer, P. (1963). The estimation of the parameters of the iterated exponential distribution from singly censored samples, *Biometrics*, **19,** 460–464.

Acknowledgements

Figures

P. 14 Adapted from Pearson, E. S. and Hartley, H. O. (Ed.) *Biometrika Tables for Statisticians*, New York: Cambridge University Press, 1958.

P. 18 Adapted from Barton, D. E. and Dennis, K. E. R. (1952). The conditions under which Gram-Charlier and Edgeworth curves are positive definite and unimodal, *Biometrika*, **39,** 425–427.

P. 24 Adapted from Johnson, N. L. (1949). Systems of frequency curves generated by methods of translation, *Biometrika*, **36,** 149–176.

P. 30 Adapted from Bhattacharyya, B. C. (1942). The use of McKay's Bessel function curves for graduating frequency distributions, *Sankhyā*, **6,** 175–182.

P. 121 Adapted from Finney, D. J. (1941). On the distribution of a variate whose logarithm is normally distributed, *Journal of the Royal Statistical Society, Series B*, **7,** 155–161.

P. 127 Adapted from Wise, M. E. (1966). The geometry of log-normal and related distributions and an application to tracer-dilution curves, *Statistica Neerlandica*, **20,** 119–142.

P. 142 & 151 Adapted from Tweedie, M. C. K. (1957). Statistical properties of inverse Gaussian distributions, I, *Annals of Mathematical Statistics*, **28,** 362–377.

P. 175 Adapted from Boyd, W. C. (1965). A nomogram for chi-square, *Journal of the American Statistical Association*, **60,** 344–346.

P. 264 Adapted from Kotel'nikov, V. P. (1964). A nomogram connecting the parameters of Weibull's distribution with probabilities, *Teoriya Veroyatnostei i ee Primeneniya*, **9,** 743–746. (English translation, **9,** 670–673.)

P. 265 Adapted from Nelson, L. S. (1967). Weibull probability paper, *Industrial Quality Control*, **23,** 452–453.

Tables

P. 27 Adapted from (i) Johnson, N. L., Nixon, E., Amos, D. E. and Pearson, E. S. (1963). Table of percentage points of Pearson curves, for given $\sqrt{\beta_1}$

and β_2, expressed in standard measure, *Biometrika*, **50,** 459–498, and (ii) Johnson, N. L. (1964). *Table of percentage points of S_U curves, for given $\sqrt{\beta_1}$ and β_2, expressed in standard measure*, Institute of Statistics, University of North Carolina, Mimeo Series No. 408.

P. 65 Adapted from Cadwell, J. H. (1953). Approximating to the distribution of measures of dispersion by a power of χ^2, *Biometrika*, **40,** 336–346.

P. 68 Adapted from Mead, R. (1966). A quick method for estimating the standard deviation, *Biometrika*, **53,** 559–564.

P. 69 Adapted from Jones, A. E. (1946). A useful method for the routine estimation of dispersion in large samples, *Biometrika*, **33,** 274–282.

P. 71 & 79 Adapted from Dixon, W. J. (1960). Simplified estimation from censored normal samples, *Annals of Mathematical Statistics*, **31,** 385–391.

P. 86 Adapted from Harter, H. L. and Moore, A. H. (1966). Iterative maximum likelihood estimation of the parameters of normal populations from singly and doubly censored samples, *Biometrika*, **53,** 205–213.

P. 130 & 131 Adapted from Harter, H. L. and Moore, A. H. (1966). Local-maximum-likelihood estimation of the parameters of three-parameter lognormal populations from complete and censored samples, *Journal of the American Statistical Association*, **61,** 842–851.

P. 178 Adapted from Pogurova, V. I. (1965). On the calculation of quantiles of the Γ-distribution, *Theory of Probability and its Applications*, **10,** 677–680.

P. 258 Adapted from Weibull, W. (1967). *Estimation of distribution parameters by a combination of the best linear order statistic method and maximum likelihood*, Technical Report AFML – TR – 67 – 105, Air Force Materials Laboratory, Wright-Patterson Air Force Base, Ohio.

P. 259 Adapted from Vogt, H. (1968). Zur Parameter- und Prozentpunktschätzung von Lebensdauerverteilungen bei kleinem Stichprobenumfang, *Metrika*, **14,** 117–131.

P. 284 Adapted from Downton, F. (1966). Linear estimates of parameters in the extreme value distribution, *Technometrics*, **8,** 3–17.

Index

This index is intended as an auxiliary to the detailed table of contents. Most references are to distributions (the word "distribution" being omitted) but there are a few other references. Distributions are only included in this index when they occur in places not easily identifiable from the table of contents.

ABCDEFGHIJ— M —76543210/69

Applied Probability and Statistics (Continued)

JAGERS · Branching Processes with Biological Applications

JOHNSON and KOTZ · Distributions in Statistics
Discrete Distributions
Continuous Univariate Distributions-1
Continuous Univariate Distributions-2
Continuous Multivariate Distributions

JOHNSON and KOTZ · Urn Models and Their Application

JOHNSON and LEONE · Statistics and Experimental Design: In Engineering and the Physical Sciences, Volumes I and II, *Second Edition*

KEENEY and RAIFFA · Decisions with Multiple Objectives

LANCASTER · The Chi Squared Distribution

LANCASTER · An Introduction to Medical Statistics

LEWIS · Stochastic Point Processes

MANN, SCHAFER and SINGPURWALLA · Methods for Statistical Analysis of Reliability and Life Data

MEYER · Data Analysis for Scientists and Engineers

OTNES and ENOCHSON · Digital Time Series Analysis

PRENTER · Splines and Variational Methods

RAO and MITRA · Generalized Inverse of Matrices and Its Applications

SARD and WEINTRAUB · A Book of Splines

SEAL · Stochastic Theory of a Risk Business

SEARLE · Linear Models

THOMAS · An Introduction to Applied Probability and Random Processes

WHITTLE · Optimization under Constraints

WONNACOTT and WONNACOTT · Econometrics

WONNACOTT and WONNACOTT · Introductory Statistics for Business and Economics, *Second Edition*

YOUDEN · Statistical Methods for Chemists

ZELLNER · An Introduction to Bayesian Inference in Econometrics

Tracts on Probability and Statistics

BHATTACHARYA and RAO · Normal Approximation and Asymptotic Expansions

BILLINGSLEY · Convergence of Probability Measures

CRAMER and LEADBETTER · Stationary and Related Stochastic Processes

JARDINE and SIBSON · Mathematical Taxonomy

RIORDAN · Combinatorial Identities